D0908790

Fundamentals of
Infrared Detector
Operation and Testing

Wiley Series on Pure and Applied Optics

The Wiley Series in Pure and Applied Optics publishes outstanding books in the field of optics. The nature of these books may be basic ("pure" optics) or practical ("applied" optics). The books are directed towards one or more of the following audiences: researchers in university, government, or industrial laboratories; practitioners of optics in industry; or graduate-level courses in universities. The emphasis is on the quality of the book and its importance to the discipline of optics.

Fundamentals of Infrared Detector Operation and Testing

JOHN DAVID VINCENT
Santa Barbara Research Center
Goleta, California

WILEY

A Wiley-Interscience Publication

John Wiley & Sons

New York / Chichester / Brisbane / Toronto / Singapore

Copyright © 1990 by Santa Barbara Research Center

All rights reserved. Published simultaneously in Canada.

Reproduction or translation of any part of this work
beyond that permitted by Section 107 or 108 of the
1976 United States Copyright Act without the permission
of the copyright owner is unlawful. Requests for
permission or further information should be addressed to
the Permissions Department, John Wiley & Sons, Inc.

Library of Congress Cataloging in Publication Data:

Vincent, John David.
    Fundamentals of infrared detector operation and testing / John
David Vincent.
        p. cm.—(Wiley series in pure and applied optics, ISSN
0277-2493)
    "A Wiley-Interscience publication."
    Bibliography: p.
    ISBN 0-471-50272-3
    1. Infrared detectors. 2. Infrared detectors—Testing.
I. Title. II. Series.

TA1570.V56 1989   89-34214
621.36'2—dc20   CIP

Printed in the United States of America

10 9 8 7 6 5 4 3 2

To Dr. George Appleton,
a gifted, energetic, and caring teacher

Foreword

It is difficult to measure the performance characteristics of infrared detectors. Detector testing has always been an arduous undertaking because of the large number of experimental variables involved. A variety of environmental, electrical, and radiometric parameters must be taken into account and carefully controlled. With the advent of large, two-dimensional detector arrays, detector performance testing has become even more complex and demanding.

The experimental methods and techniques required for infrared detector testing are quite diverse and cross several engineering disciplines. Detector testing often entails the use of vacuum equipment, cryogenic apparatus, blackbody radiators, spectrometers, very low noise amplifiers, high-speed digital data systems, and many other devices. One of the underlying difficulties is the necessity to orchestrate so much diverse experimental apparatus into a single detector test station. Obviously, the operation of such a complex array of equipment is likely to be somewhat demanding, and equipment malfunctions are likely to occur from time to time.

This diversity also places difficult requirements on the test personnel. The test engineer must be reasonably skilled in several technical areas to deal effectively with the many experimental problems that will inevitably occur. This multidisciplinary attribute may be somewhat unique to infrared detector testing. It is difficult to identify any other category of component level testing that requires such a broad range of expertise. Acquiring such a wide range of skills is not the only problem the newcomer must face. The apprentice engineer or technician must also deal with an ever-changing nomenclature and a technical literature that is often somewhat obscure and difficult to access.

The author is well aware of the problems facing newcomers to the infrared field. He has been teaching classes in detector fundamentals and testing at Santa Barbara Research Center for many years. The material for this book is an outgrowth of the notes developed for those classes. It has been selected and arranged to meet the specific needs of newcomers to infrared technology; however, it is also quite definitive since it reflects over 20 years of active engineering experience. In the past, newcomers to the field of infrared detector technology and performance testing have often been frustrated by the lack of suitable reference material. With this text, Dave Vincent has provided the newcomer with a valuable and needed resource. He has also provided a convenient reference volume that will be of significant value to the more experienced worker.

LUM EISENMAN

Del Mar, California
January 1989

Preface

This book is intended to serve as an introductory reference for the operation and testing of infrared (IR) detectors. Although I hope that it can be of help to everyone in the field, either as a reference for themselves or as a sourcebook as they teach others, it is addressed primarily to three special groups: those entering the field of IR detector design, test, or use; those who work in the peripheral areas; and those who teach and train newcomers. Formulas and examples are provided for most applications in a typical test laboratory. Most of these can be understood and applied without calculus.

There are numerous infrared texts and journals, covering many levels and specialties. Given a little time, one can find documentation on both the common topics and many esoteric ones. Experienced engineers and scientists have their own collection of notes and favorite references. Newcomers must either gather their own collections as they serve their apprenticeship, or rely on a more experienced mentor to guide them to the most convenient material. For example, a good nucleus of references for our work would include: *Optical Radiation Detectors*, by E. L. Dereniak and D. G. Crowe; *Infrared System Engineering*, by R. D. Hudson; *The Infrared Handbook*, edited by W. L. Wolfe and G. J. Zissis; *Handbook of Optics*, edited by W. G. Driscoll and W. Vaughan; *Experimental Techniques in Low Temperature Physics*, by Guy Kendell White; *Scientific Foundations of Vacuum Techniques*, by Saul Dushman; *Fundamentals of Optics*, by F. A. Jenkins and H. E. White; and *Building Scientific Apparatus*, by J. H. Moore, C. C. Davis, and M. A. Coplan. Taken together, these provide the material on detectors, radiometry, and other related laboratory skills that is essential for those who build and test IR detectors.

What I and many workers in the IR business have done is to mark selected portions of these and similar texts, and to add a handful of references from the journals and a few explanations and derivations that we have worked out ourselves or gleaned from our colleagues' work. That material, plus an index to organize the selected material in a convenient order for our purposes, provides a useful personal set of notes for our own work and for instructing others.

In the present book, I have tried to facilitate that process by providing my equivalent of that set of notes, along with references to the classic texts for those who want to pursue a topic in more detail. No textbook can replace the specialized knowledge of "old-timers" or satisfy all the requirements of different laboratories. I hope, however, that this book will serve as a convenient starting point and speed the transition to the required specialized training. I especially hope this material will be useful to those in what I refer to as the peripheral areas. The engineer who designs a dewar, the technician who assembles and blackens it, the inspector who monitors a test, and the administrator who negotiates a contract often work without much help from the detector and test engineers. This is not necessarily a deliberate choice on anyone's part, but it is still true. How much better things would work if the technician knew *why* he or she was blackening an annulus around the dewar window, or if the inspector knew why certain suspect data were adequate. I hope that this text will make it easier for them to know about and enjoy their work.

I have deliberately avoided discussion of detector architecture and fabrication methods because these are either proprietary or will soon be outdated. Similarly, I have not attempted to describe state-of-the-art data acquisition and signal processing. Instead, I have included operation and test information that is applicable to all detectors. A clear understanding of single-element detector operation, the figures of merit, and of radiometry will be useful for many years, even though the ways these techniques are implemented will change dramatically.

The choice of radiometric nomenclature was a difficult one. I have used the ANSI/CIE system, but substituted the words *incidance* for irradiance/illuminance and *sterance* for radiance/luminance. Incidance and sterance were proposed by R. Clark Jones in 1963, and I believe they are still helpful. I hope the discussion in the text will assuage the criticism of those who would have opted for other systems. I have intentionally used and mixed inches and centimeters. Many readers will find that mixture in their work, and I think it is good to face that fact and get on with it.

Chapter 1 introduces the concept of electromagnetic radiation and the other mechanisms for heat transfer; Planck's law, photons, and the definitions of the conventional figures of merit. In Chapter 2 we discuss detector types, mechanisms, and operation, and prediction of figures of merit.

Chapter 3 covers radiometric concepts and calculations, beginning with the formula for radiative transfer in an idealistic—but easy to handle— situation, and continuing with methods to extend that simple formula to real-life spectral regions, geometries, and modulation. The formulas for backgrounds and blackbody incidances in common dewar or chamber setups are motivated and examples provided.

Test sets and test setups are discussed in Chapter 4: the types of setups encountered, the issue of "calibration" of infrared test sets, and a comprehensive checklist of things to do (and not do) in designing or modifying a test setup. This chapter is the result of many years of troubleshooting and "reinventing the wheel," and will provide a helpful methodical approach to eliminating radiometric problems during the design stages or finding them in an inherited test setup.

In Chapter 5 we apply the preceding material to detector characterization. It includes a list of steps that can be taken during testing to increase the confidence and accuracy of the test results, or to identify errors if the results are suspect.

Unit 2 covers related skills, with chapters introducing measurements and error analysis, cryogenics, vacuum practices, optics, and electronics. It has been my experience that the ability to estimate uncertainties in practical measurements is very often lacking in the laboratory today; *that one skill* is probably the most useful single attribute of a competent scientist, engineer, or technician. The material in the other "ancillary skills" chapters is meant to introduce the topic, to guide students to, and help them feel comfortable browsing through the many classic texts available.

Reference material is relegated to the appendices. I have followed Hudson's example* in listing the symbols in Appendix A. Appendix B is a glossary, which may help newcomers as they learn the new vocabulary. Appendix C describes the decibel convention. Appendix D, "Characterization of Semiconductor Materials," describes tests that are not specifically detector tests, but which the practicing detector engineer or technician may be called on to specify, perform, or interpret.

I am grateful to many people for their assistance and contributions to this work. Dr. Peter Bratt's notes and memos tutored me in the prediction of detector figures of merit. Dick Chandos provided the material for Chapter 10; he was assisted by Linda Chandos. Dick brings many years of practical experience in the application of electronics to varied detector problems. I am fortunate that someone with the right experience was willing to assist me to cover a topic in which I felt inadequate. Thanks to Frank Renda and other management of Santa Barbara Research Center for their encouragement and assistance in this work. Dick Nielsen, Jim Fulton, and Bob Kvaas provided much of the material on low-background testing. My co-workers in the FPM Test Section read portions of the

* Richard D. Hudson, Jr., *Infrared System Engineering*, New York, Wiley, 1969.

manuscript, noted errors and suggested improvements. Ken Shamordola pointed out the expression for the variance of the noise, and Bob Howard introduced me to the round-off error problem and the analysis of quantization errors. Scott Daley generated the Planck's law curves. Robert Turtle and Jim Young reviewed the optics material and offered many helpful suggestions. John Sekula critiqued the manuscript and suggested several improvements.

Jim Young and Jerry Hyde have been longstanding sources of instruction and assistance, and have been patient sounding boards for discussions about radiometric and modulation transfer function calculations. The rigor and precision of their work has been a great help to me.

Hank Lay thoroughly and carefully proof read the manuscript. Fred Nicodemus provided much background material on the history and rationale of radiometric nomenclature.

Although these people provided much information and insight, I have interpreted and modified their input and take full responsibility for the results presented here.

Dr. George Appleton introduced me to physics, to cryogenics, to error analysis, and to laboratory work in general. I will always be grateful to him for his interest and support. Finally, I thank my wife, Bev, for the patience and sacrifice that allowed me the time to complete this project.

JOHN DAVID VINCENT

Goleta, California
August 1989

Contents

UNIT 2   RELATED SKILLS   273

Chapter 6.   Science and Measurements   275

Fundamentals of
Infrared Detector
Operation and Testing

Unit 1
Detectors and Testing

1

Introduction and Overview

1.1. INFRARED RADIATION

Infrared (IR) radiation is a form of radiated electromagnetic energy, obeying the same laws as those for visible light, radio waves, and x-rays. In fact, its only fundamental difference from those forms of electromagnetic radiation is its wavelength. This is shown in the chart of the electromagnetic spectrum in Figure 1.1.

The borderlines between visible, infrared, far-infrared, and millimeter waves are not absolute. These areas of the spectrum have been segregated primarily for convenience in discussions; the primary criteria are the sources used and the detectors that are available. *Light* is that portion of the spectrum to which the human eye is sensitive. *Millimeter waves* are the shortest wavelengths that can be received with the smallest microwave-like apparatus. *Infrared* and the *far-infrared* cover the wavelengths in between. Thus a statement such as "IR extends from 0.7 μm to 1000 μm" is someone's definition or convention, not a statement of a physical law.

Some characteristics of infrared radiation are that it travels basically in a straight line (although some exceptions to this are important in our work), does not penetrate metals unless they are very thin, and passes through many crystalline, plastic, and gaseous materials—including the earth's atmosphere. Most IR detectors described in this book take advantage of two atmospheric "windows"—spectral regions that transmit well: the 3- to 5-μm window, and the 8- to 12-μm window. Thus we would include the 3- to 12-μm region as the primary IR region, but we could also specify the range 1 to 50 μm, and some workers using conventional IR methods work with wavelengths as low as 0.7 μm and as high as 1000 μm.

1.2. HEAT TRANSFER

Heat is transferred in three ways:

- Radiated (electromagnetic radiation)
- Conducted (as through a hot piece of metal)
- Convected (through warm air circulating in a room)

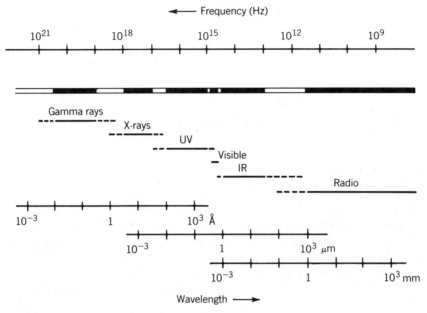

Figure 1.1. Electromagnetic spectrum.

These three methods of heat transfer are illustrated in Figure 1.2. Radiant transfer is important because our IR detectors will measure the radiant transfer of heat (or photons). We devote one chapter to the prediction of radiant transfer effects.

Warm objects radiate more IR power than do cooler ones, but all objects give off some power in the infrared. Room-temperature objects and even ice cubes emit some IR. In Chapter 3 we discuss methods to predict the power and photon flux from objects of different temperatures.

Since our eyes are not sensitive to IR, our everyday awareness of it is limited: It is generally sensed only by the heat carried by the IR radiation. If we can set up a situation in which conduction and convection are limited, it is possible to sense the infrared radiation directly: the warmth from the sun on a cold day, or the heat from a hot fire are carried through electromagnetic radiation, most of which is in the infrared.

Most detectors will be operated well below room temperature, and all three methods of heat transfer will be important to us when we consider the detector cooling problem. To minimize the heat transferred from the

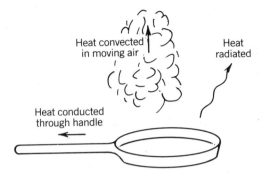

Figure 1.2.   Three methods of heat transfer.

room-temperature laboratory to the cold detectors we will need to minimize the radiated, conducted, and convected heat leaks.

1.3.   THERMAL DETECTORS

In 1800, while he was using a prism to spread sunlight into its component
colors (or wavelengths), the English astronomer Sir William Herschel
(1738–1822) made the initial discovery that there is radiant energy beyond
the visible spectrum. His experiment is described by Hudson (1969) and
illustrated in Figure 1.3. He used a thermometer to measure the temper-

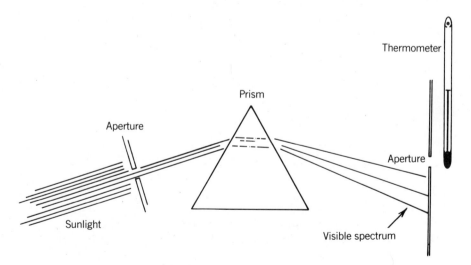

Figure 1.3.   Herschel's IR detector.

ature at places on which different portions of the spectrum fell and was surprised to find that even where he could not see any color, the thermometer registered a significant heating effect. The detectors that followed were simply more and more sensitive thermometers. Detectors that operate by sensing temperature changes are called *thermal detectors*.

1.4. PLANCK'S LAW

For any given source, some wavelengths transmit more power than others. Lummer and Pringsheim are credited with the first accurate measurements of the distribution of energy within the electromagnetic spectrum (Eisberg, 1961). Their work was done in 1899, and the measurements disagreed with predictions based on the accepted physical laws. Resolution of the disagreement became the subject of intense effort; Hudson (1969) gives a concise account of the flurry of work this discrepancy caused.

In 1900, Max Planck derived an equation (plotted in Figure 1.4) that fit the observed data. His derivation assumed that the oscillators responsible for the radiation were limited to discrete energy levels related to the frequency v of the radiation: they could have energy hv, $2hv$, $3hv$, and so on, but not $1.2hv$, or $3.7hv$, where h is a constant—now known as *Planck's constant*. His revolutionary hypothesis and the resulting solution led to modern quantum theory and earned him a Nobel Prize in 1918 (Wehr and Richards, 1960). We will use Planck's radiation law and Planck's constant to calculate the power emitted by many of our IR sources.

We look at *Planck's law* in some detail later, but some characteristics

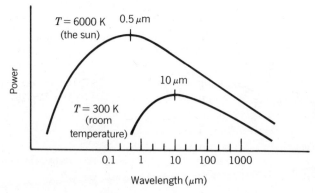

Figure 1.4. Planck's law.

are shown in Figure 1.4 and described here:

- The warmer the source, the more energy it gives off. This is a rather strong dependence: A small increase in temperature causes a large increase in emitted power.
- As the temperature of the source increases, the wavelength at which most of the energy is given off decreases. The relationship is fairly direct: Increasing the temperature by a given factor decreases the peak wavelength by the same factor.
- At short wavelengths the shape of the curve can be described very accurately by a relatively simple equation. Another simple equation works well at long wavelengths, but the equation to describe the whole curve—Planck's law—is complex.

1.5. WAVES AND PHOTONS

Many experiments with optics and other electromagnetic radiation can be explained by thinking of the radiation as a wave. For those situations we need to know the wavelength and the power transmitted by the wave. Unfortunately, other experiments do not agree with the results of calculations based on wavelike behavior, and for much of our work, we will need to visualize electromagnetic radiation in a different way.

In 1887, Hertz discovered that electromagnetic radiation striking one plate in a vacuum tube could generate a current that depended on the intensity of the "light." Subsequent experiments at first raised questions, then led to answers that helped understand the effect. In 1905, Einstein showed that the photoelectric cell was responding to light as though it were individual "packets" or "bullets" of energy instead of a continuum. This theory is now completely accepted, and we speak of *photons* or *quanta* of electromagnetic energy. Einstein was awarded the Nobel Prize in 1921 "for his contributions to mathematical physics, and especially for his discovery of the law of the photoelectrical effect" (Eisberg, 1961; Wehr and Richards, 1960).

Although the duality of light is discussed by most books on modern physics, it is a distressing situation for some students: We do not know whether light is a wave or a bunch of particles. There is a moral to be learned from this situation: There is much of which we mortals are ignorant. We can build up models, hypotheses, formulas, and laws, but they are simply mental crutches, or illustrations to help us. When they work, we think we know it all. Unfortunately, they don't always work.

Don't think of *either* the photon or wave model as being "correct."

The electromagnetic phenomena are complicated, and both models are simply analogies to help us understand and make predictions. In some situations a particular analogy may work, but in others it won't.

1.6. QUANTUM (PHOTON) DETECTORS

The photoelectric cell, and other detectors that respond to electromagnetic radiation as though it is made up of these packets of energy, are called *photon detectors*. Most of the detectors we work with are photon detectors.

1.7. IR DETECTORS AS TRANSDUCERS

A *transducer* is a device that converts one type of signal to another. We can think of the IR detector as a transducer that converts infrared to electrical signals (Figure 1.5). The incoming radiation and the electrical

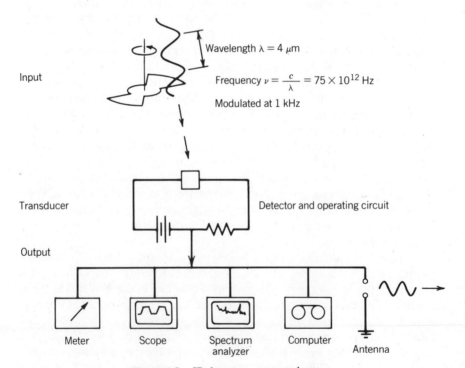

Figure 1.5.   IR detectors are transducers.

signal generated are both described in terms of wavelengths, frequencies, power, and spectral distribution.

As you begin thinking about IR detection, be careful to make a distinction in your mind between the input (IR) signal, with its wavelengths, frequencies, and power, and the output (electrical) signal, with *its* wavelengths, frequencies, and power. The infrared wavelengths have values of a few micrometers, with frequencies of about 1×10^{14} hertz (cycles per second). The electrical signals generated by our detectors (transducers) are interesting only at relatively low frequencies, from dc up to a megahertz or less.

1.8. DETECTOR PARAMETERS: DEFINITIONS

Before beginning our discussion of detectors, consider the parameters that describe how well the detectors perform. In this section we define these parameters in terms of the detector outputs and the radiometric inputs and other test conditions. Later—after we have described the various detection mechanisms—we discuss theoretical formulas that attempt to predict what these parameters will be. Later still we discuss the measurement of those parameters. The parameters used to characterize an IR detector are:

Responsivity. Electrical output for a given IR input

Noise. The "clutter" that tends to hide the true signal

Signal-to-Noise Ratio. A measure of the fidelity or "cleanliness" of a signal pattern

Noise Equivalent Power. The minimum infrared power a detector can accurately "see"

Specific Detectivity (D*). The signal-to-noise ratio that would result if the performance of your detector were scaled to a detector of a standard size, under standardized test conditions.

Linearity. How well the output signal "tracks" the infrared power

Dynamic Range. The range of IR signal levels for which the detector is useful

Frequency Response. How the responsivity changes with electrical frequency

Spectral Response. How the responsivity varies with the wavelength of the infrared power

Modulation Transfer Function (MTF). How the responsivity varies as smaller and smaller targets are focused on the detector

Crosstalk. Apparent signal from one detector due to a large signal on a nearby detector

1.8.1. Responsivity

The basic function of a detector is to convert radiant input to an output signal of some convenient type. For our purposes that output is always electrical—either a current or a voltage. The responsivity $\mathcal{R}$ is the ratio between the output signal and the radiant input (see Figure 1.6).

We will often work in terms of the *incidance E*—the flux density at our detector, exposed either in watts per square centimeter of detector area (W/cm^2) or photons per second per square centimeter [photons/($cm^2 \cdot s$)]. The radiant input is the product of the incidance and the detector area A_d.

DEFINITION OF RESPONSIVITY

$$\mathcal{R} = \frac{\text{signal output}}{\text{IR input}} \qquad (1.1)$$

$$= \frac{S}{EA_d}$$

Example: If a 30-mV signal results when a detector of area 25×10^{-6} cm^2 is exposed to an incidance of 120×10^{-6} W/cm^2, the responsivity is 10^7 V/W.

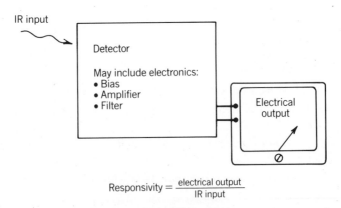

$$\text{Responsivity} = \frac{\text{electrical output}}{\text{IR input}}$$

Figure 1.6. Responsivity of a detector.

The incident power is

$$P = EA_d = (120 \times 10^{-6} \text{ W/cm}^2)(25 \times 10^{-6} \text{ cm}^2)$$

$$= 3 \times 10^{-9} \text{ W}$$

The responsivity is

$$\mathscr{R} = \frac{30 \text{ mV}}{3 \times 10^{-9} \text{ W}}$$

$$= 10^7 \text{ V/W} \qquad\qquad \blacksquare$$

Responsivity is an important parameter for a detector: It allows users to determine ahead of time how sensitive a measuring circuit they will require to "see" the expected output, or how much amplifier gain they need to get the signal levels up to a satisfactory level. Alternatively, it tells them how to determine from the output signal what the detected incidance level was.

It is most common to express the output signal in volts and the IR input in watts, so the usual units of responsivity are volts/watt (V/W). However, depending on the application and the customer's preferences, the output may be either current or voltage, and the IR input may be stated in terms of total power, power density, photon arrival rate, or photon flux density. The concept is still the same: Responsivity equals output divided by input. Watch the units carefully to make sure that you know what is intended.

In addition to the choices for measuring input and output, it is sometimes agreed to refer the output to what it would be if some different (idealized) circuit or spectral content were used. The nomenclature may warn you that some additional manipulation is meant.* Some examples are:

Short-circuit current responsivity
Open-circuit voltage responsivity
Peak spectral responsivity

---

* On the other hand, you cannot count on the use of proper or even consistent nomenclature—we often encounter vague or incorrect requirements. To be safe, get someone who knows what is meant to give you a sample calculation. If they say, "Well, you know, just the responsivity. . . ," you have a problem, because they don't know what they want.

Specific examples of this kind are discussed in the sections on circuits and spectral measurements.

Low responsivity is not itself an insurmountable problem; it is always possible to increase signal levels by adding amplifiers to the signal processing. A limitation that cannot be overcome with additional gain is the presence of noise.

1.8.2.   Noise

Noise refers to an electrical output other than the desired signal. It is unavoidable, but we strive to keep it as low as possible. Once noise enters the output, it can obscure or completely hide small signals (Figure 1.7). It is very difficult (or impossible) to find those signals again.

Some noise sources are fundamental, and for several reasons, cannot be avoided. A few of the reasons are:

- Photons do not arrive at an absolutely constant rate (the arrival rate fluctuates slightly).
- Atoms in the detector vibrate slightly, even at low temperatures.
- Electrons move randomly in the detector, not like well-drilled soldiers.

Other noise sources arise externally and can be eliminated if care is taken:

- Electrical interference: motors, ac power lines, and so on
- Temperature fluctuations
- Vibrations that cause electrical components to shift

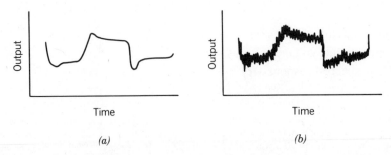

(a)   *(b)*

Figure 1.7.   Noise is "clutter" that degrades signal fidelity: (*a*) with little noise; (*b*) with more noise.

Since noise is a random deviation from the average signal output, some convention is required to decide how to assign a single number to a given noise pattern. The usual definition is the root-mean-square (rms) deviation: the square root of the mean (average) of the square of the deviation as it varies in time. (The squaring process is necessary since noise goes negative as often as it goes positive: a simple average would yield zero.) Meters and digital circuits are available that calculate rms values automatically. Formulas for calculating noise from sampled data are provided in Chapter 5.

Frequency Spectrum of Noise. Some noise components will appear at very specific frequencies: A pump will vibrate a dewar a few times every second, and an ac power line near a critical amplifier will introduce noise at the line rate—60 cycles per second (see Figure 1.8). The more fundamental noise sources, however, will add some noise more or less uniformly at all frequencies. This is referred to as *white noise*, by analogy to the fact that white light contains all wavelengths (frequencies) of light. Even if the noise is not quite white, it will generally contain a wide range of frequencies, and we can reduce the noise by eliminating unnecessary frequencies. For example, if our signal is at 200 Hz (200 cycles per second), eliminating all electrical output above 250 Hz and below 150 Hz would help improve the noise situation. The remaining 100-Hz electrical bandwidth would contain less noise than did the original wideband.

Noise Spectral Density. It can be shown that for white noise, the total noise voltage or current is proportional to the square root of the bandwidth. Even if the noise is not "pure white," the square root of bandwidth rule is often used:

$$\text{``in-band'' noise} \propto \sqrt{\Delta f} \qquad (1.2a)$$

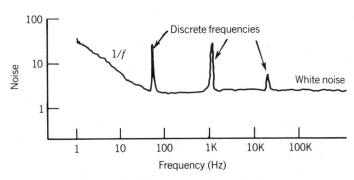

Figure 1.8. Frequency spectrum of noise.

Example: If the original bandwidth in the case mentioned above was 40 kHz, we would have reduced the noise by a factor of about 20 when we reduced the bandwidth to 100 Hz:

$$\sqrt{40 \text{ kHz}} = \sqrt{4 \times 10^4 \text{ Hz}} = 200 \text{ Hz}^{1/2}$$

$$\sqrt{100 \text{ Hz}} = 10 \text{ Hz}^{1/2}$$ ∎

To compare noise in various bandpasses, we use the noise spectral density n. It is the noise N divided by the square root of the bandpass Δf:

DEFINITION OF NOISE SPECTRAL DENSITY

$$n = \frac{N}{\sqrt{\Delta f}} \qquad\qquad (1.2b)$$

This is the noise spectral density, and it has units of volts per root hertz ($V/Hz^{1/2}$). Its numerical value is the noise that would occur if the electrical bandpass were reduced to 1 Hz.

Example: If a noise of 7.5 μV is observed in a 50-Hz bandwidth, the average noise spectral density in that bandpass is about 1.1 $\mu V/Hz^{1/2}$:

$$n = \frac{N}{\sqrt{\Delta f}}$$

$$= \frac{7.5 \ \mu V}{\sqrt{50 \text{ Hz}}}$$

$$\simeq \frac{7.5 \ \mu V}{7 \text{ Hz}^{1/2}}$$

$$\simeq 1.1 \ \mu V/Hz^{1/2}$$ ∎

1.8.3. Signal-to-Noise Ratio

The *signal-to-noise ratio* (*S/N*) is a simple way to describe the "cleanliness" of a given signal level. It is simply the signal voltage divided by the rms noise voltage. An oscilloscope trace with an *S/N* ratio of 100 or more is a very clean pattern, with negligible noise. An *S/N* ratio of 10 is a little fuzzy, but the pattern is still very clear. A ratio of 3 is pretty bad, and at 1 the signal is nearly lost. These are illustrated in Figure 1.9.

Note that *S/N* does not characterize the detector itself: We can get a

$S/N = 100$

$S/N = 10$

$S/N = 3$

$S/N = 1$

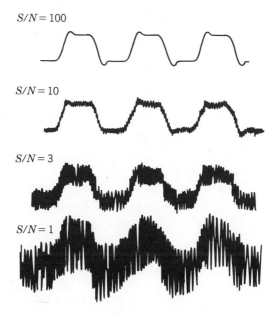

Figure 1.9. Detector output with varying signal-to-noise ratios.

better S/N for the same detector just by applying a higher incidance level. *S/N does* describe the conditions under which you are working: If you are trying to take data with a S/N ratio of 3, I would not trust your results as much as if you had a S/N of 30, or better yet, 300.

1.8.4. Noise Equivalent Power

The *noise equivalent power* (NEP) is a measure of the ultimate sensitivity of a given detector, and it is a convenient number to use to estimate what your S/N ratio will be if you know the power available. NEP is the power tha must fall on the detector to cause an S/N of 1. If you have a power available 10 times the NEP, your S/N will be 10.

NEP is determined by dividing the system noise by the responsivity (using the output per power input definition of responsivity):

DEFINITION OF NOISE EQUIVALENT POWER

$$\text{NEP} = \frac{\text{noise}}{\text{responsivity}} \tag{1.3}$$

The units of NEP are watts.

Example: For a detector with 7.5 μV of noise and responsivity 10^7 V/W, the NEP is 7.5×10^{-13} W:

$$\text{NEP} = \frac{7.5 \text{ μV}}{10^7 \text{ V/W}}$$

$$= 7.5 \times 10^{-13} \text{ W} \qquad \blacksquare$$

A variant uses the noise spectral density instead of the noise in the NEP formula:

NEP PER UNIT BANDWIDTH

$$\text{NEP}' = \frac{\text{noise spectral density}}{\text{responsivity}} \qquad (1.4a)$$

$$= \frac{\text{noise}}{\sqrt{\Delta f} \times \text{responsivity}} \qquad (1.4b)$$

This "NEP per unit bandwidth," is not a standard figure of merit, and confusion can arise if it is not clear that the standard NEP definition is not being used. The units are watts per square root hertz ($\text{W/Hz}^{1/2}$).

Example: If the noise in a 50-Hz bandwidth is 7.5 μV and the responsivity is 10^7 V/W, the NEP (per unit bandwidth) is 1.1×10^{-13} $\text{W/Hz}^{1/2}$:

$$\text{NEP}' = \frac{7.5 \text{ μV}}{\sqrt{50 \text{ Hz}} \times 10^7 \text{ V/W}} \simeq 1.1 \times 10^{-13} \text{ W/Hz}^{1/2}$$

$$\blacksquare$$

If this parameter is used, it is especially important to include the units to help minimize confusion.

1.8.5. Specific Detectivity

The NEP formula (1.3) is convenient for predicting the minimum power a given system can detect, but it has some undesirable features: (1) A good detector will have a small NEP, and (2) detectors of different sizes will have different NEPs, so we cannot say in general what a good NEP should be unless we specify the size of the detector.

The *specific detectivity* (D^*) (now normally just called *detectivity* or "D-star") eliminates those two drawbacks. A large D^* is good, and for a given environment all good detectors should have about the same D^*. D^* is the responsivity times the square root of the area, divided by the noise spectral density:

DEFINITION OF SPECIFIC DETECTIVITY

$$D^* = \frac{\text{responsivity} \times \sqrt{\text{area}}}{\text{noise}/\sqrt{\Delta f}} = \frac{\mathcal{R} \sqrt{A_d}}{N/\sqrt{\Delta f}} \qquad (1.5a)$$

$$= \frac{\text{signal} \times \sqrt{\Delta f}}{\text{noise} \times \text{incidance} \times \sqrt{\text{area}}} = \frac{S \sqrt{\Delta f}}{NE\sqrt{A_d}} \qquad (1.5b)$$

The units of D^* are cm·Hz$^{1/2}$/W.

Example: If a 25×10^{-6}-cm^2 detector has a responsivity of 10^7 V/W and a noise of 7.5 μV in a 50-Hz bandwidth, the D^* is about 4.7×10^{10} cm·Hz$^{1/2}$/W:

$$D^* = \frac{10^7 \text{ V/W} \times \sqrt{25 \times 10^{-6} \text{ cm}^2}}{7.5 \ \mu\text{V}/\sqrt{50 \text{ Hz}}}$$

$$\approx 4.7 \times 10^{10} \text{ cm·Hz}^{1/2}/\text{W} \qquad \blacksquare$$

D^* is useful in predicting S/N in a given test environment:

$$S/N = D^* \frac{\text{incidance} \times \sqrt{\text{area}}}{\sqrt{\Delta f}} \qquad (1.5c)$$

Example of the Use of D*: Suppose that we had a detector of the same material as the one in the preceding example, but with an area of 100×10^{-6} cm^2. If we used it with a circuit whose noise bandwidth was 200 Hz, and in an incidance of 10 μW/cm^2, the resulting signal-to-noise ratio would be about 335:

$$S/N = D^* \frac{\text{incidance} \times \sqrt{\text{area}}}{\sqrt{\Delta f}}$$

$$\approx 4.7 \times 10^{10} \text{ cm·Hz}^{1/2}/\text{W} \ \frac{10 \ \mu\text{W/cm}^2 \times \sqrt{100 \times 10^{-6} \text{ cm}^2}}{\sqrt{200 \text{ Hz}}}$$

$$\approx 335 \qquad \blacksquare$$

1.8.6. Linearity and Saturation

Detector outputs will increase linearly with input signal over some range of incidances, and fail to be linear at some large incidance. That is, a plot of electrical signal versus radiometric input will start out as a straight line but will eventually level off (see Figure 1.10). Linearity describes the exactness with which this is true. Linearity requirements are not always specified in the same way, but one way would be to state how far a graph of measured signal verus incidance would be allowed to deviate from the best-fit straight line.

A major source of deviation from linearity is saturation of the detector or electronic circuit. By this we mean that as the incidance and signal increase, some physical constraint is reached, and the signal cannot continue to increase. This could be due to the electrical breakdown of the detector itself, or of some electronic components. More generally, there is an amplifier in the circuit that will put out only a limited voltage. For example many operational amplifiers are powered by 6-V supplies and cannot support more than a 6-V output signal.

Dynamic Range. The dynamic range is the ratio of the highest useful signal to the lowest measurable signal. *Highest useful signal* might be defined as the point at which a given linearity specification is exceeded, and *lowest measurable signal* might be the signal at which $S/N = 1$. Other criteria are possible, however.

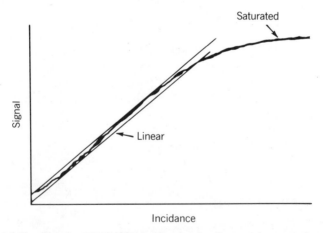

Figure 1.10.   Signal versus incidance plot showing linear and saturated regions.

1.8.7.   Frequency Response

The detector or system (detector plus electronics) will generally respond equally well to a range of low frequencies, but will not respond as well to higher frequencies. The frequency response is described by a plot of responsivity versus frequency (Figure 1.11). It can also be described by providing the corner frequency, in which case one implies that the frequency response curve obeys an equation of the form

$$S = S_0 \frac{1}{[1 + (f/f_c)^2]^{1/2}} \tag{1.6}$$

where f_c is the corner frequency and S_0 is the signal at low frequencies.

Example: Given a detector with a 1200-Hz corner frequency and a signal of 100 mV at 25 Hz, the signal at 1800 Hz is about 55.5 mV.

S_0 is 100 mV (25 Hz qualifies as a low frequency since it is well below the corner frequency), so

$$S = 100 \text{ mV} \times \frac{1}{[1 + (1800/1200)^2]^{1/2}}$$

$$= 100 \text{ mV} \times \frac{1}{(3.25)^{1/2}}$$

$$\approx 55.5 \text{ mV} \qquad \blacksquare$$

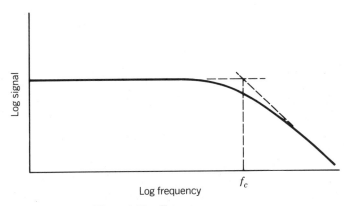

Figure 1.11.   Frequency response.

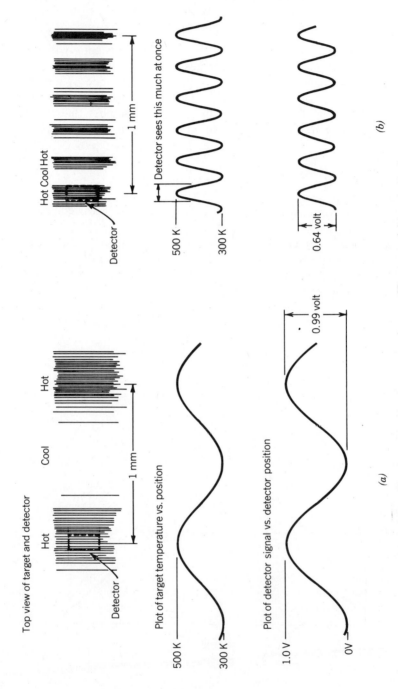

Figure 1.12. Signal generated by detector moving over IR sources: (*a*) target repeats once every millimeter (k = 1 ~/mm); (*b*) target repeats five times every millimeter (k = 5 ~/ mm); (*c*) MTF of the detector.

20

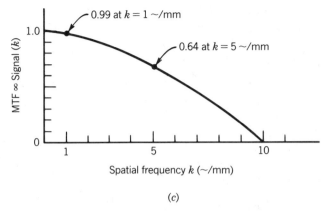

(c)

Figure 1.12.   (*Continued*)

When $f = f_c$, $S = S_0/\sqrt{2} = S_0 \times 0.707$. This is sometimes expressed by saying that at the corner frequency "the signal is down by 3 dB," and the corner frequency is sometimes referred to as the *3-dB frequency.**

1.8.8.   Spatial Considerations

Uniformity of Response. For some applications it will matter whether the detector responsivity is uniform over its surface or whether some parts are more sensitive than others. Actually, no detector is *perfectly* uniform, so the question really is how nonuniform it is, and how much nonuniformity the customer can tolerate. A response profile (a plot of responsivity versus position) is one way to report uniformity.

Modulation Transfer Function (MTF). Consider a target made up of alternating hot and cold lines, with the hot lines repeating at some frequency k every millimeter: 2 lines per millimeter, or 5 lines per millimeter, or 20 lines per millimeter. We call k the *spatial frequency* of the target, and the MTF (at spatial frequency k) is a measure of how much the detector signal will vary (be modulated) as we scan a target with that spatial frequency.†
For the example of Figure 1.12, note how the signal will fall off as k exceeds one cycle per millimeter. This detector will not "see" 0.1-mm features very well.

---

* See Appendix C for a description of the dB (decibel) system.

† Actually, the MTF definition is based on sine-wave targets; however, some MTF measurements are made with bar pairs (a hot stripe and a cold stripe), so "lines per millimeter" is an acceptable way to begin to understand spatial frequency.

MTF is sometimes called the spatial equivalent of the frequency response of a detector, or the *spatial frequency response*. It is a measure of how well the detector can sense small details. MTF depends on detector size and the uniformity of response across the detector. A small detector with uniform response will have better MTF and be able to see smaller features than will a large one, or one with a response that tapers off near the detector edges.

1.8.9. Crosstalk

If one has an array of detectors and images a spot on one detector, there should be no signal from the other detectors. In practice, some signal will be present on the others, although it should be very small. This excess signal is said to be due to *crosstalk*. Crosstalk is generally measured as a percentage of the input or driving signal. A requirement that crosstalk be less than 5% would be fairly easy to meet, and a requirement of 0.05% would be very hard to meet.

Crosstalk can be due to optical effects (e.g., reflections from the detector on which the spot was focused) and electrical effects (e.g., capacitance between the signal leads). These are illustrated in Figure 1.13.

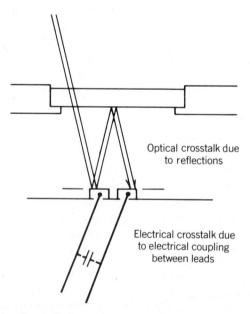

Figure 1.13.   Crosstalk due to optical reflections and electrical coupling.

REFERENCES

Eisberg, R. M. (1961). *Fundamentals of Modern Physics,* Wiley, New York.

Hudson, Richard D., Jr. (1969). *Infrared System Engineering,* Wiley, New York.

Wehr, Russell M., and James A. Richards, Jr. (1960). *Physics of the Atom,* Addison-Wesley, Reading, Mass.

SUGGESTED READINGS

Dereniak, Eustace L., and Devon G. Crowe (1984). *Optical Radiation Detectors,* Wiley, New York.   Excellent and up-to-date references; covers radiometry, detectors, and charge transfer devices.

Hudson, Richard D., Jr. (1969). *Infrared System Engineering,* Wiley, New York.   This is a standard reference in current IR labs and will be referred to often in this book. The primary limitation for our purposes is that it is directed more at the systems level than at the component level we are covering here.

RCA Electro Optics and Devices (1974). *RCA Electro Optics Handbook,* RCA, Lancaster, Pa.

Wolfe, William L., and George J. Zissis, eds. (1978). *The Infrared Handbook,* Environmental Research Institute of Michigan, Ann Arbor, Michigan.   An excellent handbook, full of details about all facets of IR.

PROBLEMS

1. Sort the following types of electromagnetic radiation into order by wavelength, with the shortest wavelength first:

Radio	Infrared
Near-infrared	Red
Gamma rays	Blue
Ultraviolet	X-rays

2. Name the three methods of heat transfer, and describe an everyday sensation or experience with each.

3. What phenomenon does Planck's law describe?

4. Describe briefly how the distribution of thermally radiated energy varies with wavelength and source temperature.

5. Invent a thermal detector of some sort: not necessarily convenient, nor repeatable, nor salable, but one that would detect electromagnetic energy.

6. What are two ways of thinking about electromagnetic energy?

7. What is the difference between a thermal and a quantum detector?

8. Name the detector parameters.

(a) Electrical output for a given IR input

(b) The signal-to-noise ratio that would result if the performance of your detector were scaled to a detector of a standard size, under standardized test conditions

(c) How the responsivity changes with the frequency at which the infrared power varies

(d) The "clutter" or unwanted electrical variation that tends to hide the true signal

(e) The minimum infrared power that a detector can accurately "see"

(f) A measure of the "cleanliness" of a signal pattern

(g) How the responsivity changes with infrared power

(h) How the responsivity varies with the spatial contrast of the infrared source

(i) How the responsivity varies with the wavelength of infrared power

9. Power input   = 100 μW
   Detector output = 30 mV
   Responsivity   = ?

10. Input   = 330 × 10^{12} photons/(cm^2·s)
    Detector area   = 1 mm × 1 mm
    Detector output = 770 mV
    Responsivity   = ?

11. Incidance   = 35 μW/cm^2
    Detector: round; diameter = 0.002 in.
    Power to the detector   = ?

2

Detector Types, Mechanisms, and Operation

In this chapter we discuss the most common types of infrared detectors, describe briefly how they operate, and show how to predict figures of merit for them. Convenient sources of more detailed information on IR detectors are Chapter 11 of the *Infrared Handbook* (Limperis, 1978) and Section 4 of the *Handbook of Optics* (Jacobs, 1978).

Some common detector types are grouped in Figure 2.1. Many of these detectors are very similar in their manufacture and operation; for example, doped germanium and doped silicon detectors are one "family," the lead salts another. The differences and similarities in these detectors are discussed in the material that follows. For now, recognize that the thermal detectors are quite different from the photon detectors, and that among the photon detectors there are two major types: photoconductors and photovoltaics.

2.1. THERMAL DETECTORS

Thermal detectors convert radiated power into a more readily measured parameter. As an example, consider how a thermometer could be used as a thermal detector: We could paint the bulb area black* so that it would absorb radiation well and allow radiant energy to fall on the bulb. The mercury would warm, expand, and we could measure the height of the mercury column to tell how much power was arriving on the "detector."

A *bolometer* is a thermal detector whose resistance depends on its temperature. Since the resistance of most semiconductors is a strong function of temperature, the resistance of a semiconductor chip can tell us how much radiant energy is falling on it.

---

* The blackening of thermal detectors determines how well they absorb radiation. An *ideal* black will absorb all wavelengths equally well, and the response of a ideal thermal detector to a given power will be the same at all wavelengths.

Thermal		Golay cell Bolometer Thermopile
Photon (quantum)	Photoconductors	Ge : Au, Ge : Hg, . . . Si : As, Si : Sb, . . . HgCdTe PbS, PbSe
	Photovoltaics	InSb HgCdTe

Figure 2.1. Types of detectors.

The *Golay cell* uses the expansion of gas when heated to sense radiated power: The gas is contained in a chamber that is closed with a reflecting membrane. When the gas is warmed, the membrane distorts and deflects a beam of light that has been focused on it. The movement of the reflected light is a measure of the incidance on the cell. See Zahl and Golay (1946), Golay (1947, 1952), Schuman (1965), and Hennerich et al. (1966) for further information.

Thermopiles are IR-detecting thermocouples. A thermocouple generates a voltage that depends on its temperature, so an increase in the IR power causes an increase in the thermopile voltage. These are discussed by Stevens (1970).

2.1.1. Bolometers

To illustrate some principles of thermal detectors, consider the carbon bolometer shown in Figure 2.2. Such devices are very sensitive, relatively rugged, and detect radiation over a very wide spectral range. They can be calibrated directly by measuring the resistance changes that result from changes in the electrical bias power.

The resistance of an ordinary carbon resistor is a strong function of temperature—in fact, a carbon resistor can be used as an inexpensive temperature sensor below about 77 K. To make a bolometer, we would mount the resistor in such a way that it is cooled to near liquid helium temperature but isolated somewhat from its cold sink. As radiation strikes it, it would warm, the resistance would decrease, and an external electrical monitor would detect the resistance change.

To make the bolometer more sensitive, we would want to make its

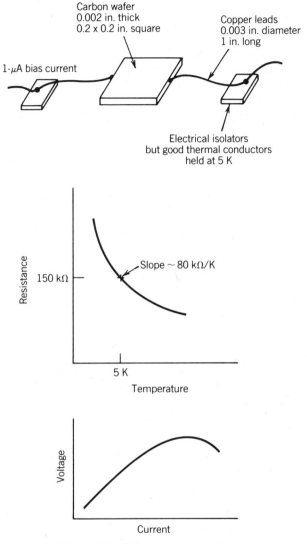

Figure 2.2. Bolometer construction.

heat capacity smaller so that a small amount of energy could heat it faster. To do this we could grind off the coating of the resistor and slice a small section of the carbon material.

Figure 2.2 shows such a slice suspended from leads that provide electrical contacts and a cooling path with a controlled heat conductance.

When no radiant signal is present, the detector would be very cold, but a small radiant load would heat the detector quickly.

Construction and performance of bolometers was described by the early users: Boyle and Rogers (1959) worked with carbon, and Low (1961, 1966) used gallium doped germanium. Bolometer theory is discussed by Strong and Lawrence (1968).

Responsivity of a Bolometer. The *responsivity* of the detector is the ratio between the output signal and the detected radiant power. If the thermal impedance of our mounting structure is 3×10^4 K/W, a 1-μW radiated input would cause a temperature change of 0.03 K: $(3 \times 10^4$ K/W$) \times (1 \times 10^{-6}$ W$)$. If the slope of the resistance temperature curve is 80 kΩ/K, an increase of 0.03 K in temperature would decrease the resistance by 2.4 kΩ: (80 kΩ/K) $\times$ (0.03 K).

If we have 1 μA running through the resistor, the change in the voltage across the resistor would be 2.4 mV. Finally, the responsivity is the ratio of that change to the input radiant power that caused it. For our detector, the responsivity is 2400 V/W:

$$\frac{2.4 \text{ mV}}{1 \text{ } \mu\text{W}} = 2400 \text{ V/W}$$

Frequency Response of a Bolometer. The *time constant* τ for such a detector is the product of the thermal impedance of its mounting structure and the heat capacity of the sensor. (This is a thermal equivalent of the electrical *RC* time of a capacitor–resistor circuit.) For our example, the thermal impedance is about 3×10^4 K/W, and the heat capacity is 1×10^{-6} J/K. The response time is the product of those two: 3×10^{-2} s or 30 ms. This detector would not respond well to changes in radiant energy that occur in much less than 30 ms.

A measure of the frequency response of the detector is the *corner frequency*, given by

$$f_c = \frac{1}{2\pi\tau} \tag{2.1}$$

Example: For our hypothetical detector, τ is 30 ms, so f_c is about 5 Hz:

$$f_c = \frac{1}{2\pi \times (30 \times 10^{-3} \text{ s})}$$

$$\approx 5 \text{ Hz} \qquad\qquad\blacksquare$$

Improvements to a Bolometer. The responsivity of a thermal detector can be increased by increasing the thermal impedance of its mounting structure, but that also makes it slower. The responsivity can be increased and the sensor made faster by decreasing the heat capacity. The primary limitation there is a mechanical one: In this case, how thin can one make the carbon chip?

A little more detailed information about the thermal performance of a bolometer is shown in Figure 2.3. The bolometer can be thought of as the thermal analog of an electrical RC circuit in which the capacitor is charged by a current source. The time constant of the bolometer is the RC time constant, for both charging (heating) and decay (cooling). When a bolometer is initially irradiated, the signal versus time appears asym-

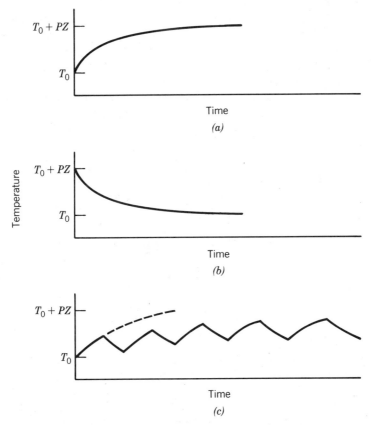

Figure 2.3. Bolometer temperature verus time: (*a*) IR power applied at $t = 0$. (*b*) IR power removed at $t = 0$; (*c*) chopped IR power applied at $t = 0$.

metric, heating more than it cools. After a few cycles, however, the steady-state symmetric pulse is obtained. Although we have discussed only one type of thermal detector, the same principles apply to all other thermal detectors.

2.2. PHOTON (QUANTUM) DETECTORS

Quantum or photon detectors generally respond to incident radiation on an electronic level. It is not necessary to heat the entire atomic structure, so these detectors are much faster than thermal detectors. The primary mechanism is dependent on the concept of semiconductor *band gaps*, as shown in Figure 2.4.

It can be shown (but it is too involved to try to do it here) that for semiconducting materials (germanium, silicon, indium antimonide, and others) the electrons can have low energies (the valance band electrons) or higher energies (the conduction band electrons) *but cannot have energies in between.* Thus there is a *gap* in the allowed energies. At low temperatures, with no incident radiation, all of the electrons are in the lower bands. It can be shown (but, once again, not here!) that when all

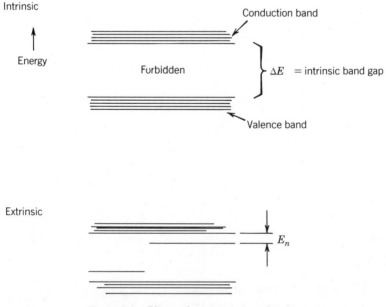

Figure 2.4. Photon detectors: energy levels.

the carriers are in the valence (lower) bands, no conduction occurs. Thus at low temperatures the electrical resistance of semiconductors is very high. At higher temperatures, motion of the atoms can give some electrons enough energy to "jump" up into the conduction bands, where they conduct electricity. The result is a characteristic decrease in electrical resistance as the temperature increases.

The energy gaps described above are characteristic of all semiconductors. Semiconductor detectors that rely on that gap for their operation are called *intrinsic detectors*. Additional energy levels can be created by deliberately adding impurities; this is referred to as *doping* the material, and the resulting detectors are called *extrinsic detectors*.

2.2.1. Carrier Generation

Carriers can be given enough energy to cross the gap into the conduction bands if they collide with an incoming "packet" of electromagnetic radiation—a photon. This generation of carriers in the conduction band is the detection mechanism for both the photoconducting (PC) and photovoltaic (PV) infrared detectors. Although the responsivity expression is different for PC and PV detectors, the rate at which carriers are generated is the same for both. The derivation of the carrier generation rate is indicated in Figure 2.5. A typical bulk detector is shown.

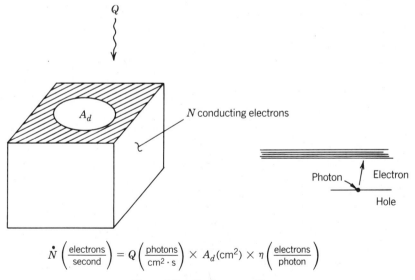

$$\dot{N}\left(\frac{\text{electrons}}{\text{second}}\right) = Q\left(\frac{\text{photons}}{\text{cm}^2 \cdot \text{s}}\right) \times A_d(\text{cm}^2) \times \eta\left(\frac{\text{electrons}}{\text{photon}}\right)$$

Figure 2.5. Carrier generation.

Quantities of importance are:

Q: The *photon incidance*—the number of photons arriving each second per square centimeter of detector area. (We will later distinguish between signal photon incidance Q_s and background photon incidance Q_{bk}, but for now there is no need to specify which we are seeing.)

A_d: The effective detector area, sometimes limited with a mechanical aperture. At this point the actual *detector* area does not matter; all we need is the area through which IR can reach the detector.

s: The spacing between the electrodes.

a: The contact area.

The photon arrival rate is the photon incidance Q times the illuminated area A_d. The *carrier* generation rate $\dot{N}$ (read as "N dot") is the photon arrival rate times the probability that a photon will create a carrier. We call this last quantity the *quantum efficiency* η, so

$$\dot{N} = \eta Q A_d \qquad (2.2)$$

2.2.2. Spectral Response of Photon Detectors

Before going on to discuss the details of photoconductors and photovoltaic detectors, it is convenient to discuss another characteristic they share— their spectral response. By *spectral response* we mean how the responsivity of the detector varies with the wavelength of the detected radiation. Thermal detectors would respond equally well to 1 mW of incident radiation, no matter what its wavelength. Photon detectors are more discriminating.

To explain the spectrum of thermal radiation, it was necessary to assume that each photon of wavelength λ carries energy E (see box with Equation 2.3a).

ENERGY OF A PHOTON (JOULES)

$$E = h\nu = \frac{hc}{\lambda} \simeq \frac{20 \times 10^{-20} \text{ J}\cdot\mu\text{m}}{\lambda} \qquad (2.3a)$$

where h = Planck's constant (about 6.6×10^{-34} J·s)
c = speed of light (about 3.0×10^{14} μm/s)
λ = wavelength (μm)

For example, a photon of 1 μm wavelength has an energy of 20×10^{-20} J; for 2 μm the energy is 10×10^{-20} J.

It is convenient to express semiconductor energy levels in terms of the *voltage* difference which would accelerate an electron to that energy: $E = e \times V$, where e is the charge on the electron—about 1.6×10^{-19} C. Doing this yields Equation 2.3b.

ENERGY OF A PHOTON (ELECTRON-VOLTS)

$$E = \frac{hc}{e\lambda} \simeq \frac{1.24 \text{ eV·μm}}{\lambda} \qquad (2.3b)$$

For example, every photon of wavelength 5.5 μm has an energy of 0.22 eV. Photons of *longer* wavelength will have *less* energy.

Cutoff Wavelength. Now if a photon does not have enough energy to "kick an electron across the band gap," the electron cannot absorb *any* energy from the photon, and the photon will pass through the material undetected. InSb has a band gap of 0.22 eV. Photons of wavelength 5.5 μm have just barely enough energy to generate a conduction electron in InSb, and photons with *longer* wavelengths will pass through InSb undetected. InSb is opaque to wavelengths less than 5.5 μm because it absorbs those photons. It is transparent beyond 5.5 μm—the photons pass through it.

Thus we expect to see a *cutoff wavelength* in our spectral response curves: a wavelength beyond which the detector is effectively blind. This is indeed seen in both photoconductors and photovoltaic detectors.

Responsivity versus Wavelength. If we define responsivity as the output (voltage, current, or whatever) per *photon* detected, the responsivity of ideal photon detectors is "flat" (independent of wavelength) up to the cutoff wavelength. Thus it is convenient to define responsivity in terms of photons.

For historical reasons, though, responsivity (including spectral response) is often reported as output per incident *watt*. When reported that way, the photon detectors show a spectral response curve that increases with wavelength (until the cutoff is reached). That is, they evidently generate a larger signal for 1 W at 5 μm than they do for 1 W at 4 μm. To see why, calculate the number of photons per second required to carry

1 W at several wavelengths. The calculation and the resulting graph are shown in Figure 2.6.

If we define *responsivity* as the output signal per *power* (watt) absorbed, we would expect to see a higher responsivity at long wavelengths because there are more photons for a given power at longer wavelengths. Note that this differs from the flat response of a thermal detector.

2.2.3. Temperature Dependence of Semiconductor Carrier Density

The carrier density or concentration (carriers per cubic centimeter) for a semiconductor material is a very predictable function of temperature. A general understanding of that dependence shows why detectors must be cooled, how the operating temperature depends on background, and how

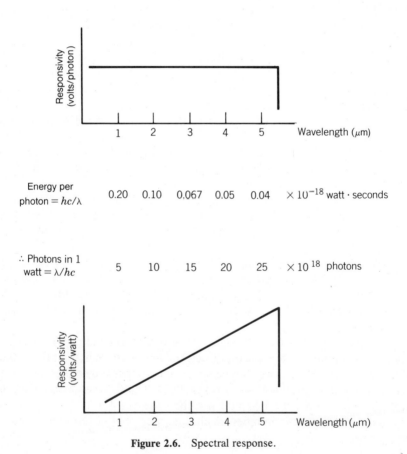

Figure 2.6. Spectral response.

resistance versus temperature is used to characterize semiconductor materials.

The temperature dependence of the carrier concentration due to thermal excitation has the form

$$n = n_0 \exp\left(-\frac{E}{kT}\right) \qquad (2.4a)$$

or

$$\log n = \log n_0 - A\frac{1}{T} \qquad (2.4b)$$

Equation $(2.4b)$ is easy to plot: It is a straight line on semilog paper if n is plotted on the vertical (log) axis and $1/T$ (or $1000/T$) is plotted on the horizontal axis.

Figure 2.7 illustrates the temperature dependence of a doped semiconductor. Begin at point A, the lowest temperature shown. Very few

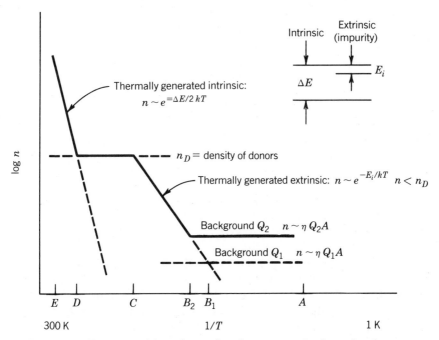

Figure 2.7. Temperature dependence of carrier concentration in semiconductors.

carriers present are due to thermal generation; the concentration is due to the photon flux and is independent of temperature. An increase in the photon flux (to Q_2) will increase the carrier concentration. This is the regime in which best detection occurs.

At background Q_1 we can increase the temperature to point B_1 without increasing the number of carriers, but above point B_1 thermally generated donors from the impurity level begin to outnumber the photon-generated carriers. The detector will no longer be *background limited*. If the background is higher (Q_2), there are more background-generated carriers, and we can operate up to a slightly higher temperature (point B_2) before thermal generation dominates. Beyond B_2 the thermally generated carriers increase rapidly with temperature until all the impurities are ionized and conducting, and there are no more to be generated—point C.

It is not until point D is reached that there are enough carriers generated thermally across the *intrinsic* energy gap to really affect the carrier concentration. Once point D is reached, however, the number of carriers increases rapidly with temperature as we move toward point E.

Experimental $n(T)$ data act as a source of information about the material: We can infer the intrinsic band gap from the slope in region D–E, the impurity concentration from n in region C–D, the impurity level from the slope of region B–C, and the photon arrival rate from n in region A–B. For detector operation, remember that we want to work in a region where the carrier concentration is dominated by the photon arrival rate, not by thermal effects. To work at reduced backgrounds will require that we work at lower temperatures—point B_1 versus point B_2.

2.2.4. Types of Photon Detectors

There are two distinct types of photon detectors: photoconductive (PC) and photovoltaic (PV) detectors. They differ in their construction and in the electrical sensing circuits with which they are used. Their names are descriptive: photoconductors are poor *conductors* whose conductivity is made better by the presence of the photon-generated carriers. (People sometimes think in terms of a *resistance* change instead of a conductance change. It is true that both resistivity and conductivity change, but it is a little easier to derive the equation for the conductivity increase than for the resistivity decrease, so keep thinking photo*conductor*.) Photoconductors are passive devices; they do not create a voltage by themselves. Photovoltaic detectors are diodes that generate an electromotive force when photons are detected. This electromotive force generates a current and voltage—photo*volt*aic.

2.2.5. Diodes and Photovoltaic Detectors

Photovoltaic (PV) detectors are diodes made from materials with a band gap less than the energy of IR photons. Before discussing PV detectors, consider briefly the construction and characteristics of an ordinary semiconductor diode.

Semiconductor Diodes. A semiconductor diode is made by joining two pieces of semiconductor material. One side is *n-type* or donor material; it contains a dopant with loosely attached electrons. This material tends to lose those electrons, donating those negative charged carriers to the conduction process. The other side is *p-type* or acceptor material; It contains a dopant that has a "need" or empty space for electrons, and tends to accept electrons from the conduction process. The "spaces" can be thought of as positively charged "holes" that move in the direction opposite to that of the electrons that fill the holes.

Immediately after the two semiconducting materials are joined, loose electrons diffuse into the *p*-type material, causing an electric field that keeps more electrons from following them. This is the diode. The construction of a diode and the electric field are shown in Figure 2.8.

For the diodes *as normally used in electronics*, the open-circuit voltage and the short-circuit current are zero. That is, if we put a voltmeter across the open leads of a diode, we see no voltage, and if we short the two leads from the diode, no current flows. (When exposed to IR photons, our PV detectors will generate an open-circuit voltage and short-circuit current.)

The normal current voltage relation for an ideal diode with no photon-generated carriers is shown in Figure 2.8 and given by the Shockley equation (Sze, 1969, p. 100):

$$I = I_s \left[\exp \left(\frac{eV}{kT} \right) - 1 \right] \tag{2.5}$$

where I_s = saturation current
   e = charge on the electron
   V = applied voltage
   T = temperature (kelvin)
   k = Boltzmann constant (about 1.38×10^{-23} J/K)

We will not need the theoretical equation for I_s. Instead, we will work in terms of the *dynamic impedance* R_0—the slope of a voltage versus current plot, measured at V equals zero. In practice we will plot current

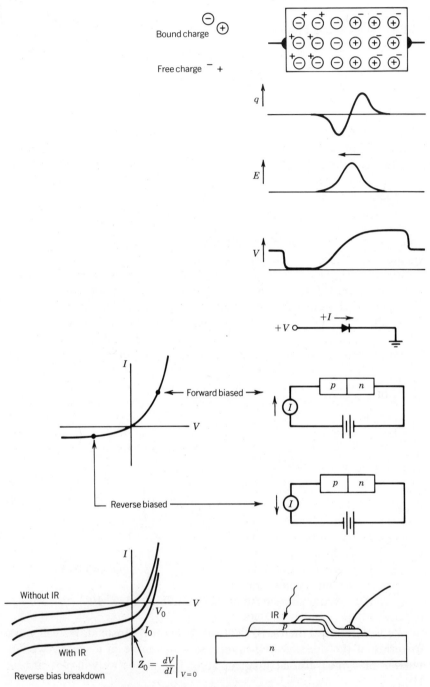

Bound charge

Free charge − +

q

E

V

$+V$ o———▶|——— $+I \longrightarrow$

I

←— Forward biased —▶

V

Reverse biased ———▶

p n

I

p n

I

I

Without IR

V

V_0

I_0

With IR

Reverse bias breakdown

$$Z_0 = \left. \frac{dV}{dI} \right|_{V=0}$$

IR

p

n

Figure 2.8. Diodes and photovoltaic detectors.

38

versus voltage, and R_0 is the reciprocal of the slope (at $V = 0$) when plotted that way:

$$R_0 = \left(\frac{dI}{dV}\right)_0^{-1} \qquad (2.6)$$

We can eliminate the saturation current by using the relationship between R_0 and I_s. Let

$$\frac{1}{R_0} = \left.\frac{dI}{dV}\right|_{V=0} = \left.\frac{I_s e}{kT} \exp\left(\frac{eV}{kT}\right)\right|_{V=0} = \frac{I_s e}{kT}$$

so

$$I_s = \frac{kT}{eR_0} \qquad (2.7)$$

We will see that the detector noise is dependent on the impedance R_0. R_0 will be measured as part of a normal diode characterization, or we can use typical data as we try to predict detector performance.

I_s is proportional to the diode area, so $R_0 A$ is independent of the diode area. Typical values of $R_0 A$ will be provided later for InSb and HgCdTe diodes. From these and the detector area we can estimate R_0 values for given background and temperatures.

Photovoltaic Detectors. Any material that can work as a photoconductive detector can, in theory, be made into a photovoltaic detector. For practical reasons Si, InSb, and HgCdTe are the common photovoltaic detector materials. As infrared radiation passes near the junction, it is absorbed and gives an electron enough energy to reach the conduction band. This generates an electron in the conduction band and leaves behind a hole that can also contribute to conduction. This process is sometimes referred to as "kicking" the electron into the conduction band, or as creating an *electron–hole pair*. The presence of an electron–hole pair changes the current–voltage relationship of the diode. That change can be monitored to provide infrared detection.

Advantages of PV detectors over PC detectors include a better theoretical limit to the signal-to-noise ratio, simpler biasing, and more accurately predictable responsivity. Photovoltaic detectors are generally more fragile than photoconductors. They are susceptible to electrostatic discharge and to physical damage during handling. Because they are so thin [10 μm (0.0004 in.)] for backside illuminated InSb], the insulating layers are susceptible to electrical punchthrough. Surface effects can allow leakage between the p and n regions that will degrade detector performance, and much of the technology for PV detectors is related to control of that surface.

2.2.6. Photoconductive Detectors

Photoconductive detectors include the doped germanium and silicon detectors (Ge:Xx, Si:Xx), and the lead salts (PbS & PbSe). The terniary compounds HgCdTe and PbSnTe can be used for PC detectors as well as PV. Responsivities of the PC detectors is greater than for the photovoltaics by a factor known as *photoconductive gain*. This complicates the formulas a little, and it requires knowledge of a few more parameters to calculate the responsivity.

More specific information for performance predictions is provided in Section 2.4. Before we consider that, let's discuss how detectors are operated.

2.3. DETECTOR OPERATION

In this section we discuss the things that must be done to make a detector work. These include the electrical stimulus that must be applied, and the mounting and cooling requirements. Figure 2.9 shows a typical setup. In this example a photoconductor is cooled by contact with a copper plate that is in contact with liquid helium. To prevent condensation of moisture on the detector and to prevent rapid boil-off of the helium, the space around the detector and the helium chamber is evacuated. A window allows infrared radiation to reach the detector.

A cold shield between the window and the detector reduces the amount of energy that can fall on the detector and the helium chamber; this improves detector performance and reduces the rate of helium boil-off. An optical filter limits the wavelengths that fall on the detector to those of interest, and reduces the detector noise.

Electrical leads from the detector pass through the vacuum wall out to the room, where they can be connected to electronic equipment. A bias circuit is used to convert resistance changes into electrical voltages. An amplifier increases the voltage levels, and an electrical filter eliminates frequencies that are not of interest, improving the signal-to-noise ratio. Finally, some sort of readout is used to measure and report the resulting signal.

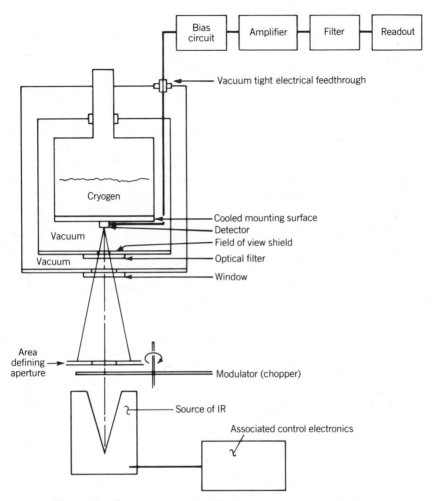

Figure 2.9.   Components required for detector operation and testing.

To provide a controlled IR input for testing the detector, we show a *blackbody*, a heated cavity whose emittance can be accurately predicted. A rotating toothed wheel (the *modulator* or *chopper*) interrupts the IR at a constant frequency, causing the electrical output to alternate for more stable amplification and to allow us to determine the frequency response of the detector. Associated electronics are provided to control the blackbody and the chopper. All of these components are discussed later in this chapter and in subsequent chapters.

2.3.1. Cooling

Requirements. Almost all detectors are cooled, either because they will not operate at room temperature or because they operate much better when cooled. This requirement complicates their use: A cooling medium must be provided, and the detector must be mounted so that it will be cold, will not be covered with condensation, but will still be accessible to the infrared. The cooling mechanism is either a cryogen (a very cold liquid) or a refrigerator of some type. The detector is usually mounted in a vacuum-tight enclosure that serves two purposes: It insulates the cold finger or the detector mounting area so that the cryogen does not boil off too rapidly, and it avoids the condensation problem by preventing moist air from contacting the detector. The normal operating temperature and cryogens used for several detectors are listed in Table 2.1.

Dewars. The technical name for a vacuum bottle (commonly called a thermos bottle) is *dewar* (named after its inventor, Sir James Dewar). Several types of dewars are available, ranging from miniature glass packages less than an inch in diameter to enormous test chambers. Metal dewars about 1 ft in diameter and 18 in. high are commonly used for experimental work with detectors (see Figure 2.10).

The use of cryogens and dewars is discussed in more detail in Chapter

Table 2.1 Typical Cooling Requirements for Common Detectors

Detector	Operating Temperature	Cryogen
Ge:Hg	20 K	He 4.2 K
Ge:Au		
Optimum	65 K	He 4.2 K
Adequate	77 K	N 77 K
PbS, PbSe[a]		
ATO	300 K	None
ITO	140 K	Freon
LTO	77 K	N 77 K
InSb	77 K	N 77 K
Bolometers	1–5 K	He 4.2 K

[a] ATO, ITO, and LTO are designations for material tailored to operate at ambient temperature, intermediate temperature, and low temperature, respectively.

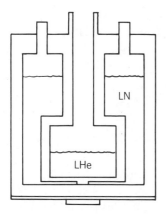

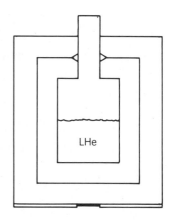

Pour filled metal dewar
with liquid nitrogen jacket
6–15 in. diameter
8–36 in. high

Pour filled metal dewar
with gas-cooled shield
6–12 in. diameter
8–16 in. high

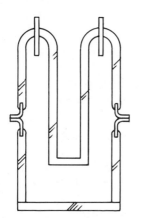

Glass dewar for use
with miniature cryostat
1–3 in. diameter
2–6 in. high

Figure 2.10.   Dewars of various sizes.

7. For now, note a few basics. The cryogens generally will boil away easily; the dewars insulate them from higher temperatures. A carefully designed dewar reduces radiated, conducted, and convected heating as much as practical. To maintain that thermal integrity requires a conscious effort, as noted below.

Mechanical Integrity. To reduce the conducted heat load, the mechanical supports are made as small as possible, and special materials are used. The following special treatment is necessary:

- Handle the dewars gently; they are often fragile.
- Any wiring from the cold stage should be handled carefully and changed only after a thermal analysis.
- Special surface preparation is essential to avoid radiated heat leaks.

Vacuum Integrity. The cryogen in the dewar is insulated thermally by a vacuum "jacket," an evacuated space. Cleanliness is essential because many materials will "outgas" over a period of time, destroying the vacuum.

- Do not handle the insides of dewars with bare hands.
- Use only approved materials in the vacuum space.

Refrigerators. A second way to obtain the required temperature is with refrigerators. These generally incorporate the same kind of insulating techniques as those used for dewars, and require the same precautions. In addition, the refrigerator and its refrigerant must be maintained. Refrigerators are of two principal types: *closed-cycle* refrigerators, which recycle the same fluid (often helium) over and over; and *open-cycle* refrigerators, which draw from a high-pressure tank of gas, with the gas vented to the atmosphere after it performs its cooling task.

Other Cooling Methods. *Radiation coolers* are used in space applications: They face the sky (which in space is very cold) and radiate energy away into space. They require no operating fluid and have no moving parts except the mechanisms required to keep them pointing away from the sun.

Thermoelectric (TE) *coolers* provide cooling by forcing current through a junction of dissimilar metals. These coolers are small, simple, reliable, generate no audible noise, and operate in any orientation. Their primary disadvantages are the relatively large amount of electrical power required to achieve a given cooling level, and the limited temperatures that can be reached. Chapter 15 in *The Infrared Handbook* (Donabedian, 1978) and Hudson (1969) provide convenient overviews of TE coolers and other cooling systems.

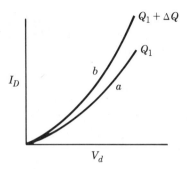

Figure 2.11. Detector V–I curves.

2.3.2. Photoconductor Biasing

The detector element is characterized by its voltage–current (V–I) curve, shown by curve a in Figure 2.11. As the incidance increases, the conductance increases, and the curve shifts (curve b in the figure). This shift is a measurable response to the IR. To sense this change, a bias battery and a resistor are generally connected to the detector, as shown in Figure 2.12. The resistor is called the *load resistor*. An increase in the detector conductance both increases the current and decreases the voltage across the detector. If the radiation is modulated, an ac voltage will result at point A; this signal may be filtered and amplified.

The load resistor is necessary if a signal voltage is desired: If the load resistance were zero, the voltage across the detector would be exactly equal to the dc battery voltage and signal voltage would not be possible.

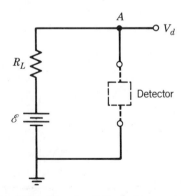

Figure 2.12. Bias circuit for a photoconductor.

Operating Point from the *V–I* Curve and Load Line. The voltage and current that will flow depend on both the detector and the bias circuit. The detector is characterized by its *V–I* curve. The circuit is characterized by its *load line*: the graph of the current that flows in terms of the voltage across the detector. In the circuit illustrated, the voltage is given by

$$V_d + IR = \mathscr{E} \qquad (2.8a)$$

so

$$I = \frac{\mathscr{E}}{R} - V_d \frac{1}{R} \qquad \cdot \quad (2.8b)$$

This is shown in Figure 2.13. The current is $\mathscr{E}/R$ when V_d = zero, is zero if V equals $\mathscr{E}$, and is a linear function of V_d—with slope $1/R$—in between.

If we had an algebraic expression for the *V–I* curve, we could calculate the voltage across the detector and the current in the circuit algebraicaly. We can arrive at the same answer and visualize the interaction of detector and the bias circuit better, by superimposing the detector *V–I* curves and the load line of the bias circuit (Figure 2.14). The operating point (current and voltage) is the intersection of the load line and the *V–I* curve.

Short-Circuit Signal Current and Open-Circuit Signal Voltage. It turns out to be easy to derive the equations for the detector signal *current* in a circuit with zero load resistance. Unfortunately, many applications rely on *voltage* amplifiers, and these require a load resistor. Thus we will want to predict the signal voltage once we know the load resistance and the

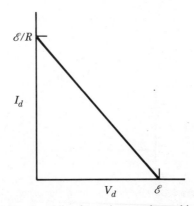

Figure 2.13. Load line for a photoconductor bias circuit.

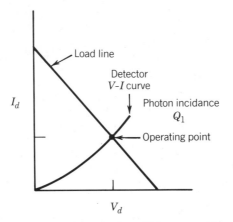

Figure 2.14. Operating point from $V–I$ curve and load line.

ideal signal current. We will also want to know the signal voltage that could be developed if the load resistor were much larger than the detector resistance. This ideal voltage is the open-circuit voltage, v_{oc}. The desired equations are Equation 2.9 and 2.10.

SIGNAL VOLTAGE v FROM SHORT-CIRCUIT SIGNAL CURRENT i_{sc}

$$v = i_{sc}R_{parallel} \tag{2.9}$$

OPEN-CIRCUIT SIGNAL VOLTAGE V_{oc} FROM OBSERVED SIGNAL VOLTAGE v

$$v_{oc} = v\,\frac{R_d + R_L}{R_L} \tag{2.10}$$

where i_{sc} is the ideal, or short-circuit current (which we derive later) and $R_{parallel}$ is the parallel resistance of the detector and load:

$$\frac{1}{R_{parallel}} = \frac{1}{R_{detector}} + \frac{1}{R_{load}}$$

Figure 2.15 is helpful in deriving those relationships. It shows two $V–I$ curves for a hypothetical detector: one for a photon incidance Q_{bk} and

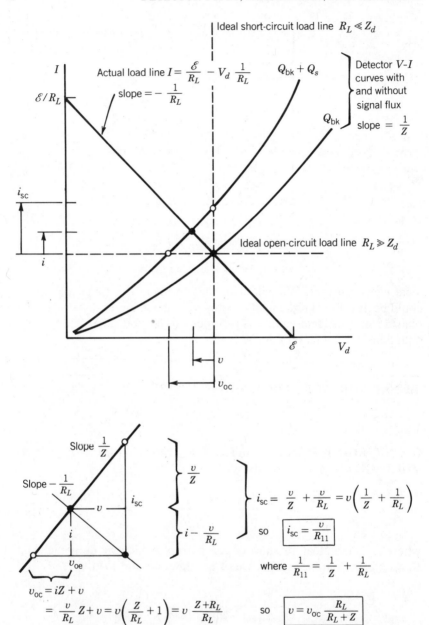

Figure 2.15. Visualizing the effect of bias circuit on signal.

another where the arrival rate has increased by the photon signal incidance Q_s. It also shows load lines for three bias circuits. The line labeled "actual load line" represents a realistic circuit. The other two represent idealizations, one for the short-circuit limit (a load resistor much less than the detector resistance), the other for the open-circuit limit (load resistance much greater than the detector resistance).

The operating point for a given background and bias circuit is the intersection of the corresponding *V–I* curve and load line. The *signal voltage v* is the difference between the operating voltage for $Q = Q_{bk}$ and the operating voltage when $Q = Q_{bk} + Q_s$. The *signal current i* can be found in a similar way. Note that v is a maximum in the open-circuit limit, is smaller for the actual circuit, and is zero for the short-circuit limit. Similarly, the signal current is a maximum in the short-circuit limit, is smaller for the actual circuit, and is zero in the open-circuit limit.

Choice of Load Resistor. The load resistor (or feedback resistor in an op-amp circuit) affects the detector performance in several ways. The system requirements or goals listed below will determine the optimum resistance value.

1. *Required Frequency Response.* Capacitance in the detector circuit will cause the signal and noise to "roll off" at high frequencies. The frequency at which roll-off begins can be moved to high frequencies by the use of a load resistor that is small relative to the detector dynamic impedance. The corner frequency f_c is given by

$$f_c = \frac{1}{2\pi RC} \tag{2.11}$$

2. *Responsivity.* For any given bias voltage across the detector, the voltage responsivity can be maximized by the use of a load resistance much greater than the detector resistance.

3. *Available Bias Voltage.* If the available bias supply voltage is less than the optimum or recommended bias voltage, large voltage drops across the load cannot be tolerated. The maximum responsivity in this case will be obtained when the load resistor equals the detector resistance.

4. *Johnson Noise.* A small load resistor will generate noise that may be large compared to the detector noise, resulting in a degradation of the overall detector performance. (The ratio of Johnson noise to generation–recombination noise varies as 1 over the square root

of R, so a large load resistance is best.) Specific Johnson noise formulas are presented in Section 2.4.3.

2.3.3. Photovoltaic Circuits

A photovoltaic detector is a diode that generates a current that is proportional to the photon arrival rate. There is no fundamental reason why the bias circuit described for a PC detector cannot be used with a PV detector. (The bias battery would be very low voltage, or even zero.) In practice this is never done because a much more convenient scheme is available.

Current Mode Preamplifier. The operational amplifier feedback circuit shown in Figure 2.16 is generally used with these detectors. Such a circuit is called a *current mode preamplifier*. This circuit holds the voltage across the detector close to zero, where the detector performs best. The signal current generated by the detector times the feedback resistance value yields a signal voltage that may be filtered and amplified by conventional voltage amplifiers:

$$V_{\text{out}} = -i_{\text{sc}}R_f \left(1 + \frac{1}{A} + \frac{R_F}{Az} \right)^{-1} \approx -i_{\text{sc}}R_f \qquad (2.12)$$

Benefits of Using an Operational Amplifier with PV Detectors:
*   The detector voltage V_{det} is held near zero
*   Large feedback resistances R_f can be used for low noise:

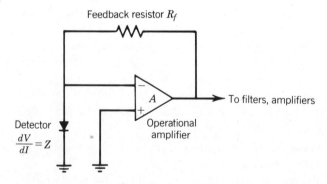

Figure 2.16.   Use of operational amplifier in the feedback mode for PV detectors.

$$i_N = \left(\frac{4kT}{R_f}\right)^{1/2}$$

• Frequency response is limited by R_f/A instead of R_L

Effect of Feedback Resistance. The criteria for selecting a feedback resistor for a PV circuit are similar to those for the PC load resistor.

1. Frequency Response. If the feedback resistance is too large, the circuit response will "roll off" at frequencies that may be of interest. The corner frequency f_c is given by

$$f_c = \frac{A}{2\pi R_f C} \tag{2.13}$$

   where A is the open-loop gain of the operational amplifier (A is typically greater than 100) and C is the capacitance on the input to the amplifier.
2. Amplifier Saturation. Operational amplifiers will saturate if the product of the feedback resistance and the short-circuit current of a photovoltaic detector exceeds the supply voltage of the amplifier—typically, 2 to 9 V.
3. Johnson Noise. The feedback resistance should be large enough that its Johnson noise will be small compared to the detector noise.

It can be shown that a feedback resistance such that $I_{dc}R_f = 1$ V will generally satisfy both the saturation and the Johnson noise requirements. (This can be derived easily from equations 2.25 and 2.31 and the saturation criterion given in this section.)

2.4. DETECTOR PARAMETERS: MODELS AND PREDICTIONS

In Section 1.8 we defined several figures of merit for detectors. In this section we obtain equations with which to predict (using some inherent detector characteristics) values of those figures of merit. Later in the chapter we provide typical material characteristics and calculate predicted figures of merit for several detectors. We begin with a fundamental parameter, current: first for a photovoltaic detector, then for a photoconductive detector. We can then derive the detector responsivity.

2.4.1.   Current Equations

Consider first the current for a photovoltaic detector, as shown in Figure 2.17. Carriers are generated by incoming photons and reach the junction at a rate (in carriers/second)

$$\dot{N} = \eta Q A_d \qquad (2.14a)$$

in which η is the probability that a photon will create a carrier; Q is the photon incidance; and A is measured in cm². In a photon detecting diode, the carriers generated are moved (by the internal field of the diode) across the junction just as fast as they reach the junction. The current is the arrival rate $(\eta Q A_d)$ times the charge e on each carrier:

$$I = \eta Q A_d e \qquad (2.14b)$$

This is the *photovoltaic current equation.*

Example: Consider a 0.002-in. (2-mil) square PV detector with a quantum efficiency of 60% and an incidance of 4×10^{15} photons/(cm²·s). The resulting current is 9.6 nA:

$A_d = (0.002 \text{ in.} \times 2.54 \text{ cm/in.})^2 \approx 25 \times 10^{-6} \text{ cm}^2$

$I = \eta Q A_d e$

$\quad = 0.60[4 \times 10^{15} \text{ photons/(cm²·s)}](25 \times 10^{-6} \text{ cm}^2)(1.6 \times 10^{-19} \text{ C})$

$\quad = 96 \times 10^{-10} \text{ C/s}$

$\quad = 9.6 \text{ nA}$ ∎

For a photoconductive detector the current equation is a little more involved. The current I is the product of (1) the carrier concentration n, (2) the mean drift velocity v, (3) the charge e on the carriers, and (4) the

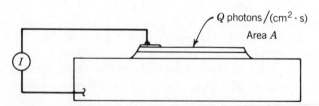

Figure 2.17. Notation for photovoltaic detector current equation.

coss-sectional area a:

$$I = nvea \qquad (2.15)$$

We can predict the current if we can provide expressions for the carrier concentration in terms of the photon incidence and for the drift velocity.

Figure 2.18 shows a representative photoconductive detector and motivates the derivation. The notation is the same as that used for the PV detector, and the carrier generation rate is also the same:

$$\dot{N} = \eta Q A_d \qquad (2.16)$$

The total number of carriers "alive" at any time is their generation rate $\dot{N}$ times their mean lifetime τ. The carrier concentration n is the total number of carriers alive at any time, divided by the volume of the sample:

$$n = \frac{(\text{generation rate})(\text{lifetime})}{as}$$
$$= \frac{(\eta Q A_d)\tau}{as} \qquad (2.17)$$

The drift velocity is the mobility μ times the electric field E (the electric field E is the voltage V across the detector divided by the interelectrode spacing s):

$$v = \mu E \qquad (2.18)$$

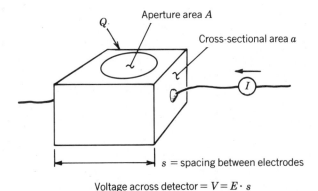

Aperture area A

Cross-sectional area a

s = spacing between electrodes

Voltage across detector = $V = E \cdot s$

Figure 2.18. Notation for photoconductor detector current equation.

Combining the equations above and rearranging the order of terms yields the photoconductor current equation:

$$I = [\eta Q A_d e] \frac{\mu \tau E}{s} \qquad (2.19)$$

The particular grouping of the parameters in the photoconductive current equation was chosen for two reasons:

1. The expression in brackets is identical to the *photovoltaic* detector current.
2. The quantity $\mu \tau E/s$ is dimensionless and can be thought of as a "gain." It is called the *photoconductive gain G*; depending on the detector type and the electric field, it can be as large as several thousand.

Example: Consider a square detector 2 mils on a side, with a $\mu \tau$ product of 0.5×10^{-3} cm^2/V, biased to an electric field of 300 V/cm. The photoconductive gain is about 30:

$$G = \frac{\mu \tau E}{s}$$

$$= \frac{(0.5 \times 10^{-3} \text{ cm}^2/\text{V}) (300 \text{ V/cm})}{(0.002 \text{ in.})(2.54 \text{ cm/in.})}$$

$$\simeq 30 \qquad \blacksquare$$

By writing the PC current equation in terms of the photoconductive gain, only one equation is needed for the photon-induced current in photon detectors, the generalized current equation (Eq. 2.20). This equation is a key expression, and its derivation helps teach the roles of various parameters in detector performance.

GENERALIZED CURRENT EQUATION

$$I = \eta Q A_d e G \qquad (2.20)$$

where for photovoltaics $G = 1$ and for photoconductors $G = \mu \tau E/s$.

Using this formula as a starting point, one could derive a formula for the resistance of a PC detector. An alternative is to assume a typical electric field and calculate the resulting photoconductive gain, the current, and the voltage across the detector (E/s). To obtain the resistance, divide the voltage by the current. Although it requires two steps, this method has the advantage of providing both the current and the resistance, and eliminates the need to remember one more formula.

In Eq. 2.20 Q is the background photon flux density and I is the resulting dc current. If Q is replaced by the *signal* flux density, then the same equation yields the *short circuit signal current*. To obtain signal voltages (or voltage responsivities) combine the short circuit signal current with the detector resistance and load resistance as described in Eq. 2.9 and 2.10, and Fig. 2.15.

2.4.2. Responsivity

Remember that the responsivity is the signal output divided by the radiant input, but that the output may be in current or voltage, and the input may be in watts or in photons per second (or sometimes W/cm^2 or photons/$(cm^2 \cdot s)$). Consider the simplest of the common definitions: output current per input photon flux (QA_d)(photons/s). The responsivity is then easily deduced from the signal current equation:

$$\mathscr{R} = \eta eG \qquad (2.21)$$

The units are amperes per (photon/second). If the responsivity is defined in terms of output voltage, we multiply the signal current by the effective circuit impedance. For a photoconductor this is the resistance of the detector and load in parallel, or in the idealized open-circuit voltage case, it is just the detector resistance. For a photovoltaic detector, the effective circuit impedance is the feedback resistance, or for more complex circuits it is a value referred to as Z, the *transfer impedance* (often shortened to *transimpedance*). If we use the symbol Z for the effective impedance, then

$$\mathscr{R} = \eta eZG \qquad (2.22)$$

In the more common definition of output voltage per watt of input power, we note that the input power is the photon arrival rate times the energy per photon ($E = hc/\lambda$), so that

$$\text{power} = \frac{QA_d hc}{\lambda} \qquad (2.23)$$

and

$$\mathcal{R} = \frac{\eta QA_d eZG}{QA_d hc/\lambda}$$

$$= \eta e \frac{\lambda}{hc} ZG \qquad (2.24)$$

The formulas for the other possible definition of responsivity are given in Figure 2.19, along with responsivity values at 5 μm for a 2-mil square PC detector with a 60% quantum efficiency, a PC gain of 30, in a circuit such that the effective impedance Z is 1 MΩ.

At very low backgrounds the resistance of doped Ge and Si PC detectors becomes very high, and a modification of the responsivity prediction is required. The PC gain calculated earlier will still apply at low frequencies, but at higher frequencies the gain will saturate to a value of 0.5. This is described in Section 2.5.4.

2.4.3. Noise

Noise comes about because of the random nature of the motion of the charge carriers. The total noise is the rss (root of the sum of the squares) of all the contributing noises. There are numerous sources of noise, but

		Current (A)	Voltage (V)
QA	photons/s	ηeG $= 1.44 \times 10^{-18}$ A/(ph/s)	ηeGz $= 1.44 \times 10^{-12}$ V/(ph/s)
Q	photon/s/cm^2	ηAeG $= 36 \times 10^{-24}$ A/[ph/(cm^2·s)]	$\eta AeGz$ $= 36 \times 10^{-18}$ V/[ph/(cm^2·s)]
EA	W	$\eta eG\lambda/(hc)$ $= 36$ A/W	$\eta eGz\lambda/(hc)$ $= 36 \times 10^6$ V/W
E	W/cm^2	$\eta AeG\lambda/(hc)$ $= 900$ μA/(W/cm^2)	$\eta AeGz\lambda/(hc)$ $= 900$ V/(W/cm^2)

Figure 2.19. Responsivity equations and values for one example. The example assumes η = 0.60, A = 25 × 10^{-6} cm^2, G = 30, λ = 5 μm, and Z = 1 MΩ. $\lambda/(hc)$ = 25 × 10^{18} photon/(w·s).

for most applications only a few noise sources dominate the performance of IR detectors. Although the noise mechanisms are not exactly the same for both detector types, the resulting formulas are very similar, so only two equations are needed to describe most of the cases of interest. We deliberately exclude from consideration at this time sources of "external" noise—sources that can be eliminated by careful setup and shielding. Noise due to amplifiers is deferred to Chapter 10.

Johnson Noise. *Johnson noise* is generated in all resistors. The expressions are shown in Equations 2.25 and 2.26.

JOHNSON NOISE CURRENT AND VOLTAGE

$$i_N^2 = \frac{4kT}{R} \Delta f \qquad (2.25)$$

$$v_N^2 = 4kTR \, \Delta f \qquad (2.26)$$

where　　k = Boltzmann constant (about 1.38×10^{-23} J/K)
　　　　　T = temperature of the resistor (kelvin)
　　　　　R = resistance (ohms)
　　　　　Δf = noise bandwidth (hertz)

Example: A 1-MΩ resistor at room temperature generates a Johnson noise spectral density of 130×10^{-15} A/Hz$^{1/2}$ or 130 nV/Hz$^{1/2}$:

$$i_N^2 = \frac{(4 \times 1.38 \times 10^{-23} \text{ J/K})(300\text{K})}{10^6 \, \Omega} \Delta f$$

$$= (16{,}560 \times 10^{-30} \text{ J/}\Omega) \, \Delta f$$

$$i_N = (130 \times 10^{-15} \text{ A/Hz}^{1/2}) \sqrt{\Delta f}$$

$$v_N^2 = (4 \times 1.38 \times 10^{-23} \text{ J/K})(300\text{K} \times 10^6 \, \Omega) \, \Delta f$$

$$= (16{,}560 \times 10^{-18} \text{ J·}\Omega) \, \Delta f$$

$$v_N \simeq (130 \times 10^{-9} \text{ V/Hz}^{1/2}) \sqrt{\Delta f} \qquad \blacksquare$$

It is generally most convenient to work with noise currents. In a typical PC bias circuit the load resistor and the detector resistor are in parallel across the output. In that case the net noise current is the rss of the two noise currents:

$$i_N^2 = i_{N,1}^2 + i_{N,2}^2 \qquad (2.27a)$$

To minimize the Johnson noise contributed by one resistor in parallel with another, the resistance should be as large as possible. Thus the load resistance and the feedback resistance should be as large as possible in order to minimize their contribution of Johnson noise.

To obtain the net Johnson noise voltage of two resistors in parallel, multiply the net noise current (as described above) by the parallel resistance of the two:

$$v_N^2 = (i_{N,1}^2 + i_{N,2}^2)\frac{R_1 R_2}{R_1 + R_2} \qquad (2.27b)$$

For resistors in *series*, the net noise voltage is the rss of the individual Johnson noise voltages, and the net noise current is the net noise voltage divided by the sum of the resistances.

$$v_N^2 = v_{N,1}^2 + v_{N,2}^2 \qquad (2.28a)$$

$$i_N^2 = \frac{v_{N,1}^2 + v_{N,2}^2}{R_1 + R_2} \qquad (2.28b)$$

Johnson noise is also known as *thermal noise* or *Nyquist noise*. It applies only to resistive elements: a feedback resistor, a load resistor, or a PC detector. It does not apply to ideal diodes, although we shall see that a noise expression similar to that for Johnson noise does result.

Shot Noise in PV Detectors. An ideal PV diode is not resistive—there is no I^2R power lost—so Johnson noise does not apply to ideal PV detectors. The dominant noise is called *shot noise*; it is due to the random arrival rate of carriers at the PV junction.

When viewed at the atomic or electron level, the current in our detectors does not flow in a smooth, uniform way. It is the result of discrete bundles of charge crossing a surface at a random rate. The *average* rate—and the resulting current—are predictable. We have calculated that rate and current. The current we called the dc current; if it is modulated slowly (by atomic or electron standards), it is the signal current. We now want to estimate the *rms variation* in the current about its average value due to fluctuations at the electron level. To do that we need an expression for the variation in the rate at which carriers reach the diode junction.

Noise is a statistical phenomenon, and statistical concepts are not easy to condense into short summaries. So, once again, we resort to "it can be shown that": . . . the *variation* $\delta\dot{N}$ in the arrival rate $\dot{N}$ is proportional to the square root of the arrival rate $\dot{N}$ and to the square root of the bandwidth being observed. The actual formula is

$$\delta\dot{N} = \sqrt{2\dot{N}\,\Delta f} \qquad (2.29)$$

Units of $\delta\dot{N}$ are (carriers) per second; these are the same units as for $\dot{N}$. This variation in arrival rate causes a variation in current; this is a *noise current*:

$$i_N = \delta\dot{N}e = \sqrt{2\dot{N}\,\Delta f}\;e \qquad (2.30)$$

We can write this in a convenient way if we use $I = \dot{N}e$; see Equation 2.31. This formula is true whether the carriers are generated thermally or by photons. To apply it, one must know or be able to predict the actual currents. A subtlety must be observed when applying this formula: If the current is due to two or more independent phenomena, the noise from each must be calculated, using the absolute value of the current. The shot noise due to thermally generated carriers will provide an example of such a case.

SHOT NOISE IN PV DETECTORS

$$i_N^2 = 2Ie\,\Delta f \qquad (2.31)$$

Shot Noise in the Photon Limit. We can limit ourselves to variations in the rate of arrival of carriers generated by arriving photons, assuming that the detector is cold enough that thermally generated carriers are negligible. In that case we decided earlier that the current I was given by

$$I = \eta Q A_d e$$

Substituting this in the equations above yields

$$i_N^2 = 2\eta Q A_d\,\Delta f\;e \qquad (2.32)$$

Jones (1953, 1957) points out that the statistics for noise are not actually quite as simple as stated here. Photons obey Bose–Einstein statistics, which have a larger fluctuation rate than we have assumed. Kittel and Kroemer (1980) mentions this phenomenon: "It has been said that 'bosons travel in flocks'." To account for this, the photon variance should contain a *Bose factor*. This can be included in our noise formulas if we multiply the spectral photon incidence by the Bose factor before integrating (Eq. 2.33).

USE OF BOSE FACTOR

$$Q(\lambda) \rightarrow Q(\lambda) \left[\frac{e^x}{e^x - 1} \right] \qquad (2.33)$$

where

$$x = \frac{hc}{\lambda kT}$$

The factor in brackets above is the Bose factor. For an ideal detector viewing a room-temperature background, the factor for most of the background incidence is only slightly greater than unity, and the effect is normally not considered in noise and D^* predictions. For a detector whose spectral range is limited to long wavelengths, the factor could be significant.

Shot Noise in the Thermal Limit. We can also apply the shot noise formula to the thermally induced current. The diode current due to thermal generation is

$$I = I_s \left[\exp\left(\frac{eV}{kT}\right) - 1 \right] \qquad (2.34)$$

The current equation above is actually the combination of two independent currents:

$$I_1 = I_s \exp\left(\frac{eV}{kT}\right) \qquad (2.35a)$$

$$I_2 = -I_s \qquad (2.35b)$$

As mentioned earlier, we must apply the noise formula to each of these separately, using the absolute value of the current. We then calculate the rss of the two noise terms, yielding

$$i_N^2 = 2I_s \left[\exp\left(\frac{eV}{kT}\right) + 1 \right] e \, \Delta f \qquad (2.36a)$$

Even though there is zero net current at zero bias voltage, the noise is not zero:

$$i_N^2 \big|_{V=0} = 4I_s e \, \Delta f \qquad (2.36b)$$

Using the relation

$$\frac{1}{R_0} = \frac{dI}{dV}\bigg|_{V=0} = \frac{I_s e}{kT} \qquad (2.37)$$

allows us to eliminate the saturation current I_s; the resulting formula is very similar to the Johnson noise formula:

$$i_N^2 = \frac{4kT}{R_0} \Delta f \qquad (2.38)$$

Here R_0 is dV/dI—the slope of a V versus I graph, or the inverse of the slope of an I versus V graph, measured at zero detector voltage, for an ideal diode (no leakage).

Total PV Noise. If there is leakage associated with the diode, dV/dI is lower than that expected from thermal current equation, and it will contribute Johnson noise consistent with the leakage resistance. The observed slope dI/dV is the sum of the thermal and leakage slopes, so

$$\frac{1}{R_{observed}} = \frac{1}{R_{thermal}} + \frac{1}{R_{leakage}} \qquad (2.39)$$

and the expressions for the Johnson noise and the thermally generated shot noise can be combined:

$$i_N^2 = \frac{4kT}{R_{thermal}} + \frac{4kT}{R_{leakage}} = \frac{4kT}{R_{observed}} \qquad (2.40)$$

The total PV detector noise (at zero bias) due to photon-generated shot noise, thermal-generated shot noise, and Johnson noise from any leakage present is shown in Equation 2.41.

TOTAL PV DETECTOR NOISE

$$i_N^2 = \left(2\eta Q A_d e + \frac{4kT}{R_0} \right) \Delta f \qquad (2.41)$$

Here R_0 is the observed value of dV/dI; it includes both any leakage present and the slope due to an ideal diode.

The relationship of the various noise expressions discussed so far is shown in Figure 2.20. The expression above does not include the so-called $1/f$ *noise*, nor does it include amplifier noise. Prediction of $1/f$ noise is generally empirical. Amplifier noise is discussed in Chapter 10.

GR Noise in Photoconductors. For a photoconductor the noise mechanisms are Johnson noise (discussed earlier) and *generation–recombination noise* (GR noise). The PC current is dependent on the average carrier density. This depends on the carrier generation rate and lifetime. Thus the randomness comes about from two independent mechanisms: The carriers are generated at a random rate, and the carriers have a random lifetime. (Alternatively, we can think of random generation and random recombination.)

The fluctuation in carrier concentration due to generation is the square root of $2\dot{N}\,\Delta f$, as is that due to recombination. The total fluctuation is the RSS sum of the two:

$$\delta\dot{N} = \sqrt{4\dot{N}\,\Delta f} \qquad (2.42)$$

We derive the resulting PC noise current much as we did the PC signal current. This derivation results in an expression much like the PC shot noise, but it includes a factor of 4 (instead of 2) and it includes the photoconductive gain term G:

$$i_N^2 = 4IeG\,\Delta f \qquad (2.43)$$

GR Noise in the Photon Limit. We have already derived the current due to photon generation; inserting it into the equation above yields

Equations for $i_N^2/\Delta f$

	Shot Noise 2Ie		Johnson Noise $\dfrac{4kT}{R}$		
	Due to Photon-Generated Current ηQAe	Due to Thermally Generated Current	Due to Shunt R_s	Due to Diode $R_{o,\,ideal}$	Due to Feedback R_f
General expressions	$2\eta QAe$	$I_s\left[\exp\left(\dfrac{eV}{kT_d}\right) - 1\right]$	$\dfrac{4kT_d}{R_s}$	0	$\dfrac{4kT_f}{R_f}$
	$\longrightarrow$	$2I_s\left[\exp\left(\dfrac{eV}{kT_d}\right) + 1\right]$	$\longrightarrow$		$\longrightarrow$
		$\dfrac{2kT_d}{R_{o,\,ideal}}\left[\exp\left(\dfrac{eV}{kT}\right) + 1\right]$			
		$\downarrow$ if $V = 0$			
		$\dfrac{4kT_d}{R_{o,\,ideal}}$	$\dfrac{4kT_d}{R_s}$		
Total noise^2 at $V = 0$	$2\eta QAe$	$+\,\dfrac{4kT_d}{R_{o,\,actual}}$		$+$	$\dfrac{4kT_f}{R_f}$

Figure 2.20. Relationship of photovoltaic noise expressions.

PC GENERATION—RECOMBINATION NOISE CURRENT

$$i_N^2 = 4\eta Q A_d\, e^2 G^2\, \Delta f \qquad (2.44)$$

For long wavelength detectors, apply the Bose factor (Eq. 2.33) to the photon incidance associated with generation noise, but not the recombination noise: replace $4Q(\lambda)$ with $2Q(\lambda) \times$ Bose factor $+ 2Q(\lambda)$ before integrating over wavelength.

GR Noise in the Thermal Limit. Although we could insert the equations for thermally generated currents into the noise equation (2.42), the result is cumbersome. For theoretical purposes, it is enough to note that the current is dominated by the number of carriers and that the noise varies as the square root of the number of carriers. Examination of Figure 2.7 (temperature dependence of the carrier concentration) shows that carrier concentration increases rapidly as temperature increases beyond the background limited value. We will generally arrange the temperatures so that the thermally generated noise is negligible.

To determine the expected GR noise from experimental data, insert measured current in equation (2.42). This is valid whether the current is thermally generated, or due to photons, or due to a combination of the two.

Total PC Noise. Combining the Johnson and GR noise expressions yields Equations 2.45*a* and 2.45*b*.

TOTAL NOISE FOR PHOTOCONDUCTORS

$$i_N^2 = \left(4IeG + \frac{4kT}{R} \right) \Delta f \qquad (2.45a)$$

TOTAL NOISE FOR PHOTOCONDUCTORS WITH NEGLIGIBLE THERMAL GENERATION

$$i_N^2 = \left(4\eta Q A_d e^2 G^2 + \frac{4kT}{R} \right) \Delta f \qquad (2.45b)$$

2.4.4. Noise Equivalent Power

The noise equivalent power (NEP) was defined as the noise divided by the responsivity, so to predict NEP, we only have to apply the applicable noise and responsivity expressions derived earlier. A detector that is limited by photon-generated noise is termed a *background-limited infrared photon detector* (BLIP). Since no detector can have noise less than the BLIP limit, the BLIP NEP is a convenient figure of merit. Using the photon-generated noise expressions yields Equations 2.46a and 2.46b.

BLIP NEP (PV)

$$NEP = \frac{hc}{\lambda} \sqrt{\frac{2QA_d}{\eta}} \sqrt{\Delta f} \qquad (2.46a)$$

BLIP NEP (PC)

$$NEP = \frac{hc}{\lambda} \sqrt{\frac{4QA_d}{\eta}} \sqrt{\Delta f} \qquad (2.46b)$$

Note that the PC gain canceled out of the NEP expression, so that the PC and the PV NEP expressions are the same except for a square root of 2. Remember that this expression applies only to BLIP detectors—it is the best NEP that can be achieved.

2.4.5. Specific Detectivity

The specific detectivity D^* was defined as

$$D^* = \frac{\text{responsivity} \times \sqrt{A_d}}{\text{noise}/\sqrt{\Delta f}} \qquad (2.47)$$

For background-limited photon detectors, at wavelengths less than the cutoff, substitution of the responsivity and photon-generated noise equations derived earlier yields the BLIP D^* equations 2.48a and 2.48b.

D^* FOR BLIP LIMIT (PV)

$$D^* = \frac{\lambda}{hc} \sqrt{\frac{\eta}{2Q}} \qquad (2.48a)$$

D^* FOR BLIP LIMIT (PC)

$$D^* = \frac{\lambda}{hc} \sqrt{\frac{\eta}{4Q}} \qquad (2.48b)$$

Dependence of $D*$ on Cutoff Wavelength. The dark lines in Figure 2.21 show how the BLIP $D*$ varies with wavelength for two hypothetical PC detectors. One detector has a 4-μm cutoff wavelength and the other cuts off at 5 μm. Both detectors see a room-temperature background through a full hemisphere. The 5-μm detector sees more photons than does the 4-μm detector, so its noise is higher and its $D*$ is less. This effect is partially compensated by the higher responsivity at 5 μm than at 4 μm. The light lines indicate the BLIP $D*$ (at the cutoff wavelength) versus wavelength for ideal detectors of other cutoff wavelengths.

If we begin with a detector whose cutoff is at very short wavelengths, the responsivity is low, but very few background photons are seen, so the noise is also very low and the $D*$ is good. As the cutoff wavelength increases toward 10 μm (where the background peaks), the effective background increases rapidly—far faster than the responsivity, so $D*$ gets worse.

Beyond 30 μm, very few additional background photons are collected,

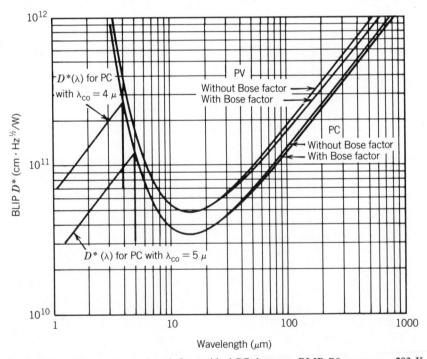

Figure 2.21. $D*$ versus wavelength for an ideal PC detector. BLIP $D*$ assumes a 293 K background seen through 180° field of view.

but the responsivity continues to increase proportional to wavelength, so D^* increases linearly with wavelength. Inclusion of the Bose factor (see photon noise in Section 2.4.3) affects D^* only at the longest wavelengths, as shown in Figure 2.21. This effect on D^* is seldom considered.

2.4.6. Modulation Transfer Function

Consider a detector of width w whose responsivity may vary across its surface: $\mathcal{R} = \mathcal{R}(x)$. Imagine that detector moving across targets whose incidance varies with position as shown in Figure 2.22: $E_0 \cos(2\pi kx)$. As that detector moves across a target, the signal S will vary sinusoidally— we say the signal is *modulated*. The signal extremes S_0 will occur when the detector is centered over one of the incidance peaks and minima, and will have a value given by

$$S_0 = \int_{-\infty}^{\infty} \mathcal{R}(x)\, E_0 \cos(2\pi kx)\, dx$$

When k is zero (or small enough so that kw is much less than one) the target incidance varies slowly compared to the detector size. In that case the incidance is nearly the same at all points on the detector and the signal amplitude is as large as it can ever be for that detector:

$$S_{0,\text{max}} = E_0 \int_{-\infty}^{\infty} \mathcal{R}(x)\, dx$$

The modulation transfer function is the ratio of the signal amplitude for targets of spatial frequency k to that with zero spatial frequency, as shown in Equation 2.49.

MTF FROM DETECTOR RESPONSE PROFILE

$$\text{MTF}(k) = \frac{\displaystyle\int_{-\infty}^{\infty} \mathcal{R}(x) \cos(2\pi kx)\, dx}{\displaystyle\int_{-\infty}^{\infty} \mathcal{R}(x)\, dx} \qquad (2.49)$$

For an ideal detector the responsivity is a constant $\mathcal{R}_0$ inside some width w, but zero outside that width. In that case the integrals can be done in closed form, as seen in Equation 2.50.

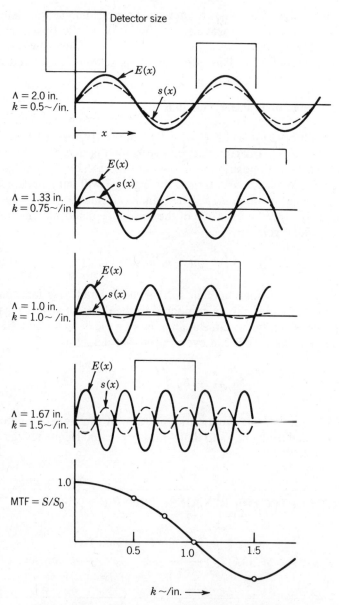

Figure 2.22. MTF peak-to-peak signal versus target size.

MTF FOR AN IDEAL DETECTOR OF WIDTH *w*

$$\text{MTF}(k) = \frac{\sin(\pi k w)}{\pi k w} \tag{2.50}$$

Example: For an ideal 10 mil (0.010 inch, or 0.25 mm) detector, the MTF at 2 ~/mm is 0.637 and drops to zero at 4 ~/mm:

$$\text{MTF}(2 \sim/\text{mm}) = \frac{\sin(\pi 2 \sim/\text{mm} \times 0.25 \text{ mm})}{2 \sim/\text{mm} \times 0.25 \text{ mm}}$$

$$= \frac{\sin(\pi/2)}{\pi/2}$$

$$= 0.637$$

For $\text{MTF}(k) = 0$, $\sin(\pi k w) = 0$

so $\pi k w = \pi$

or $k = 1/w = 1/(0.25 \text{ mm}) = 4 \sim/\text{mm}$ ∎

If the detector nonuniformities can be predicted, the MTF can be predicted using the method outlined above.

2.5. DETECTOR MATERIALS AND PERFORMANCE PREDICTIONS

In this section we discuss some of the characteristics of detectors that are unique to the particular types. Topics include handling precautions, material parameters, and performance predictions. The handling precautions are meant only as a reminder of some of the problems that may be encountered. For specific guidance refer to the manufacturer. Allowable environmental exposure (temperature, humidity, solvents, and contamination), biasing limits, and static protection need to be considered.

Similarly, the material parameters and prediction methods are intended only as a starting point. Actual performance depends on the material parameters for the material available, and there is no substitute for measurements to determine those parameters for *your* material. The information provided here should allow you to make an estimate of the performance to be expected; this should then be confirmed by consultation with others or by direct measurement. Once the methods are understood, they can be combined with experimental data to extrapolate performance to other backgrounds, bias conditions, and detector sizes. Because they are based on timely data, those extrapolations are more reliable than predictions from this or other handbook data.

Unless more specific data are available, the spectral response of the photon detectors should be assumed to be that of an ideal detector (proportional to wavelength up to the cutoff, zero beyond). A certain amount of rounding near the peak and some sensitivity beyond the cutoff are always present but difficult to predict. Convenient sources of more detailed information on IR detectors are Chapter 11 of *The Infrared Handbook* (Limperis, 1978) and Section 4 of the *Handbook of Optics* (Jacobs, 1978).

All detector performance predictions require a knowledge of detector size, cutoff wavelength, background photon incidance and spectral content, spectral content of the signal, biasing conditions, and electronics. Information that is unique to each detector type is presented in the material that follows.

2.5.1. Photovoltaic Indium Antimonide (InSb)

The cutoff wavelength for InSb is about 5.5 μm and its spectral response curve peaks at about 5.0 μm. The current is given by equation (2.14); the responsivity can be calculated from the equations of Figure 2.19. For uncoated detectors the reflectance is about 36%, so assume a quantum efficiency of 64%. If the detector is anti-reflective coated, the quantum efficiency can be very nearly unity for some wavelengths.

Noise can be calculated from equations (2.40); R_0A values are given in Figure 2.23. Capacitance of InSb detectors is about 450 picofarad per square millimeter.

Like other photovoltaic detectors PV InSb is static sensitive; precautions must be taken to prevent the discharge of static electricity through the device. They are also fragile. Unless specific direction is received from the manufacturer:

- Do not apply voltages greater than ± 100 mV.
- Do not apply currents greater than 150 μA per square millimeter of detector area.
- Do not attempt to clean them.

Ultraviolet (UV) and visible radiation can convert the surface of any exposed *n*-type base material to *p*-type, enlarging the junction area. This

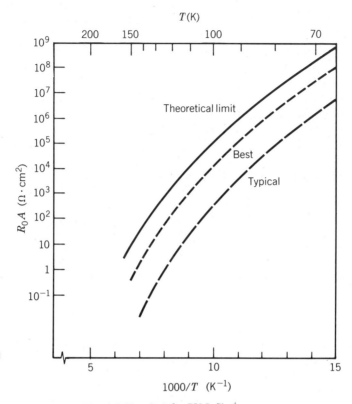

Figure 2.23. R_0A for PV InSb detectors.

can increase the effective area of the detectors, lower the impedance, increase the signals, and cause excessive crosstalk. This is a temporary effect that dissipates when the detector is at room temperature. To avoid this "flashing" effect, do not expose the detectors to direct visible or UV radiation, especially when cold.

2.5.2.   PV Mercury–Cadmium–Telluride (HgCdTe*)

HgCdTe is actually a combination of mercury telluride (HgTe) and cadmium telluride (CdTe). The relative concentrations of the two molocules are deliberately adjusted in the growth process to obtain the desired mix-

---

* The abbreviations HCT, or MCT are also sometimes used. In speech, "merc–cad–tel-luride" often replaces "mercury–cadmium–telluride."

ture. If we let x indicate the relative concentration of CdTe ($x = 0.2$, for example), the concentration of HgTe would be $1 - x$. Thus HgCdTe is sometimes (more properly) written as

$$(HgTe)_{1-x} (CdTe)_x \qquad \text{or sometimes} \qquad Hg_{1-x} Cd_x Te$$

The variable concentration allows one to adjust the cutoff wavelength and operating temperature of the detectors as shown in Figure 2.24. See Reine (1983), Hansen et al (1982), and Schmit and Stelzer (1969) for additional information.

HgCdTe can be used as either a photovoltaic detector or a photocon-

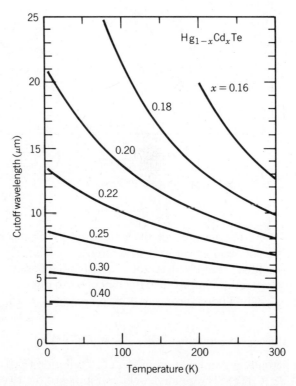

Figure 2.24. Variation of HgCdTe cutoff wavelength with operating temperature for several values of alloy composition x. Plotted from an expression published by Hansen et al. (1982). [From M. B. Reine, "Status of HgCdTe Technology," Infrared Detectors, William L. Wolfe, ed., *Proc SPIE* 443, 2 (1983). Used by permission.]

ductor; both are common. The material can be cut, lapped, and polished to the desired thickness, or it can be grown from vapor or liquid onto a substrate. In either case the detectors are very thin, and therefore fragile. Definition of the detector shape, and lead attachment, are done by sequential plating and photolithography.

Responsivity and noise of PV HgCdTe can be predicted using formulas of Figure 2.19 and Equation 2.41. Assume a quantum efficiency of 60% (uncoated) or unity (for an optimized coating). Theoretical R_0A products are given in Figure 2.25.

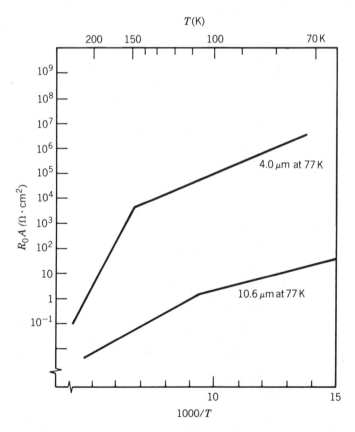

Figure 2.25. Theoretical R_0A for PV HgCdTe detectors. [From Broudy et al. (1978). "Infrared/Charge Coupled Devices (IR/CCD) Hybrid Focal Planes," *Proc. SPIE* 132, 10.

2.5.3. Schottky Barrier Detectors; Platinum Silicide

Platinum silicide (PtSi) detectors are one of a family of Schottky barrier devices. These are relatively new. They are made on silicon wafers, so they take advantage of a well-established Si technology and material availability. They are low in cost, have a high yield, and can be made in very closely packed two-dimensional arrays. They operate at 60 to 70 K. The detection mechanism is distinct from that of the PV and PC detectors discussed earlier. Photons with high-enough energy tunnel through a potential barrier E_b.

The quantum efficiency drops off over the entire spectral range, reaching zero when the cutoff wavelength is reached. These detectors are described by Shepherd (1983) and Aguilera (1987). They give an expression for the quantum efficiency in terms of the barrier energy E_b and a constant for the material called the *Schottky quantum yield*. For Pt:Si they give a barrier energy of 0.22 eV, a cutoff wavelength of about 5.6 μm, and a Schottky quantum yield of about 0.35 eV^{-1}.

2.5.4. Doped Germanium and Silicon

These detectors are made from bulk, predoped germanium and silicon. The simplest detectors are single elements, cut and lapped to size, with

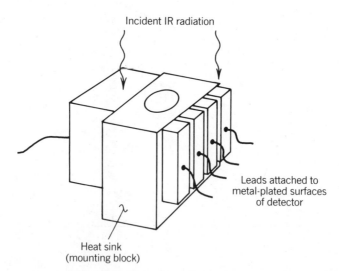

Figure 2.26. Construction of typical Ge:xx and Si:xx detectors, showing a single element and a four-element array.

contacts added, then glued or soldered to a heat sink for ease in mounting. Multielement arrays can be made by separating individual elements after the precut material is mounted to a heat sink (see Figure 2.26).

Detection is dominated by bulk properties; the surface plays little effect. The result is that the detectors are very rugged and stable. The incident surface of an uncoated detector can be cleaned with a sharpened Q-tip, or the entire detector can be cleaned by rinsing it in solvents. Detectors with AR coatings must be treated more carefully.

The thickness of the material used is determined by the optical absorption coefficient for the particular doping. In order to absorb most of the radiation, the path length must be from one to three times the absorption length. Typical path lengths (detector thickness) are 3 mm for most of the doped germanium and silicon detectors. Review articles by

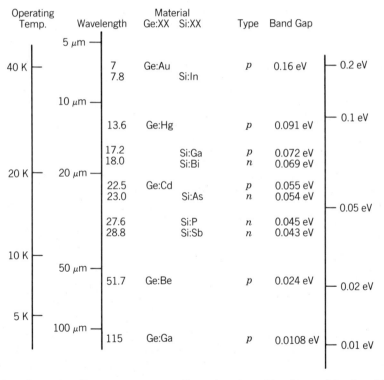

Figure 2.27. Operating temperature, cutoff wavelength, and band gap of the doped germanium and silicon detectors.

Bratt (1977), Bratt, Lewis, and Nielsen (1978), and Sclar (1983) describe the fabrication and operating characteristics of these detectors; they provide a good starting point for more detailed study. Numerous other articles describing specific systems or applications may be found in *Proceedings of SPIE*.

Figure 2.27 shows the operating temperature, band gap, and cutoff wavelengths for doped Ge and Si detectors. The BLIP NEP and D^* predictions require only a knowledge of the background and the quantum efficiency. For high background applications, prediction of noise and responsivity require a knowledge of the quantum efficiency, mobility–lifetime product, and the optimum electric field.

The lifetime of doped germanium and silicon can be deliberately varied: fast response is obtained with high resistivity, short lifetime material, and high responsivity is obtained from low resistance, long lifetime material.

For representative or first cut estimates, use the following values:

Quantum efficiency: 30%
Optimum electric field: 300 V/cm
$\mu\tau$ Product
  High responsivity: 5×10^{-3} cm^2/V
  Standard 1×10^{-3} cm^2/V
  High speed 1×10^{-6} cm^2/V

A great deal of variation is possible: quantum efficiency could be 20% or 40%, the optimum electric field could be 100 or 500 V/cm, and the $\mu\tau$ product could be 3 times higher or lower than the values given. A better estimate for your situation would require measurements made on detectors from ingots grown like the one you plan to use and preferably from material close to the place from which your detector will be cut.

Saturation of the PC Gain. At very low backgrounds the resistance of doped Ge and Si PC detectors becomes very high. This limits the frequency response of the responsivity. The PC gain calculated earlier will still apply at low frequencies and is referred to as the *dc photoconductive gain*. At higher frequencies the gain will saturate to a value of 0.5.

The transition between the dc gain and the saturated gain occurs when the electrical period equals the dielectric relaxation time τ_{DR}. Stated in terms of frequency, this is seen in Equations 2.51 and 2.52.

PC GAIN ROLL-OFF DUE TO DIELECTRIC RELAXATION TIME

$$f_{DR} = \frac{1}{2\pi\tau_{DR}} \qquad (2.51)$$

where

$\tau_{DR} = \epsilon\epsilon_0\rho$

$\quad \epsilon = $ dielectric constant of the detector material $\qquad (2.52)$

$\quad\quad = 16.6$ for germanium, 13.0 for silicon

$\quad \epsilon_0 = $ permeability of free space

$\quad\quad = 8.85 \times 10^{-14}$ F/cm

$\quad \rho = $ resistivity of the detector material ($\Omega\cdot$cm)

Formula (2.51) predicts the point at which the gain and responsivity have dropped by 3 dB, as shown in Figure 2.28. See Blouke et al (1972) and Bratt (1977) for a more detailed discussion.

To predict detector performance when dielectric relaxation time effects may be important, it helps to begin with an assumed electric field, then calculate the following parameters in the order indicated:

Dc PC gain: $\mu\tau E/S$

The corresponding dc current: $\eta QAeG$

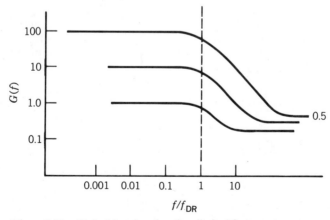

Figure 2.28. Dielectric relaxation-time-limited frequency response.

Resistance and resistivity: $R = V/I$, $\rho = Ra/s$

Dielectric relaxation time: $\tau = \epsilon\epsilon_0\rho$

Frequency at which the gain rolls off: $f_{DR} = (2\pi\tau)^{-1}$

At this point you can sketch the photoconductive gain versus frequency using Figure 2.28 as a guide. Using the gain appropriate to the frequencies of interest, calculate the responsivity and noise values from the equations of Figure 2.19 and (2.43).

2.5.5. PC Mercury-Cadmium-Telluride (HgCdTe)

The composition of mercury-cadmium-telluride detector material, and the effect on cutoff wavelength and its temperature dependence is the same for PC detectors as for PV detectors; this is discussed in section 2.5.2.

When used as a photoconductor, these detectors have a low-resistance surface that provides an electrical short around the bulk, and has a strong effect on the overall detector resistance and voltage responsivity. The actual resistance depends on the surface preparation and control, so— unlike the doped Ge and Si detectors—it cannot be predicted from the background and mobility lifetime product, and must be determined experimentaly. Typical values are 30 to 100 Ω for a square detector.

The prediction of detector performance is similar to that of the doped germanium and silicon detectors. The voltage responsivity is

$$\mathcal{R} = \frac{\lambda}{hc}\, \eta e G R \qquad (2.53)$$

where R is the parallel resistance of the load and detector (bulk and shunt). Assume a quantum efficiency of 0.70 (not coated) or near unity if the coating is optimum for the wavelengths of interest, a mobility lifetime product of 0.15 cm²/V, and an electric field of 5 to 10 V/cm.

Example: The voltage responsivity of a square HgCdTe PC detector 0.004-in. on a side at a wavelength of 10 μm is about 20 kV/W:

$$\mu\tau = (1 \times 10^5 \text{ cm}^2/\text{V}\cdot\text{s})(1.5 \ \mu\text{s}) = 0.15 \text{ cm}^2/\text{V}$$

$$E = 5 \text{ V/cm}$$

$$G = \frac{\mu\tau E}{s} = 75$$

$$A = 100 \times 10^{-6} \text{ cm}^2$$

$$\mathcal{R} = \frac{\lambda}{hc} \eta e GR$$

$$= \frac{10 \times 10^{-4} \text{ cm}}{20 \times 10^{-24} \text{ J·cm}} \, 0.70 \times 1.6 \times 10^{-19} \text{ C} \times 75 \times 50 \, \Omega$$

$$= 21 \times 10^3 \text{ V/W} \qquad \blacksquare$$

Calculate GR noise from the usual PC noise current expression (2.43), then multiply by the actual detector resistance (surface shunt in parallel with bulk) to obtain the GR noise voltage. Johnson noise voltage must also use the actual detector resistance. BLIP D^* and NEP may be calculated from the normal PC BLIP D^* equations, but the decreased detector resistance increases the chances that Johnson noise will affect performance.

Another complication associated with the HgCdTe detectors is the dependence of mobility and lifetime on background and temperature. For a doped germanium detector, the mobility–lifetime product is relatively insensitive to background and temperature—at least in the normal temperature operating range. This means that the responsivity is independent of background and temperature, the resistance varies as 1 over the background, and the noise current decreases as the square root of background.

For HgCdTe this is not true. Reducing the background tends to increase the mobility–lifetime product, so responsivity goes up, and noise will not decrease as fast as the square root of background—it may even increase. This makes diagnostics and predictions difficult.

2.5.6. Lead Salts

Lead sulfide (PbS) and lead selenide (PbSe) are deposited from a chemical solution onto substrates. The resulting film is 1 to 2 μm thick. It is etched to delineate detectors of the desired size and shape. Plated metal is etched to provide electrical contacts (see Figure 2.29). The chemistry of the lead salt detectors can be adjusted to favor operation at any desired temperature range. The designations LTO, ITO, and ATO stand for "low temperature operation," "intermediate temperature operation," and "ambient temperature operation." Because the detectors are thin, the heat associated with large currents will damage them permanently. Do not exceed the recomended bias voltage or current by more than 50%.

Electrostatic discharge is a concern for only the very smalllest of these detectors. Environmental exposure is critical for these detectors, so they are safest in a sealed vacuum package. In general, they should not be exposed to humidities in excess of 50%. The detectors are very sensitive to acids. Even the fine saliva mist or spray from the human mouth contains

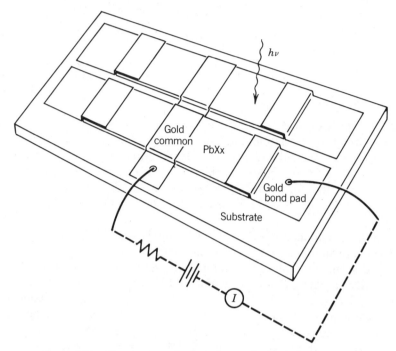

Figure 2.29.   Construction of a four-element lead salt detector array.

enough acid to etch pinholes in the thin films. Although they can be rinsed in some solvents, this should be done only after careful research and planning. A protective overglass is sometimes used on PbS, but there is no protective coat (except its own natural oxide) available for PbSe.

Bright visible light degrades the performance of these detectors, so they are normally handled in light levels of 15 footcandles or less. PbS is more sensitive to light but can be rejuvenated by a special baking operation. The number of times this "light recovery bake" can be done is limited.

Still one more complication associated with the lead salt detectors is the prediction of their performance. Detectivity D^* can be predicted reasonably well by equation (2.47), using quantum efficiencies of 0.10 to 0.30. Models for predicting responsivity and noise are more complicated and less general than for either the doped germanium detectors or HgCdTe detectors. Instead of attempting to predict responsivity the user is advised to consult with a vendor or someone who has operated detectors of the type, size, and operating conditions being considered. Johnson (1983) provides a comprehensive review of lead salt detector history, construction, and performance.

REFERENCES

Aguilera, R. (1987). "256 × 256 Hybrid Schottky Focal Plane Arrays," *Proc. SPIE* 782, 108–113.

Blouke, M. M., E. E. Harp, C. R. Jeffus, and R. L. Williams (1972). "Gain saturation in extrinsic germanium photoconductors operating at low temperatures" *J. Appl. Phys.* 43, 188.

Boyle, W. F., and K. F. Rogers, Jr. (1959). "Resistance Characteristics of a New Low Temperature Bolometer," *J. Opt. Soc. Am.* 49, 66.

Bratt, Peter R. (1977). "Impurity Germanium and Silicon Infrared Detectors," in *Semiconductors and Semimetals,* Vol. 12, *Infrared Detectors II*, Academic Press, New York.

Bratt, P. R., N. N. Lewis, and R. L. Nielsen (1978). "Ge:Ga and Ge:Be Photoconductive Detectors for Far Infrared Astronomy from a Space Platform," in Utilization of Infrared Detectors, Irving J. Spiro, ed., *Proc. SPIE* 132, 97.

Broudy, R., M. Gurnee, S. Iwasa, T. Tredwell, and J. White (1978). "Infrared/Charge Coupled Devices (IR/CCD) Hybrid Focal Planes," *Proc. SPIE* 132, 10.

Donabedian, Martin (1978). "Cooling Systems," Chapter 15 in *The Infrared Handbook*, William L. Wolfe and George J. Zissis, eds., Environmental Research Institute of Michigan, Ann Arbor, Michigan.

Golay, M. J. E. (1947). "A Pneumatic Infra-Red Detector," *Rev. Sci. Instrum.,* 18, 357.

Golay, Marcel J. E. (1952). "Bridges across the Infrared-Radio Gap," *Proc. IRE* 40, 1161.

Hansen, G. L., J. L. Schmit, and T. N. Casselman (1982). *J. Appl. Phys.* 53, 7099.

Hennerich, K., et al. (1966). "The Linearity of Golay Detectors," *Infrared Phys.* 6, 123.

Hudson, Richard D., Jr. (1969). *Infrared System Engineering,* Wiley, New York.

Jacobs, Stephen F. (1978). "Nonimaging Detectors," Section 4 in *Handbook of Optics*, Walter G. Driscoll, ed., William Vaughan, assoc. ed., McGraw-Hill, New York.

Johnson, T. H. (1983). "Lead Salt Detectors and Arrays—PbS and PbSe," *Proc. SPIE* 443, 60.

Jones, R. Clark (1953). "Performance of Detectors for Visible and Infrared Radiation," in *Advances in Electronics*, Vol. 5, Academic Press, New York.

Jones, R. Clark (1957). "Quantum Efficiency of Photoconductors," *Proc. IRIS* 2, 13.

Kittel, Charles, and Herbert Kroemer (1980). *Thermal Physics,* W.H. Freeman, San Francisco.

Limperis, Thomas (1978). "Detectors," Chapter 11 in *The Infrared Handbook*, Williams L. Wolfe and George J. Zissis, eds., Environmental Research Institute of Michigan, Ann Arbor, Michigan.

Low, Frank J. (1961). "Low Temperature Germanium Bolometer," *J. Opt. Soc. Am.* 51, 1300–1304.

Low, Frank J. (1966). "Thermal Detection Radiometry at Short Millimeter Wavelengths," *Proc. IEEE* 54, 477–484.

Reine, M. B. (1983). "Status of HgCdTe Technology," *Proc. SPIE* 443, 2.

Schmit, J. L., and E. L. Stelzer (1969). "Temperature and Alloy Compositional Dependences of the Energy Gap of $Hg_{1-x}Cd_xTe$," *J. Appl. Phys.* 40, 4865.

Schuman, Mark (1965). "Sensitivity of a Selective Pneumatic Detector," *Appl. Opt.* 4, 1442.

Sclar, N. (1984). "Development Status of Silicon Extrinsic IR Detectors—II," *Infrared Detectors*, William L. Wolfe, ed., *Proc. SPIE* 443, 11.

Shepherd, F. D. (1983). "Schottky Diode Based Infrared Sensors," *Proc. SPIE* 443, 42.

Stevens, Norman B. (1970). "Radiation Thermopiles," Chapter 7 in *Semiconductors and Semimetals*, Vol. 5, Academic Press, New York, p. 287.

Strong, John, and Paul W. Lawrence, Jr. (1968). "Bolometer Theory," *Appl. Opt.* 7, 49.

Sze, S. M. (1969). *Physics of Semiconductor Devices*, Wiley-Interscience, New York.

Zahl, Harold A., and Marcel J. E. Golay (1946). "Pneumatic Heat Detector," *Rev. Sci. Instrum.* 17, 511.

SUGGESTED READING

Boyd, Robert W. (1983). *Radiometry and the Detection of Optical Radiation*, Wiley, New York.

Bratt, Peter R. (1977). "Impurity Germanium and Silicon Infrared Detectors," in *Semiconductors and Semimetals*, Vol. 12, *Infrared Detectors II*, Academic Press, New York.

Dereniak, Eustace L., and Devon G. Crowe (1984). *Optical Radiation Detectors*, Wiley, New York. Excellent and up-to-date reference; covers radiometry, detectors, and charge transfer devices. Recommended.

Hudson, Richard D., Jr. (1969). *Infrared System Engineering*, Wiley, New York.

Jones, R. C., D. Goodwin, and G. Pullan (1960). *Standard Procedure for Testing Infrared Detectors and Describing Their Performance*, Office of Director of Defense Research and Engineering, Washington, D.C.

Kingston, R. H. (1978). *Detection of Optical and Infrared Radiation*, Springer-Verlag, Berlin.

Limperis, Thomas (1978). "Detectors," Chapter 11 in *The Infrared Handbook*, William L. Wolfe and George J. Zissis, eds., Environmental Research Institute of Michigan, Ann Arbor, Michigan.

Spiro, Irving J., ed. (1978). *Utilization of Infrared Detectors, Proc. SPIE* 132.

Stevens, Norman B. (1970). "Radiation Thermopiles," in *Semiconductors and Semimetals,* Vol. 5, Academic Press, New York, p. 287.

Van Vliet, K. M. (1958). "Noise in Semiconductors and Photoconductors," *Proc. Inst. Radio Engr.* 46, 1004–1018.

Willardson, R. K., and Albert C. Beer, eds. (1981). *Semiconductors and Semimetals,* Vol. 18, *Mercury Cadmium Telluride,* Academic Press, New York.

Wolfe, William L., ed. (1984). *Infrared Detectors, Proc. SPIE* 443.

PROBLEMS

Detector Types, Mechanisms

1. What is the effect on the frequency response of a thermopile if we choose to increase the size of the junction but otherwise do not change the device?

2. What is the cutoff wavelength for a semiconductor with a band gap of 0.2 eV? 0.02 eV?

3. Which of the two materials in Problem 2 would operate at the warmer temperature?

4. Describe briefly how thermal detectors work.

5. What specific property changes with temperature in the following thermal detectors?

   (a) Mercury-in-glass thermometer

   (b) Carbon bolometer

   (c) Thermopile

   (d) Golay cell

6. What precautions should you observe in handling the following detectors?

   (a) PV InSb

   (b) PbSe

   (c) Ge:Hg

7. Describe briefly the spectral characteristics of the following detectors.

   (a) PV InSb

   (b) Ge:Hg

   (c) Bolometer

8. Mention a few advantages and disadvantages of PV and PC detectors.

Detector Operation

1. List a few ways that detectors can be cooled.
2. Review or list precautions that one must use in handling dewars, and why these precautions are necessary.
3. Why is a load resistor used in a photoconductive detector bias circuit? Would it be necessary if we used a current meter to measure the signal?
4. Sketch the $V-I$ curve for a hypothetical PV detector and a PC detector.
5. (a) Sketch the $V-I$ curve for a PC detector whose resistance is 2 MΩ. Show voltages from 0 to 100 V, and label both axis.

    (b) On the same graph as part (a), sketch the load line for a PC circuit with a 100-V bias battery and a 1- MΩ load resistor.

    (c) From the intersection of the curves of parts (a) and (b), determine the voltage and current that will result if the detector of part (a) is used with the bias circuit of part (b).

    (d) Add the $V-I$ curve for the detector after its resistance is driven down to 1.8 MΩ by increasing the photon flux falling on it.

    (e) Using the graphical method, determine the new detector voltage, and the signal voltage—the change in voltage—due to the increased photon flux.

6. Consider a PV detector of dc current I operated in a feedback amplifier with saturation voltage of 15 V. Show that use of a feedback resistance R_f such that IR_f is about 1 V will provide Johnson noise much less than the detector shot noise, without saturating the amplifier. (Use Equations 2.25 and 2.31.)

Detector Figures of Merit

DETECTOR A

PV InSb
The operating temperature is 77 K.
The RA product is 1×10^6 Ω·cm^2.
The detector is circular, with a 2-mm diameter.
The effective background is 1.0×10^{14} photons/(s·cm^2).
Assume a quantum efficiency of 60%.
The feedback resistor is 1 MΩ, uncooled.
The electrical bandwidth is not given (call it Δf).

---

1. (a) For detector A (PV InSb, shown in Figure 2.30), find:
       (i) The expected current
       (ii) The photoconductive gain
       (iii) The voltage of required bias battery
       (iv) The responsivity [A/(photon/s)]
       (v) The responsivity (V/W)
       (vi) The Johnson noise from the feedback resistor
       (vii) The photon shot noise
    (b) Imagine that you set up detector A and measure a current of 600 nA. Predict:
       (i) The total shot noise

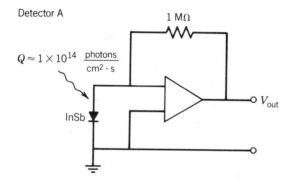

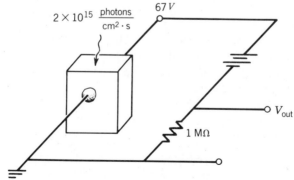

Figure 2.30. Detectors for problems.

      (ii) The NEP

      (iii) The D^*

2. For detector B (shown in Figure 2.30), find:

    (a) The expected current

    (b) The photoconductive gain

    (c) The voltage of required bias battery

    (d) The electric field

    (e) The responsivity [A/(photon/s)]

    (f) The responsivity (V/W)

    (g) The Johnson noise from the load resistor

    (h) The photon GR noise

    (i) The NEP

    (j) The D^*

DETECTOR B

Ge:Hg (PC)

The operating temperature is 5 K.

The size is: 3 mm "deep," 1 mm $\times$ 2 mm exposed to IR, leads on the 2 mm $\times$ 3 mm faces.

The effective background is 2×10^{15} photons/(s·cm^2).

The quantum efficiency is about 65%.

The mobility–lifetime product is 0.5×10^{-3} cm^2/V.

The load resistor is at room temperature, 1 MΩ.

The voltage across the detector is 67 V.

The electrical bandwidth is 50 Hz.

---

3. For detector B, determine the spatial frequency at which the MTF will be zero when scanned in the direction of the current flow. Calculate the expected MTF at four spatial frequencies below or slightly above that limiting frequency.

3

Radiometry

In its strictest sense the word *radiometry* relates to the detection and measurement of radiated electromagnetic energy, but it is often used to describe the prediction or calculation of the power transferred from one object or surface to another. In addition, since many of the concepts of radiometry are common to those of photometry (related to vision and detection by the human eye) and to the transfer of photons, we will use the word "radiometry" to describe all three. (The word *phluometry* has been coined to cover this generalization, but "radiometry" is still a little better known.) The calculations described in this chapter are a necessary part of the characterization of detector performance and the prediction of signal and noise levels.

Throughout this book we concentrate on blackbody sources and test setups used in detector testing. This allows some simplifications that are not always possible in other applications of radiometry. Once the student is familiar with the approach presented here it would be wise to review a more general approach; this is especially necessary if measurement of "real" targets or high precision work is contemplated. The *National Bureau of Standards Self Study Manual on Optical Radiation Measurements* (Nicodemus, 1976) is one source of complete and rigorous instruction. Unfortunately, completeness and rigor generally are achieved at the expense of brevity and ease of reading.

3.1. THE GENERAL METHOD

For a real-life situation, the computations can be very complex. We first describe a simplified case which we can solve easily, then show how to extend it to more realistic cases. We also introduce the vocabulary of radiometry.

To begin the discussion of power or photon transfer by radiation, consider the problem shown in Figure 3.1*a*. We want to know how much energy is radiated from the source to the detector. (The formulas we will

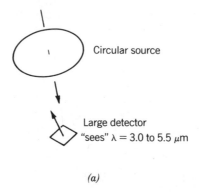

Circular source

Large detector
"sees" $\lambda = 3.0$ to 5.5 μm

(a)

Very small source

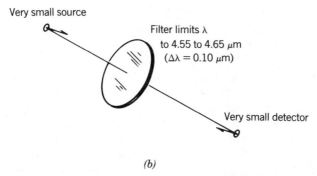

Filter limits λ
to 4.55 to 4.65 μm
$(\Delta\lambda = 0.10$ μm$)$

Very small detector

(b)

Figure 3.1. Two cases of photon transfer: (a) typical situation; (b) incremental case.

arrive at here are not restricted to detector and source situations, but it is helpful to have a specific example in mind.)

We will eventually also want to consider the radiation going the other way—from the detector to the emitter. The formulas we develop will let us do that. (But in many problems one surface will be much warmer than the other, and the emission from the colder surface to the warmer will be negligible.)

It would be nice if we could write a general equation that would work for all situations. That gets too complicated for us at this point. Instead, we start out with the much simpler case shown in Figure 3.1*b*. It can be extended later to more complicated situations.

3.1.1. Incremental Limit

To keep the geometry simple, we begin with the power transferred between two very small surface areas, as shown in Figure 3.1*b*. To remind us that they are very small, we use the notation ΔA_e and ΔA_d. To simplify the spectral considerations we consider only the power in the narrow bandwidth interval $\lambda = \lambda_0 \pm \Delta\lambda/2$. The resulting formula is

INCREMENTAL POWER TRANSFER

$$P = \frac{(\epsilon\ \Delta A_s\ \cos\ \theta_s)\ (\alpha\ \Delta A_r\ \cos\ \theta_r)}{\pi r^2}\ \tau M(\lambda,\ T) \qquad (3.1)$$

The symbols involved are listed in the next paragraph. Don't be alarmed by the length of the formula—you don't have to memorize it, nor must you use it in this form! It merely serves as an outline for some ideas that will be useful later and to introduce the factors that affect us.

Relevant Parameters
  Source

  ϵ: emissivity: a measure of how well it radiates

$$\epsilon = \begin{cases} 1.000 & \text{for a perfect (ideal) emitter} \\ 0.995 & \text{for specially treated black cavities} \\ 0.02 & \text{for some specially selected gold-plated surfaces} \end{cases}$$

ΔA_s: incremental source area, usually in units of cm^2
θ_s: angle between the line connecting our two surfaces and a perpendicular to the source surface

Receiver

α: absorptivity—a measure of how well it absorbs

$$\alpha = \begin{cases} 1.000 & \text{for a perfect (ideal) absorber} \\ 0.995 & \text{for specially treated black cavities} \\ 0.02 & \text{for some specially selected gold-plated surfaces} \end{cases}$$

ΔA_r: incremental receiver area, usually in units of cm^2
θ_r: angle between the line connecting our two surfaces and a perpendicular to the receiver surface

Other

$M(\lambda,\ T)$: spectral exitance of the source at temperature T, wavelength λ
$\Delta\lambda$: spectral bandwidth of interest
r: distance between source and detector

τ: composite transmittance of all the optical components between the source and receiver

Motivation for the Incremental Formula. The factors in the incremental power interchange formula make intuitive sense:

Increasing the areas involved, or the ability of the source to emit, or of the receiver to absorb, all increase the power transferred.

The transmittance of anything in the path between the source and receiver will affect the power transferred.

The cosine factors correct for the fact that the surfaces may not be facing each other directly: unless θ_s and θ_r are zero, the effective area will be less than the actual area (see Figure 3.2). (In all of our work we will assume that the emitter is *Lambertian*—the power emitted falls off with angle as predicted by our incremental law. This is valid for a blackbody. Other behavior is possible but will not be important for most detector testing applications.)

The r^2 in the denominator makes sense if you think about a sphere of radius r surrounding the emitter (see Figure 3.3). The area of the surface of the sphere is $4\pi r^2$. If we increase the source-to-surface distance r, the surface area of the sphere increases, so the density of radiation (the power per unit area, or incidance) will be decreased as $1/r^2$. We do not need the factor of 4 in our equation: A real surface emits in only one direction, so we deal only with half a sphere, and the average over all angles will provide another factor of 2.

All of the above are geometrical or material parameters. The spectral exitance M contains the physical laws about how radiated power depends on the wavelengths and temperatures involved: M is a function of wavelength and emitter temperature. The formula for exitance in terms of wavelength and temperature is called *Planck's law*.

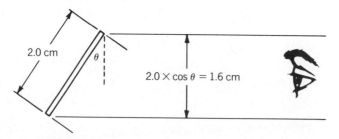

Figure 3.2.   Cosine factors account for foreshortening.

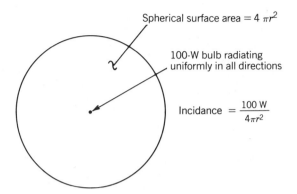

Figure 3.3. $1/\pi r^2$ accounts for expanding surface area.

3.1.2. Extensions

The incremental formula is pretty easy to use: just multiplication and division processes. Everything needed is defined. It applies, however, only to very special situations: infinitesimally small detectors, emitters, and spectral bandwidths. What do we do if we find ourselves with a large detector, or a wide field of view, or a bandpass that is not limited? Our equation requires extension in four ways, which can often be treated separately. These are described here and illustrated in Figure 3.4.

1. We add together (or integrate) several expressions such as the one above until we have included all the *wavelengths* of interest to our problem. This part of the operation is *spectral* integration.
2. We add together (or integrate) similar expressions to cover the surface *area* of a real detector and real emitter instead of just the very small detector and emitter areas we used for our starting equation. This is *spatial* or *geometrical* integration.
3. We consider the way in which the situation changes with *time*. This is *temporal* variation. The change in signal with time is referred to as *modulation*.
4. Electromagnetic radiation, including IR, involves sinusoidaly varying electric fields. These fields are oriented perpendicular to the direction of travel, so any wave can be broken into two possible components. For example, if the wave is headed straight up, one component would include those whose electric fields were oriented east–west and the other would include the north–south fields. These two components are called the two polarizations. We discuss polarization only in the case of off-axis reflections, in Chapter 9.

Spectral: break wide bandpass into many narrow bands

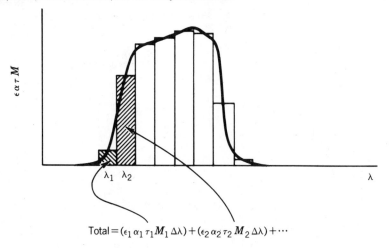

$$\text{Total} = (\epsilon_1\,\alpha_1\,\tau_1 M_1\,\Delta\lambda) + (\epsilon_2\,\alpha_2\,\tau_2\,M_2\,\Delta\lambda) + \cdots$$

Spatial: break large area into many small ones

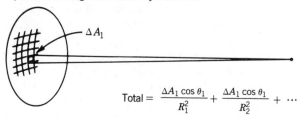

$$\text{Total} = \frac{\Delta A_1 \cos\theta_1}{R_1^2} + \frac{\Delta A_1 \cos\theta_1}{R_2^2} + \cdots$$

Temporal effects: break waveform into many sine waves

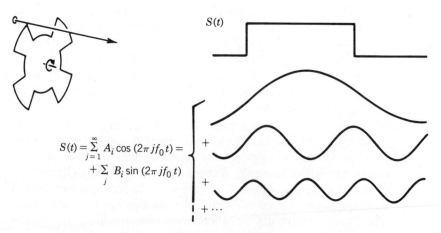

$$S(t) = \sum_{j=1}^{\infty} A_i \cos\left(2\pi j f_0 t\right) = \\ + \sum_j B_i \sin\left(2\pi j f_0 t\right)$$

Figure 3.4. Extensions of the incremental case to real situations.

Before proceding with the first three extensions, we show how different users group some of the factors together in a way that is convenient for their purposes, and give that group of factors a name.

3.1.3. Nomenclature and Notation

For many applications it is convenient to group the geometrical factors in the Planck's law calculations in different ways, resulting in physical quantities with different names but the same spectral distribution. For detector work the flux density at the detector plane is handy. It is called the *incidance, E*. Other combinations of the factors in the radiometric formulas are the *sterance*, the *exitance*, the *intensity*, and the *flux*. Unfortunately, there is not a complete agreement on the names and symbols for these quantities, so the student may encounter two or more words and symbols for the same quantity.

The nomenclature, symbols, and units used in this book are listed in Table 3.1; a brief discussion of the motivation for these and other choices follows. Make sure that you know which is intended when you see radiometric symbols in formulas. The units are often helpful in resolving ambiguities, but even the units do not always prevent ambiguity.

Evolution of the Nomenclature. The nomenclature and notation for radiometric, photometric, and quantum calculations evolved independently, and different words were used to describe analogous parameters. In radiometry, the nomenclature and notation described by Bell (1959) were well established. Photometry had its own nomenclature, so someone familiar with one field had to learn a new set of words when trying to work in another field, even though the geometry was identical. Neither vocabulary lent itself to use in the other field because words such as "radiance" and "irradiance" were clearly radiometric in origin, while "luminance" and "illuminance" were related to the lumen—the unit of photometric flux.

Jones (1963), Nicodemus, Geist, Zolewski and others have attempted to take advantage of the commonality of radiometric, photometric, and photon-flux calculations; see Spiro's column "Radiometry and Photometry" (1974) for some of the history and debate on this subject. The intention was to use just one vocabulary—with the appropriate modifiers—to describe radiometric, photometric, and photon problems.

By 1967 several international committees and organizations had agreed on a new scheme, described by McAdam (1967) and Hudson (1969, Appendix 2). They used the symbols and much of the nomenclature of Table 3.1, but they also referenced the older symbols and nomenclature. Despite

Table 3.1   Radiometric Nomenclature

Word Used in this Book	CIE, ANSI Name	Symbol	Units
Flux	Flux	Φ	
Radiant flux	Radiant flux	Φ_e	Watts (W)
Photon flux	—[a]	Φ_q	Photons/s
Luminous flux	Luminous flux	Φ_v	Lumens (lm)
Exitance	Exitance	M	
Radiant exitance	Radiant exitance	M_e	W/cm^2
Photon exitance	—	M_q	$Photons/(cm^2 \cdot s)$
Luminous exitance	Luminous exitance	M_v	lm/cm^2
Intensity	Intensity	I	
Radiant intensity	Radiant intensity	I_e	W/sr
Photon intensity	—	I_q	$Photons/(sr \cdot s)$
Luminous intensity	Luminous intensity	I_v	lm/sr
Sterance	—	L	
Radiant sterance	Radiance	L_e	$W/(cm^2 \cdot sr)$
Photon sterance	—	L_q	$Photons/(cm^2 \cdot sr \cdot s)$
Luminous sterance	Luminance	L_v	$lm/(cm^2 \cdot sr)$
Incidance	—	E	
Radiant incidance	Irradiance	E_e	W/cm^2
Photon incidance	—	E_q, Q	$Photons/(cm^2 \cdot s)$
Luminous incidance	Illuminance	E_v	lm/cm^2

[a] — indicates that term is not specified in ANSI or CIE.

its "official" adoption, this system was not accepted completely. For example, R. D. Hudson, one of the members of the Nomenclature Committee of the Optical Society of America, provides this system as an appendix to his book *IR Systems Engineering* (Hudson, 1969) but uses the older system, stating that he does not favor use of the adopted system.

Even today the older system of names and symbols is often encountered. This is due in part to reluctance to change, to the influence of Hudson's book, and because the word "exitance," seems contrived or artificial. The choice of the symbol E for incidance is a problem for people working with detectors since E is the common symbol for electric field.

Current Nomenclature. The most "official" radiometric nomenclature is now that of the International Commission on Illumination (CIE 1970) and the American National Standards Institute (ANSI/IES RP-16-1986). Those standards do not address the photon-flux quantities and they are only partially successful in achieving a common vocabulary for radiant and luminous fluxes.

As shown in Table 3.1, the nomenclature used in this book is consistent with the CIE and ANSI/IES standards, except that *incidance* is substituted for *irradiance/illuminance*, and *sterance* is substituted for *radiance/ luminance*. *Incidance* and *sterance* were proposed by Jones (1963). The resulting set of terms has the desired commonality: we can speak of radiant incidance, or photon-flux incidance, or luminous incidance. I find Jones's (1963) reasoning very convincing:

> The words radiance and irradiance (and luminance and illuminance) are not very different and are easily confused, particularly in oral discourse. Yet the concepts to which they refer are really quite different. . . . The real advantage of the proposal is that it facilitates generalization. Photometry and radiometry share a geometrical structure that is more general than both.

One field which could use the proposed notation has been mentioned— *photon-flux*. Jones cites three other fields.

Incidance and *sterance* were included as acceptable alternates by the Nomenclature Committee of the American Optical Society (McAdam 1967, Hudson 1969, Appendix 2), but have not found general use. The word *exitance* was proposed for similar reasons and it has been accepted by CIE and ANSI.

Nicodemus endorses the idea of a common vocabulary, but feels that the word *incidance* will always be troublesome because it is so similar to incidence. He suggests the words *areance* (instead of *incidance* and *exitance*), *pointance* (instead of intensity), and *sterance* (see Spiro, 1974).

When working with the detector figures of merit we will make one exception out of deference to tradition and to avoid the confusion between incidance and electric field: the letter Q will be used for the photon incidance: $Q = E_q$. If the energy is spread over the entire Planck distribution, we refer to the "total" incidance, and if the incidance is in a finite spectral band, we talk about the "in-band" incidance. No special symbols are standard for these cases.

To state that the energy is limited to a narrow spectral bandpass, we use the word *spectral* as a prefix: "spectral incidance." This is indicated symbolically by adding the subscript λ to the usual symbol. The units of a spectral parameter are those for the total or in-band quantity *per unit wavelength*. The units of exitance, for example, are W/cm^2; for spectral exitance the units are $W/(cm^2 \cdot \mu m)$. To get in-band exitance, we multiply or integrate the spectral value times a spectral bandpass, in micrometers.

All of these expressions have counterparts or analogs in terms of photons exchanged, and we can work in either photons or watts. It is more convenient to use photons when working with photon detectors, but several figures of merit are written in terms of watts for historical reasons.

The CIE/ANSI notation uses the same basic symbols for the radiometric (watts) and photometric (weighted for the spectral response of the human eye) equivalents. The subscript e is used to identify radiometric terms, and v is used for photometric terms. We use the method proposed by Murray, Nicodemus and Wunderman (1971) to extend this convention by using the subscript q for photon equivalents. Where the context makes the meaning clear, or where the general case applies, the subscript is dropped. Be aware however, that the IR community is sometimes not consistent in the nomenclature for the photon analogs.

The five different words used in radiometry are illustrated in Figure 3.5.

Flux. The radiant flux Φ is the power (rate at which energy is transferred). The units are watts (W). One watt is 1 joule per second.

Exitance. We define the radiant exitance M as the power into a hemisphere per unit emitter area. The units are watts per square centimeter (W/cm^2) for the total exitance, and W/cm$^2 \cdot \mu$m for the spectral exitance.

Intensity. Describing the geometry of effective emitter size when viewed from a distance away (or collector size when radiated from a distance) will be more convenient if we work with *solid angles*. These are discussed in section 3.3.1, but for now note that they are an analog or extension of our conventional plane angles (360° or π radians to a circle) to three dimensions, and that the units are steradians: 4π steradians to a sphere. We define the intensity I as the power from a given source per unit projected solid angle:

$$I = \frac{\Phi}{\Omega} \qquad (3.2a)$$

The units are watts/steradian (W/sr) for total radiant intensity, and W/(sr$\cdot\mu$m) for spectral radiant intensity.

The intensity is useful for describing the radiation from a target or source. We do not really care (and may not know) the area or temperature of a satellite or a star. Given the intensity and the solid angle subtended by the collector, we can calculate the total received power:

$$\Phi = I\Omega \qquad (3.2b)$$

Power Φ

Φ watts collected

Exitance M

Source area A_s

$\Phi_{180°}$ watts emitted into hemisphere

$$M = \frac{\Phi_{180°}}{A_s}$$

Intensity I

Receiver

$$\Omega_r \simeq \frac{A_r}{d^2}$$

$$I = \frac{\text{power } \Phi \text{ to receiver}}{\Omega_r}$$

Sterance L

Source area A_s

Ω_r

$$L = \frac{\text{power } \Phi \text{ to receiver}}{A_s \cdot \Omega_r}$$

Incidance E

Receiver area A_r

$$E = \frac{\text{power } \Phi \text{ to receiver}}{A_r}$$

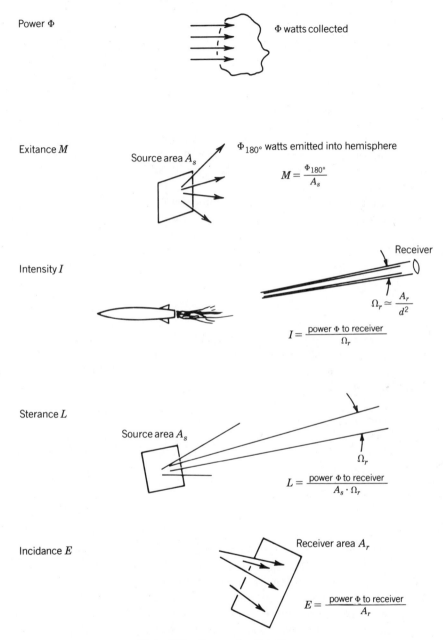

Figure 3.5.   Terms of radiometry.

Sterance. The sterance L is the power transferred per unit emitter area and unit receiver projected solid angle:

$$L = \frac{d^2\phi}{\cos\theta\ dw\ dA} = \frac{d^2\theta}{d\Omega\ dA} \tag{3.3}$$

The units of radiant sterance are W/(cm²·sr) for the total sterance, and W/(cm²·sr·μm) for the spectral sterance. There is a nice symmetry about the power exchange formula when expressed in terms of sterance or exitance: The area of the emitter, the area of the collector, and the angles between the normals and the connecting path all enter the formula in the same way.

For a blackbody or other Lambertian source, the exitance M and the sterance L differ only by a factor of π, the effective solid angle of the 180° field of view:

$$M = L\pi \tag{3.4a}$$

$$L = \frac{M}{\pi} \tag{3.4b}$$

Many people are more accustomed to using the exitance than the sterance, and some graphs show the exitance instead of the sterance. Both are acceptable and accurate if used correctly. We use "exitance" in the remainder of this book.

Incidance. The three previous parameters (exitance, intensity, and sterance) characterize the source. The incidance characterizes the radiation at some point in space—usually at the detector plane. We define radiant incidance E as the power per unit collector area from a given source:

$$E = \frac{\Phi}{A_c} \tag{3.5a}$$

The units are watts per square centimeter (W/cm²) for the total incidance and (W/cm²·μm) for the spectral incidance. Note that both the exitance and incidance are flux densities: powers per unit area. The exitance can be calculated given only the temperature of the blackbody, but the incidance depends also on the solid angle subtended by the blackbody.

Incidance is convenient for describing the radiant environment at a detector test station. It can be precalculated and posted at the station;

the power Φ to the detector is then easy to calculate:

$$\Phi = EA_d \tag{3.5b}$$

(If the detector is in a dewar with a window or filter, they will reduce the total power to the detector, and we will need to include their transmittances in our calculation.) For most of our detector testing we will refer to the incidance—either in W/cm^2 or in photons/(s·cm^2).

3.1.4. Blackbodies, Graybodies, and Emissivity

Consider the emission from disks made of a variety of materials, but all of the same size, and all at the same temperature. If we plot the emitted power versus wavelength, we will find that each material has a characteristic graph—the solid lines on Figure 3.6a—but that they all fall at or below a smooth limiting curve—the dashed curve. Some of these radiators will come very close to the smooth curve, but none will exceed it.

Now construct cavities of these materials, making the area of the opening the same size as our original disks, but with inside surface areas much larger than the opening. Heat those cavities to the same temperature as our original disks. The resulting emission curves will match the smooth limiting curve very well. [Some cavity designs work better than others, but we need not dwell on that subtlety now. Papers by Chandos and Chandos (1974) and Bartell and Wolfe (1976) may be consulted for rigorous derivations of the effect of cavity design on emitting properties.]

When an ideal cavity is cool, the opening appears perfectly black to the eye. Because of this we call these ideal emitters *blackbodies*, and the ideal curve is the "blackbody radiation curve." Objects with emissive curves similar to that of a blackbody, but lower, are sometimes called *graybodies*, and the ratio between the actual exitance of an object and that of the blackbody value is the emissivity ϵ:

$$\text{emissivity} = \epsilon = \frac{M_{\text{actual}}}{M_{\text{ideal}}} \tag{3.6}$$

For an ideal blackbody, $\epsilon = 1$. For most of our laboratory "blackbodies," ϵ is within a few percent of unity.

For opaque materials the emissivity ϵ at a given wavelength λ is unity minus the reflectance ρ at that wavelength.

$$\epsilon(\lambda) = 1 - \rho(\lambda) \tag{3.7a}$$

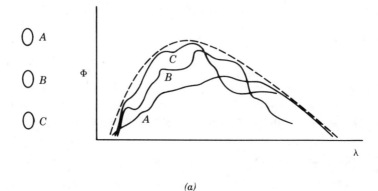

(a)

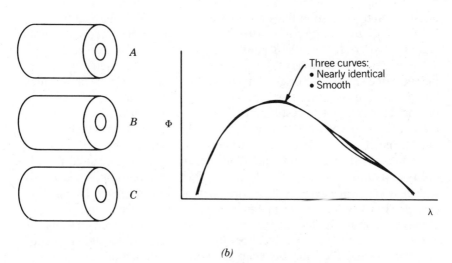

(b)

Figure 3.6. Spectral content of power depends on source material unless cavity shape is used. Emissivity curves. (a) Consider spectral power from 3-disks: same area, same temperature, different materials, A, B, C. Consider spectral power from 3 cavities: same geometry, same temperatures, different materials A, B, C.

If the cavity and surrounding are at the same temperature, (3.7a) also applies to the average emissivity $\bar{\epsilon}$ and average reflectance $\bar{\rho}$:

$$\bar{\epsilon} = 1 - \bar{\rho} \qquad (3.7b)$$

Good emitters are poor reflectors, and vice versa. Cavities are good emitters—they trap incoming light (low reflectivity) and have emissivity near unity.

In the article "Blackbody, blackbody simulator, blackbody simulator cavity, blackbody simulator cavity aperture, and blackbody simulator aperture are each different from one another," Bartell (1989b) points out important conceptual differences which should be recognized but are generally overlooked when we speak of blackbodies. A (true) blackbody is an idealization. The devices in our laboratories that we *call* blackbodies are actually *blackbody simulators*. They use approximately isothermal cavities and a separated aperture. For a well-designed simulator, accurate radiometric calculations can be done by treating that separated aperture (not the cavity itself, nor the aperture of the cavity) as the IR source. The IR incidence depends on the distance from the separated aperture.

3.2. SPECTRAL CONTENT

In this section we talk about the distribution of energy (and photon rate) over the spectrum (wavelengths) of interest. We also describe methods of calculating the energy (and photon rate) in narrow bandwidths, in wide bandwidths, and in the entire spectral region (zero to infinite wavelengths). The spectral distribution is defined by Planck's law, so we start there.

3.2.1. Planck's Law for Spectral Exitance

Planck's law is a statement of how the radiant heat transfer depends on temperature and wavelength. It can be in terms of either power or photons per second, and is usually tailored to describe the sterance L or the exitance M. We will work in terms of exitance.

The exitance formulas are most conveniently used in a compressed form which will be described a little later, but it may be of interest to have the equations in their complete form:

$$M_q(\lambda, T) = \frac{2\pi c}{\lambda^4(e^{hc/\lambda kT} - 1)} \qquad (3.8a)$$

$$M_e(\lambda, T) = \frac{2\pi hc^2}{\lambda^5(e^{hc/\lambda kT} - 1)} \qquad (3.8b)$$

where c = speed of light in a vacuum
  $\simeq 2.998 \times 10^{10}$ cm/s
 h = Planck's constant
  $\simeq 6.626 \times 10^{-34}$ J·s
 k = Boltzmann's constant
  $\simeq 1.381 \times 10^{-23}$ J/K

Note that

$$M_e(\lambda, T) = M_q(\lambda, T) \frac{hc}{\lambda}$$

Since we will normally use mixed length units (wavelengths in μm, areas in cm^2), some conversion of units is necessary to use the formulas. The more compact form described next eliminates the need for this conversion.

To simplify writing the formulas, a single letter (x) is sometimes substituted for the exponent, and the formulas and the computations are further simplified by lumping the required fundamental constants into three new constants: c_1, c_1', and c_2. This form is recommended for practical spectral exitance calculations, as described in Equations 3.9a and 3.9b.

PLANCK'S LAW FOR SPECTRAL EXITANCE

$$M_q(\lambda, T) = \frac{c_1'}{\lambda^4(e^x - 1)} \qquad \text{photons/(cm}^2\text{·s·}\mu\text{m)} \qquad (3.9a)$$

$$M_e(\lambda, T) = \frac{c_1}{\lambda^5(e^x - 1)} \qquad \text{W/(cm}^2\text{·}\mu\text{m)} \qquad (3.9b)$$

where $\quad x = \dfrac{c_2}{\lambda T}$ $\hfill (3.9c)$

λ = wavelength (μm)
T = temperature of the source (kelvin)
c_1' = first radiation constant for photon exitance
$\quad = 2\pi c$
$\quad = 1.8836515 \times 10^{23}$ photons·μm^3/(cm^2·s)
$\quad \simeq 1.884 \times 10^{23}$ photons·μm^3/(cm^2·s)
c_1 = first radiation constant for radiant exitance
$\quad = 2\pi hc^2$
$\quad = 3.7417749 \times 10^4$ W·μm^4/cm^2
$\quad \simeq 3.74 \times 10^4$ W·μm^4/cm^2
c_2 = second radiation constant
$\quad = hc/k$
$\quad = 14387.69$ μm·K
$\quad \simeq 14388$ μm K

The first value listed for the radiation constants is a current best estimate (Cohen and Taylor, 1987); the second is an approximation that will yield results accurate to 0.05% or better; this will be discussed in a later section.

Limiting Cases. Two limiting cases are of interest for historical reasons and because they are sometimes useful for approximate calculations:

1. Wien's law applies if $x \gg 1$ ($\lambda \leq \lambda_{pk}$)

$$M_q(\lambda, T) \simeq \frac{c_1'}{\lambda^4 e^x} \qquad (3.10a)$$

$$M_e(\lambda, T) \simeq \frac{c_1}{\lambda^5 e^x} \qquad (3.10b)$$

Wien's law is in error by less than 1.5% at λ_{pk}, and improves rapidly in accuracy for wavelengths less than λ_{pk}.

2. The Rayleigh–Jeans law applies if $x \ll 1$ ($\lambda \gg \lambda_{pk}$)

$$M_q(\lambda, T) \simeq \frac{c_1' kT}{\lambda^3 hc} = \frac{c_1' T}{c_2 \lambda^3} \qquad (3.11a)$$

$$M_e(\lambda, T) \simeq \frac{c_1 kT}{\lambda^4 hc} = \frac{c_1 T}{c_2 \lambda^4} \qquad (3.11b)$$

As we mentioned in Section 3.13, we could just as well use Planck's law to give the blackbody sterance L:

$$L(\lambda, T) = \frac{M(\lambda, T)}{\pi} \qquad (3.12)$$

The spectral exitance can be calculated directly from Planck's law, read directly or interpolated from graphs or tables of exitance versus wavelength (see Figures 3.7 and 3.8, and Tables 3.2 and 3.3). These can be read with enough accuracy to be useful for quick calculations and checks on other sources.

Still a third method is to use a radiation slide rule. In the precomputer era, radiation slide rules were common. Now they are hard to find, but even the simplest and least accurate is very useful, and accurate enough

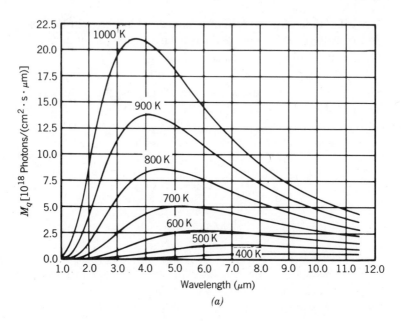

(a)

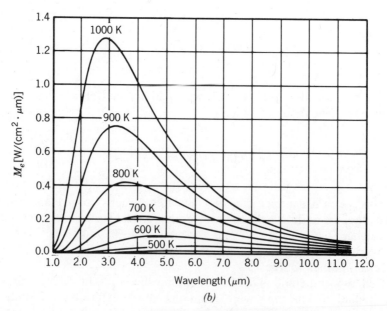

(b)

Figure 3.7. Planck's law: spectral exitance versus wavelength: (a) M_q linear plot; M_e linear plot.

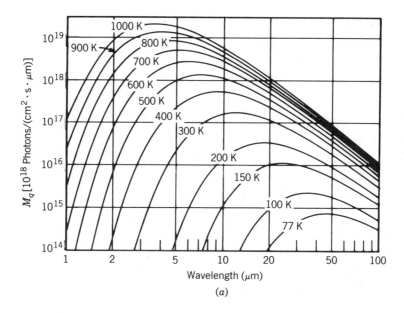

(a)

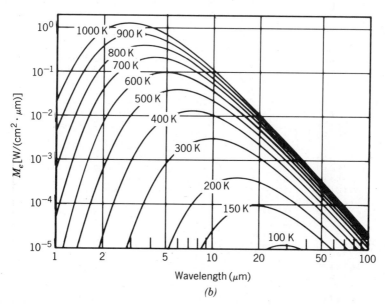

(b)

Figure 3.8. Planck's law: spectral exitance versus wavelength: (a) M_q log plot; (b) M_e log plot.

Table 3.2   Spectral Photon Exitance [photons/(cm·s·μm)]

Wavelength (microns)	Temperature (K)					
	200	273	293	300	400	500
1.00	1.08E-08	2.43E+00	8.89E+01	2.79E+02	4.50E+07	6.00E+10
1.50	5.52E+01	2.05E+07	2.25E+08	4.84E+08	1.43E+12	1.73E+14
2.00	2.81E+06	4.23E+10	2.56E+11	4.54E+11	1.82E+14	6.64E+15
2.50	1.53E+09	3.37E+12	1.42E+13	2.25E+13	2.72E+15	4.84E+16
3.00	8.96E+10	5.46E+13	1.81E+14	2.65E+14	1.44E+16	1.59E+17
3.50	1.49E+12	3.62E+14	1.01E+15	1.41E+15	4.32E+16	3.38E+17
4.00	1.14E+13	1.40E+15	3.43E+15	4.57E+15	9.15E+16	5.53E+17
4.50	5.24E+13	3.77E+15	8.38E+15	1.08E+16	1.55E+17	7.69E+17
5.00	1.70E+14	7.97E+15	1.64E+16	2.06E+16	2.27E+17	9.57E+17
5.50	4.30E+14	1.42E+16	2.73E+16	3.36E+16	2.98E+17	1.11E+18
6.00	9.02E+14	2.23E+16	4.06E+16	4.91E+16	3.63E+17	1.21E+18
6.50	1.65E+15	3.18E+16	5.53E+16	6.60E+16	4.19E+17	1.28E+18
7.00	2.70E+15	4.22E+16	7.05E+16	8.31E+16	4.63E+17	1.31E+18
7.50	4.07E+15	5.29E+16	8.55E+16	9.96E+16	4.96E+17	1.31E+18
8.00	5.72E+15	6.34E+16	9.95E+16	1.15E+17	5.19E+17	1.30E+18
8.50	7.62E+15	7.34E+16	1.12E+17	1.28E+17	5.32E+17	1.27E+18
9.00	9.70E+15	8.24E+16	1.23E+17	1.40E+17	5.38E+17	1.22E+18
9.50	1.19E+16	9.05E+16	1.32E+17	1.49E+17	5.37E+17	1.18E+18
10.00	1.42E+16	9.74E+16	1.40E+17	1.57E+17	5.31E+17	1.12E+18
10.50	1.64E+16	1.03E+17	1.46E+17	1.63E+17	5.21E+17	1.07E+18
11.00	1.86E+16	1.08E+17	1.50E+17	1.67E+17	5.08E+17	1.01E+18
11.50	2.07E+16	1.11E+17	1.53E+17	1.69E+17	4.94E+17	9.61E+17
12.00	2.27E+16	1.14E+17	1.54E+17	1.70E+17	4.77E+17	9.08E+17
12.50	2.45E+16	1.16E+17	1.55E+17	1.70E+17	4.60E+17	8.58E+17
13.00	2.62E+16	1.16E+17	1.54E+17	1.69E+17	4.42E+17	8.10E+17
13.50	2.76E+16	1.17E+17	1.53E+17	1.67E+17	4.25E+17	7.64E+17
14.00	2.89E+16	1.16E+17	1.52E+17	1.65E+17	4.07E+17	7.20E+17
15.00	3.10E+16	1.14E+17	1.46E+17	1.59E+17	3.72E+17	6.41E+17
16.00	3.24E+16	1.11E+17	1.40E+17	1.51E+17	3.39E+17	5.70E+17
17.00	3.33E+16	1.06E+17	1.33E+17	1.43E+17	3.09E+17	5.09E+17
18.00	3.36E+16	1.01E+17	1.25E+17	1.34E+17	2.81E+17	4.55E+17
19.00	3.35E+16	9.62E+16	1.18E+17	1.26E+17	2.56E+17	4.08E+17
20.00	3.32E+16	9.10E+16	1.11E+17	1.18E+17	2.34E+17	3.66E+17
21.00	3.26E+16	8.57E+16	1.03E+17	1.10E+17	2.13E+17	3.30E+17
22.00	3.18E+16	8.06E+16	9.67E+16	1.03E+17	1.95E+17	2.98E+17
13.00	2.62E+16	1.16E+17	1.54E+17	1.69E+17	4.42E+17	8.10E+17
24.00	2.98E+16	7.11E+16	8.43E+16	8.91E+16	1.63E+17	2.45E+17
25.00	2.88E+16	6.67E+16	7.87E+16	8.30E+16	1.50E+17	2.23E+17
26.00	2.77E+16	6.25E+16	7.35E+16	7.74E+16	1.38E+17	2.04E+17
27.00	2.65E+16	5.87E+16	6.86E+16	7.22E+16	1.27E+17	1.86E+17
28.00	2.54E+16	5.50E+16	6.42E+16	6.74E+16	1.17E+17	1.71E+17
29.00	2.43E+16	5.17E+16	6.00E+16	6.30E+16	1.08E+17	1.57E+17
30.00	2.33E+16	4.85E+16	5.62E+16	5.89E+16	1.00E+17	1.45E+17
35.00	1.84E+16	3.58E+16	4.09E+16	4.28E+16	7.00E+16	9.84E+16
40.00	1.46E+16	2.69E+16	3.05E+16	3.18E+16	5.05E+16	6.99E+16
45.00	1.16E+16	2.06E+16	2.32E+16	2.41E+16	3.75E+16	5.13E+16
50.00	9.37E+15	1.61E+16	1.80E+16	1.87E+16	2.86E+16	3.87E+16
55.00	7.63E+15	1.28E+16	1.43E+16	1.48E+16	2.23E+16	3.00E+16
60.00	6.27E+15	1.03E+16	1.15E+16	1.19E+16	1.77E+16	2.36E+16
70.00	4.37E+15	6.99E+15	7.72E+15	7.97E+15	1.17E+16	1.54E+16
80.00	3.16E+15	4.93E+15	5.43E+15	5.60E+15	8.10E+15	1.06E+16
90.00	2.35E+15	3.61E+15	3.96E+15	4.08E+15	5.84E+15	7.62E+15
100.00	1.79E+15	2.72E+15	2.97E+15	3.06E+15	4.35E+15	5.65E+15

600	700	800	900	1000	2000
7.26E+12	2.23E+14	2.91E+15	2.15E+16	1.06E+17	1.42E+20
4.24E+15	4.17E+16	2.31E+17	8.75E+17	2.54E+18	3.10E+20
7.31E+16	4.05E+17	1.46E+18	3.98E+18	8.85E+18	3.32E+20
3.29E+17	1.30E+18	3.63E+18	8.07E+18	1.53E+19	2.88E+20
7.86E+17	2.46E+18	5.81E+18	1.13E+19	1.94E+19	2.33E+20
1.33E+18	3.54E+18	7.41E+18	1.32E+19	2.09E+19	1.84E+20
1.84E+18	4.34E+18	8.30E+18	1.38E+19	2.07E+19	1.46E+20
2.24E+18	4.82E+18	8.60E+18	1.36E+19	1.96E+19	1.16E+20
2.51E+18	5.02E+18	8.49E+18	1.28E+19	1.80E+19	9.37E+19
2.66E+18	5.02E+18	8.13E+18	1.19E+19	1.62E+19	7.63E+19
2.72E+18	4.89E+18	7.64E+18	1.09E+19	1.45E+19	6.27E+19
2.71E+18	4.67E+18	7.08E+18	9.86E+18	1.30E+19	5.21E+19
2.64E+18	4.40E+18	6.51E+18	8.90E+18	1.15E+19	4.37E+19
2.54E+18	4.11E+18	5.95E+18	8.02E+18	1.02E+19	3.70E+19
2.42E+18	3.82E+18	5.43E+18	7.21E+18	9.13E+18	3.16E+19
2.28E+18	3.53E+18	4.95E+18	6.49E+18	8.14E+18	2.71E+19
2.15E+18	3.26E+18	4.50E+18	5.85E+18	7.28E+18	2.35E+19
2.01E+18	3.00E+18	4.10E+18	5.28E+18	6.52E+18	2.04E+19
1.88E+18	2.77E+18	3.74E+18	4.77E+18	5.86E+18	1.79E+19
1.76E+18	2.55E+18	3.41E+18	4.32E+18	5.28E+18	1.58E+19
1.64E+18	2.35E+18	3.12E+18	3.93E+18	4.77E+18	1.39E+19
1.53E+18	2.17E+18	2.85E+18	3.57E+18	4.32E+18	1.24E+19
1.42E+18	2.00E+18	2.61E+18	3.26E+18	3.92E+18	1.11E+19
1.33E+18	1.85E+18	2.40E+18	2.98E+18	3.57E+18	9.92E+18
1.24E+18	1.71E+18	2.21E+18	2.73E+18	3.26E+18	8.92E+18
1.16E+18	1.58E+18	2.03E+18	2.50E+18	2.98E+18	8.06E+18
1.08E+18	1.47E+18	1.88E+18	2.30E+18	2.73E+18	7.30E+18
9.43E+17	1.27E+18	1.61E+18	1.96E+18	2.31E+18	6.05E+18
8.27E+17	1.10E+18	1.38E+18	1.68E+18	1.97E+18	5.06E+18
7.28E+17	9.60E+17	1.20E+18	1.45E+18	1.69E+18	4.28E+18
6.43E+17	8.42E+17	1.05E+18	1.25E+18	1.47E+18	3.65E+18
5.71E+17	7.41E+17	9.17E+17	1.10E+18	1.28E+18	3.14E+18
5.08E+17	6.56E+17	8.08E+17	9.62E+17	1.12E+18	2.72E+18
4.54E+17	5.83E+17	7.15E+17	8.49E+17	9.84E+17	2.37E+18
4.07E+17	5.20E+17	6.36E+17	7.53E+17	8.71E+17	2.08E+18
1.24E+18	1.71E+18	2.21E+18	2.73E+18	3.26E+18	8.92E+18
3.31E+17	4.19E+17	5.09E+17	6.00E+17	6.91E+17	1.62E+18
3.00E+17	3.78E+17	4.58E+17	5.39E+17	6.20E+17	1.45E+18
2.72E+17	3.42E+17	4.13E+17	4.85E+17	5.58E+17	1.29E+18
2.48E+17	3.11E+17	3.74E+17	4.39E+17	5.04E+17	1.16E+18
2.26E+17	2.83E+17	3.40E+17	3.98E+17	4.56E+17	1.05E+18
2.07E+17	2.58E+17	3.10E+17	3.62E+17	4.15E+17	9.46E+17
1.90E+17	2.36E+17	2.83E+17	3.30E+17	3.78E+17	8.58E+17
1.28E+17	1.57E+17	1.87E+17	2.17E+17	2.47E+17	5.50E+17
8.96E+16	1.10E+17	1.30E+17	1.50E+17	1.70E+17	3.74E+17
6.53E+16	7.94E+16	9.35E+16	1.08E+17	1.22E+17	2.65E+17
4.90E+16	5.93E+16	6.96E+16	8.00E+16	9.04E+16	1.95E+17
3.77E+16	4.54E+16	5.32E+16	6.10E+16	6.89E+16	1.47E+17
2.96E+16	3.56E+16	4.16E+16	4.76E+16	5.36E+16	1.14E+17
1.92E+16	2.30E+16	2.68E+16	3.06E+16	3.44E+16	7.25E+16
1.32E+16	1.57E+16	1.82E+16	2.08E+16	2.33E+16	4.89E+16
9.41E+15	1.12E+16	1.30E+16	1.48E+16	1.66E+16	3.45E+16
6.95E+15	8.26E+15	9.56E+15	1.09E+16	1.22E+16	2.53E+16

Table 3.3   Spectral Radiant Exitance [W/(cm$^2 \cdot \mu$m)]

					Temperature (K)	
Wavelength (microns)	200	273	293	300	400	500
1.00	2.12E-27	4.80E-19	1.75E-17	5.51E-17	8.88E-12	1.18E-08
1.50	7.25E-18	2.69E-12	2.96E-11	6.36E-11	1.88E-07	2.28E-05
2.00	2.77E-13	4.17E-09	2.52E-08	4.47E-08	1.79E-05	6.55E-04
2.50	1.21E-10	2.66E-07	1.12E-06	1.77E-06	2.15E-04	3.81E-03
3.00	5.89E-09	3.59E-06	1.19E-05	1.74E-05	9.49E-04	1.04E-02
3.50	8.37E-08	2.04E-05	5.70E-05	7.91E-05	2.43E-03	1.90E-02
4.00	5.61E-07	6.88E-05	1.69E-04	2.25E-04	4.51E-03	2.73E-02
4.50	2.30E-06	1.65E-04	3.67E-04	4.73E-04	6.80E-03	3.37E-02
5.00	6.70E-06	3.14E-04	6.45E-04	8.11E-04	8.93E-03	3.78E-02
5.50	1.54E-05	5.09E-04	9.79E-04	1.21E-03	1.07E-02	3.96E-02
6.00	2.96E-05	7.32E-04	1.33E-03	1.61E-03	1.19E-02	3.98E-02
6.50	5.00E-05	9.64E-04	1.68E-03	2.00E-03	1.27E-02	3.87E-02
7.00	7.60E-05	1.19E-03	1.99E-03	2.34E-03	1.30E-02	3.68E-02
7.50	1.07E-04	1.39E-03	2.25E-03	2.62E-03	1.30E-02	3.45E-02
8.00	1.41E-04	1.56E-03	2.45E-03	2.83E-03	1.28E-02	3.19E-02
8.50	1.77E-04	1.70E-03	2.60E-03	2.98E-03	1.23E-02	2.93E-02
9.00	2.12E-04	1.81E-03	2.70E-03	3.07E-03	1.18E-02	2.68E-02
9.50	2.47E-04	1.88E-03	2.75E-03	3.10E-03	1.11E-02	2.44E-02
10.00	2.79E-04	1.92E-03	2.76E-03	3.09E-03	1.05E-02	2.21E-02
10.50	3.08E-04	1.94E-03	2.73E-03	3.05E-03	9.78E-03	2.01E-02
11.00	3.34E-04	1.93E-03	2.69E-03	2.98E-03	9.11E-03	1.82E-02
11.50	3.55E-04	1.91E-03	2.62E-03	2.90E-03	8.46E-03	1.65E-02
12.00	3.73E-04	1.87E-03	2.54E-03	2.79E-03	7.84E-03	1.49E-02
12.50	3.87E-04	1.82E-03	2.44E-03	2.68E-03	7.26E-03	1.35E-02
13.00	3.97E-04	1.77E-03	2.34E-03	2.56E-03	6.71E-03	1.23E-02
13.50	4.04E-04	1.70E-03	2.24E-03	2.44E-03	6.20E-03	1.12E-02
14.00	4.07E-04	1.64E-03	2.13E-03	2.32E-03	5.73E-03	1.01E-02
15.00	4.07E-04	1.50E-03	1.92E-03	2.08E-03	4.89E-03	8.42E-03
16.00	3.99E-04	1.36E-03	1.73E-03	1.86E-03	4.18E-03	7.03E-03
17.00	3.86E-04	1.23E-03	1.54E-03	1.66E-03	3.58E-03	5.90E-03
18.00	3.68E-04	1.11E-03	1.37E-03	1.47E-03	3.08E-03	4.98E-03
19.00	3.48E-04	9.99E-04	1.22E-03	1.31E-03	2.66E-03	4.23E-03
20.00	3.27E-04	8.97E-04	1.09E-03	1.16E-03	2.30E-03	3.61E-03
21.00	3.06E-04	8.05E-04	9.71E-04	1.03E-03	2.00E-03	3.10E-03
22.00	2.85E-04	7.22E-04	8.66E-04	9.18E-04	1.75E-03	2.67E-03
13.00	3.97E-04	1.77E-03	2.34E-03	2.56E-03	6.71E-03	1.23E-02
24.00	2.45E-04	5.84E-04	6.92E-04	7.31E-04	1.34E-03	2.01E-03
25.00	2.27E-04	5.26E-04	6.20E-04	6.55E-04	1.18E-03	1.76E-03
26.00	2.10E-04	4.74E-04	5.57E-04	5.87E-04	1.05E-03	1.54E-03
27.00	1.94E-04	4.28E-04	5.01E-04	5.27E-04	9.28E-04	1.36E-03
28.00	1.79E-04	3.88E-04	4.52E-04	4.75E-04	8.26E-04	1.20E-03
29.00	1.65E-04	3.51E-04	4.08E-04	4.28E-04	7.37E-04	1.07E-03
30.00	1.53E-04	3.19E-04	3.69E-04	3.87E-04	6.60E-04	9.50E-04
35.00	1.04E-04	2.02E-04	2.31E-04	2.41E-04	3.94E-04	5.54E-04
40.00	7.20E-05	1.33E-04	1.50E-04	1.57E-04	2.49E-04	3.44E-04
45.00	5.10E-05	9.04E-05	1.02E-04	1.06E-04	1.64E-04	2.25E-04
50.00	3.70E-05	6.36E-05	7.12E-05	7.38E-05	1.13E-04	1.53E-04
55.00	2.73E-05	4.59E-05	5.12E-05	5.30E-05	7.99E-05	1.07E-04
60.00	2.06E-05	3.39E-05	3.77E-05	3.90E-05	5.82E-05	7.76E-05
70.00	1.23E-05	1.97E-05	2.17E-05	2.25E-05	3.29E-05	4.35E-05
80.00	7.78E-06	1.22E-05	1.34E-05	1.38E-05	2.00E-05	2.62E-05
90.00	5.14E-06	7.90E-06	8.67E-06	8.94E-06	1.28E-05	1.67E-05
100.00	3.53E-06	5.35E-06	5.86E-06	6.03E-06	8.58E-06	1.11E-05

600	700	800	900	1000	2000
1.43E−06	4.40E−05	5.74E−04	4.24E−03	2.10E−02	2.79E+01
5.58E−04	5.47E−03	3.04E−02	1.15E−01	3.34E−01	4.07E+01
7.20E−03	3.99E−02	1.44E−01	3.92E−01	8.72E−01	3.27E+01
2.60E−02	1.02E−01	2.86E−01	6.36E−01	1.21E+00	2.27E+01
5.16E−02	1.62E−01	3.82E−01	7.45E−01	1.27E+00	1.53E+01
7.49E−02	2.00E−01	4.17E−01	7.42E−01	1.18E+00	1.04E+01
9.06E−02	2.14E−01	4.09E−01	6.79E−01	1.02E+00	7.20E+00
9.81E−02	2.11E−01	3.77E−01	5.94E−01	8.58E−01	5.10E+00
9.90E−02	1.98E−01	3.35E−01	5.06E−01	7.09E−01	3.70E+00
9.55E−02	1.80E−01	2.92E−01	4.27E−01	5.82E−01	2.73E+00
8.94E−02	1.61E−01	2.51E−01	3.58E−01	4.78E−01	2.06E+00
8.20E−02	1.41E−01	2.15E−01	2.99E−01	3.93E−01	1.58E+00
7.43E−02	1.24E−01	1.83E−01	2.51E−01	3.24E−01	1.23E+00
6.67E−02	1.08E−01	1.56E−01	2.11E−01	2.69E−01	9.72E−01
5.95E−02	9.40E−02	1.34E−01	1.78E−01	2.25E−01	7.78E−01
5.30E−02	8.19E−02	1.15E−01	1.51E−01	1.89E−01	6.29E−01
4.71E−02	7.14E−02	9.86E−02	1.28E−01	1.59E−01	5.14E−01
4.18E−02	6.23E−02	8.51E−02	1.10E−01	1.35E−01	4.24E−01
3.71E−02	5.45E−02	7.37E−02	9.41E−02	1.15E−01	3.53E−01
3.30E−02	4.78E−02	6.40E−02	8.12E−02	9.91E−02	2.96E−01
2.94E−02	4.21E−02	5.58E−02	7.04E−02	8.55E−02	2.50E−01
2.62E−02	3.71E−02	4.89E−02	6.12E−02	7.40E−02	2.12E−01
2.34E−02	3.28E−02	4.29E−02	5.35E−02	6.44E−02	1.82E−01
2.09E−02	2.91E−02	3.78E−02	4.69E−02	5.63E−02	1.56E−01
1.88E−02	2.59E−02	3.35E−02	4.13E−02	4.94E−02	1.35E−01
1.69E−02	2.31E−02	2.97E−02	3.65E−02	4.35E−02	1.18E−01
1.52E−02	2.07E−02	2.64E−02	3.24E−02	3.85E−02	1.03E−01
1.24E−02	1.67E−02	2.11E−02	2.57E−02	3.04E−02	7.95E−02
1.02E−02	1.36E−02	1.71E−02	2.06E−02	2.43E−02	6.24E−02
8.44E−03	1.11E−02	1.39E−02	1.68E−02	1.97E−02	4.97E−02
7.05E−03	9.22E−03	1.15E−02	1.37E−02	1.61E−02	4.00E−02
5.92E−03	7.69E−03	9.51E−03	1.14E−02	1.32E−02	3.26E−02
5.01E−03	6.47E−03	7.96E−03	9.48E−03	1.10E−02	2.68E−02
4.26E−03	5.47E−03	6.71E−03	7.97E−03	9.24E−03	2.23E−02
3.65E−03	4.66E−03	5.70E−03	6.75E−03	7.81E−03	1.86E−02
1.88E−02	2.59E−02	3.35E−02	4.13E−02	4.94E−02	1.35E−01
2.72E−03	3.44E−03	4.18E−03	4.93E−03	5.68E−03	1.33E−02
2.36E−03	2.98E−03	3.61E−03	4.25E−03	4.89E−03	1.14E−02
2.06E−03	2.59E−03	3.13E−03	3.68E−03	4.23E−03	9.81E−03
1.81E−03	2.27E−03	2.73E−03	3.20E−03	3.68E−03	8.48E−03
1.59E−03	1.99E−03	2.40E−03	2.80E−03	3.21E−03	7.37E−03
1.41E−03	1.76E−03	2.11E−03	2.46E−03	2.82E−03	6.43E−03
1.25E−03	1.55E−03	1.86E−03	2.17E−03	2.48E−03	5.64E−03
7.19E−04	8.85E−04	1.05E−03	1.22E−03	1.39E−03	3.10E−03
4.42E−04	5.40E−04	6.39E−04	7.38E−04	8.38E−04	1.84E−03
2.86E−04	3.48E−04	4.10E−04	4.72E−04	5.34E−04	1.16E−03
1.93E−04	2.34E−04	2.75E−04	3.15E−04	3.56E−04	7.68E−04
1.35E−04	1.63E−04	1.91E−04	2.19E−04	2.47E−04	5.28E−04
9.72E−05	1.17E−04	1.37E−04	1.56E−04	1.76E−04	3.75E−04
5.41E−05	6.47E−05	7.54E−05	8.61E−05	9.68E−05	2.04E−04
3.24E−05	3.87E−05	4.50E−05	5.12E−05	5.75E−05	1.20E−04
2.06E−05	2.45E−05	2.84E−05	3.24E−05	3.63E−05	7.56E−05
1.37E−05	1.63E−05	1.88E−05	2.14E−05	2.40E−05	4.98E−05

for most background calculations and as a "sanity check" on computed values. (It is always worthwhile to do such a sanity check when first using incidance values provided by someone else. It is somehow common to find a value reported to four significant figures, but incorrect by a power of 10, or π!).

The notation on radiation slide rules is very compact—often just a single letter symbol. Make sure you know whether the rule provides the exitance or the sterance. These will give values accurate to 20% or better. They require three steps:

1. Find M_{pk}, the exitance at the peak wavelength.
2. Find the ratio of $M(\lambda)/M_{pk}$.
3. Multiply the two together to obtain $M(\lambda)$.

Temperature Dependence. Figures 3.7 and 3.8 shows spectral exitance curves for blackbody emitters at various temperatures. Note that when plotted on log paper they all have the same shape; they are merely shifted horizontaly and vertically. Tables 3.2 and 3.3 provide numerical values for several temperatures and wavelengths.

3.2.2. Spectral Integral

Consider a detector that is limited with a spectral filter to "see" only those wavelengths between 10 and 11 μm. What fraction of the total emitted radiation can it see? The problem is illustrated in Figure 3.9. The figure shows the spectral exitance, the exitance as a function of wavelength, as well as the spectral range over which the detector responds.

The curve itself is the Planck radiation function. The *total* exitance—

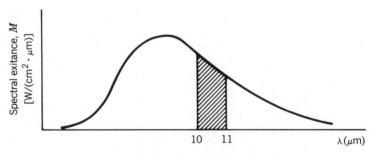

Figure 3.9.   Spectral integral.

all wavelengths—can be thought of as the "area" under the entire curve, and the exitance of interest is the area under selected portions of the curve. If our detector sees only those photons or watts in the crosshatched area, the effective exitance is proportional to the crosshatched area. The necessary calculation is split into three types: the total (all-wavelengths) case, the wideband case, and the narrowband case. The total case and the narrowband case both turn out to be easy, but the wideband case is more difficult.

Total Exitance. The total exitance (all wavelengths) is easily calculated since the integral over all wavelengths of the Planck function has been done in closed form. The result is given in terms of the Stefan–Boltzmann constant (watts), or an analogous constant σ' if photons are desired. The necessary formulas (Equations 3.13a and 3.13b) are given in several texts, including Pivovonsky and Nagle (1961) and Hudson (1969).

TOTAL BLACKBODY EXITANCE

$$M_{\text{total}} = \sigma_q T^3 \quad \text{(photons/(cm}^2\text{·s)}) \qquad (3.13a)$$

$$M_{\text{total}} = \sigma T^4 \qquad \text{W/cm}^2 \qquad (3.13b)$$

where T = temperature of the source (kelvin)

σ_q = photon analog of the Stefan–Boltzmann constant

$$= \frac{4\pi^4 k^3}{25.79436 h^3 c^2}$$

$$= 1.520487 \times 10^{11} \text{ photons/(s·cm}^2\text{·K}^3)$$

$$\simeq 1.52 \times 10^{11} \text{ photons/(s·cm}^2\text{·K}^3)$$

σ = Stefan–Boltzmann constant

$$= \frac{2\pi^5 k^4}{15 h^3 c^2}$$

$$= 5.67051 \times 10^{-12} \text{ W/(cm}^2\text{·K}^4)$$

$$\simeq 5.67 \times 10^{-12} \text{ W/(cm}^2\text{·K}^4)$$

The first value listed for the constants is a current best estimate (Cohen and Taylor, 1987); the second is a convenient approximation that will yield results accurate to 0.05% or better. The accuracy of Planck's law calculations will be discussed in a later section.

Narrowband Calculation. If a plot of exitance versus wavelength on linear paper (not log paper) shows little curvature over the spectral band of interest, the narrowband approximation will probably be adequate. In that case, the calculations are easily done with a hand-held calculator using the Planck exitance function (3.9) directly. The method is to approximate the area under the curve by a rectangle (see Figure 3.10) so that the desired (in band) exitance is the spectral exitance at the center wavelength times the bandpass, as shown in Equation 3.14.

EXITANCE IN A NARROW SPECTRAL BAND

$$M_{\text{in band}} \simeq M(\lambda_c, T)(\lambda_2 - \lambda_1) \qquad (3.14)$$

The error introduced by using formulas (3.14) can be estimated by comparing the exitance at the center of the band with the average of its values at the two endpoints:

$$\text{relative error} = \frac{M(\lambda_c) - [M(\lambda_1) + M(\lambda_2)]/2}{3M(\lambda_c)} \qquad (3.15)$$

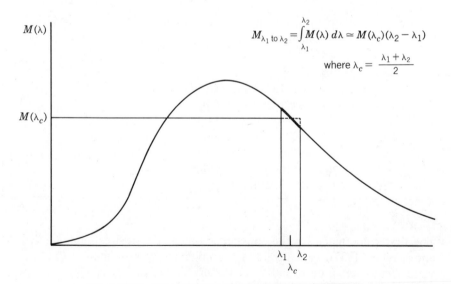

Figure 3.10. Narrowband spectral integral approximation.

On the other hand, if you go to the trouble of calculating the exitance at the center and the two endpoints, it is very little additional effort to use a simple Simpson's rule integration formula to calculate a more accurate in-band value:

$$M_{\text{in band}} \simeq \frac{M(\lambda_1) + 4M(\lambda_c) + M(\lambda_2)}{6} (\lambda_2 - \lambda_1) \qquad (3.16)$$

Wideband Calculation. The integration of Planck's radiation law over a wide spectral band is more difficult than for a narrow band, or for the total integral. It can be done in several ways. The choice depends on the facilities available, the desired accuracy, and the width of the spectral band.

If the band is not too broad, the narrowband approximation may be useful, at least as a quick check on other methods.

If any great accuracy is desired, the calculations must be done with a computer or looping calculator routine. Such a program includes or calls a routine to calculate the spectral exitance, and an integration routine. Chapter 1 of the *Infrared Handbook* (Wolfe and Zissis, 1978) provides programs for Hewlett-Packard and Texas Instruments calculators.

Tables such as Tables 3.4 and 3.5 can be used; these provide the total exitance from zero to the specified wavelength. To obtain the exitance between two wavelengths, subtract the corresponding entries:

$$M(\lambda_1 \text{ to } \lambda_2) = M(0 \text{ to } \lambda_2) - M(0 \text{ to } \lambda_1) \qquad (3.17a)$$

This is illustrated in Figure 3.11.

Radiation slide rules use a similar method; they can be used to give values accurate within 20%. The method requires four steps:

1. Find the total exitance M_{total}.
2. Find the fraction from zero to λ_2.
3. Find the fraction from zero to λ_1.
4. Find the in-band exitance from the relation

$$M(\lambda_1 \text{ to } \lambda_2) = \left[\frac{M(0 \text{ to } \lambda_2)}{M_{\text{total}}} - \frac{M(0 \text{ to } \lambda_1)}{M_{\text{total}}} \right] M_{\text{total}} \qquad (3.17b)$$

For wavelengths less than the peak, the Wien's law approximation can

Table 3.4 Integrated Photon Exitance [photons/(cm^2·s)]

Wavelength (microns)	Temperature (K)					
	200	273	293	300	400	500
1.00	1.53E-10	4.78E-02	1.88E+00	6.05E+00	1.32E+06	2.22E+09
1.50	4.60E+00	1.72E+06	1.90E+07	4.09E+07	1.28E+11	1.70E+13
2.00	2.44E+05	4.09E+09	2.57E+10	4.63E+10	2.33E+13	1.07E+15
2.50	1.60E+08	4.53E+11	2.05E+12	3.31E+12	5.45E+14	1.25E+16
3.00	1.25E+10	1.05E+13	3.76E+13	5.66E+13	4.27E+15	6.11E+16
3.50	2.81E+11	9.63E+13	2.91E+14	4.15E+14	1.79E+16	1.83E+17
4.00	2.83E+12	4.93E+14	1.31E+15	1.80E+15	5.08E+16	4.05E+17
4.50	1.67E+13	1.72E+15	4.15E+15	5.50E+15	1.12E+17	7.36E+17
5.00	6.79E+13	4.57E+15	1.02E+16	1.32E+16	2.07E+17	1.17E+18
5.50	2.10E+14	1.00E+16	2.10E+16	2.66E+16	3.38E+17	1.69E+18
6.00	5.33E+14	1.91E+16	3.79E+16	4.72E+16	5.04E+17	2.27E+18
6.50	1.16E+15	3.25E+16	6.18E+16	7.60E+16	7.00E+17	2.89E+18
7.00	2.23E+15	5.10E+16	9.33E+16	1.13E+17	9.21E+17	3.54E+18
7.50	3.91E+15	7.48E+16	1.32E+17	1.59E+17	1.16E+18	4.19E+18
8.00	6.35E+15	1.04E+17	1.79E+17	2.13E+17	1.42E+18	4.85E+18
8.50	9.67E+15	1.38E+17	2.32E+17	2.74E+17	1.68E+18	5.49E+18
9.00	1.40E+16	1.77E+17	2.90E+17	3.41E+17	1.95E+18	6.11E+18
9.50	1.94E+16	2.20E+17	3.54E+17	4.13E+17	2.21E+18	6.71E+18
10.00	2.59E+16	2.67E+17	4.23E+17	4.90E+17	2.48E+18	7.29E+18
10.50	3.36E+16	3.18E+17	4.94E+17	5.70E+17	2.74E+18	7.83E+18
11.00	4.23E+16	3.70E+17	5.68E+17	6.52E+17	3.00E+18	8.35E+18
11.50	5.22E+16	4.25E+17	6.44E+17	7.36E+17	3.25E+18	8.85E+18
12.00	6.30E+16	4.81E+17	7.20E+17	8.21E+17	3.50E+18	9.32E+18
12.50	7.48E+16	5.39E+17	7.98E+17	9.06E+17	3.73E+18	9.76E+18
13.00	8.75E+16	5.97E+17	8.75E+17	9.91E+17	3.96E+18	1.02E+19
13.50	1.01E+17	6.55E+17	9.52E+17	1.07E+18	4.17E+18	1.06E+19
14.00	1.15E+17	7.13E+17	1.03E+18	1.16E+18	4.38E+18	1.09E+19
15.00	1.45E+17	8.29E+17	1.18E+18	1.32E+18	4.77E+18	1.16E+19
16.00	1.77E+17	9.42E+17	1.32E+18	1.47E+18	5.13E+18	1.22E+19
17.00	2.10E+17	1.05E+18	1.46E+18	1.62E+18	5.45E+18	1.28E+19
18.00	2.43E+17	1.15E+18	1.59E+18	1.76E+18	5.74E+18	1.32E+19
19.00	2.77E+17	1.25E+18	1.71E+18	1.89E+18	6.01E+18	1.37E+19
20.00	3.10E+17	1.35E+18	1.82E+18	2.01E+18	6.26E+18	1.41E+19
21.00	3.43E+17	1.43E+18	1.93E+18	2.13E+18	6.48E+18	1.44E+19
22.00	3.75E+17	1.52E+18	2.03E+18	2.23E+18	6.68E+18	1.47E+19
13.00	8.75E+16	5.99E+17	8.78E+17	9.93E+17	3.96E+18	1.02E+19
24.00	4.37E+17	1.67E+18	2.21E+18	2.42E+18	7.04E+18	1.53E+19
25.00	4.66E+17	1.74E+18	2.29E+18	2.51E+18	7.20E+18	1.55E+19
26.00	4.94E+17	1.80E+18	2.37E+18	2.59E+18	7.34E+18	1.57E+19
27.00	5.21E+17	1.86E+18	2.44E+18	2.66E+18	7.47E+18	1.59E+19
28.00	5.47E+17	1.92E+18	2.50E+18	2.73E+18	7.59E+18	1.61E+19
29.00	5.72E+17	1.97E+18	2.56E+18	2.80E+18	7.71E+18	1.62E+19
30.00	5.96E+17	2.02E+18	2.62E+18	2.86E+18	7.81E+18	1.64E+19
35.00	7.00E+17	2.23E+18	2.86E+18	3.11E+18	8.23E+18	1.70E+19
40.00	7.82E+17	2.39E+18	3.04E+18	3.29E+18	8.53E+18	1.74E+19
45.00	8.47E+17	2.50E+18	3.17E+18	3.43E+18	8.75E+18	1.77E+19
50.00	9.00E+17	2.60E+18	3.28E+18	3.54E+18	8.91E+18	1.79E+19
55.00	9.42E+17	2.67E+18	3.36E+18	3.62E+18	9.04E+18	1.81E+19
60.00	9.76E+17	2.72E+18	3.42E+18	3.69E+18	9.14E+18	1.82E+19
70.00	1.03E+18	2.81E+18	3.51E+18	3.79E+18	9.28E+18	1.84E+19
80.00	1.07E+18	2.87E+18	3.58E+18	3.85E+18	9.38E+18	1.86E+19
90.00	1.09E+18	2.91E+18	3.63E+18	3.90E+18	9.45E+18	1.86E+19
100.00	1.11E+18	2.94E+18	3.66E+18	3.94E+18	9.50E+18	1.87E+19
all	1.21E+18	3.08E+18	3.81E+18	4.09E+18	9.69E+18	1.89E+19

600	700	800	900	1000	2000
3.27E+11	1.19E+13	1.80E+14	1.52E+15	8.45E+15	2.58E+19
4.75E+14	5.36E+15	3.42E+16	1.48E+17	4.87E+17	1.45E+20
1.44E+16	9.57E+16	4.06E+17	1.27E+18	3.24E+18	3.10E+20
1.06E+17	5.02E+17	1.66E+18	4.29E+18	9.35E+18	4.66E+20
3.78E+17	1.44E+18	4.03E+18	9.20E+18	1.81E+19	5.96E+20
9.06E+17	2.95E+18	7.37E+18	1.54E+19	2.83E+19	7.00E+20
1.70E+18	4.94E+18	1.13E+19	2.22E+19	3.88E+19	7.82E+20
2.73E+18	7.24E+18	1.56E+19	2.90E+19	4.89E+19	8.47E+20
3.92E+18	9.71E+18	1.99E+19	3.56E+19	5.83E+19	9.00E+20
5.22E+18	1.22E+19	2.40E+19	4.18E+19	6.68E+19	9.42E+20
6.57E+18	1.47E+19	2.80E+19	4.75E+19	7.45E+19	9.77E+20
7.93E+18	1.71E+19	3.16E+19	5.27E+19	8.14E+19	1.01E+21
9.26E+18	1.94E+19	3.50E+19	5.74E+19	8.75E+19	1.03E+21
1.06E+19	2.15E+19	3.82E+19	6.16E+19	9.29E+19	1.05E+21
1.18E+19	2.35E+19	4.10E+19	6.54E+19	9.78E+19	1.07E+21
1.30E+19	2.53E+19	4.36E+19	6.89E+19	1.02E+20	1.08E+21
1.41E+19	2.70E+19	4.60E+19	7.19E+19	1.06E+20	1.09E+21
1.51E+19	2.86E+19	4.81E+19	7.47E+19	1.09E+20	1.10E+21
1.61E+19	3.00E+19	5.01E+19	7.72E+19	1.12E+20	1.11E+21
1.70E+19	3.13E+19	5.18E+19	7.95E+19	1.15E+20	1.12E+21
1.79E+19	3.26E+19	5.35E+19	8.16E+19	1.18E+20	1.13E+21
1.86E+19	3.37E+19	5.50E+19	8.34E+19	1.20E+20	1.14E+21
1.94E+19	3.47E+19	5.63E+19	8.51E+19	1.22E+20	1.14E+21
2.01E+19	3.57E+19	5.76E+19	8.67E+19	1.24E+20	1.15E+21
2.07E+19	3.66E+19	5.87E+19	8.81E+19	1.26E+20	1.15E+21
2.13E+19	3.74E+19	5.98E+19	8.94E+19	1.27E+20	1.16E+21
2.19E+19	3.82E+19	6.08E+19	9.06E+19	1.29E+20	1.16E+21
2.29E+19	3.95E+19	6.25E+19	9.27E+19	1.31E+20	1.17E+21
2.38E+19	4.07E+19	6.40E+19	9.46E+19	1.33E+20	1.17E+21
2.45E+19	4.17E+19	6.53E+19	9.61E+19	1.35E+20	1.18E+21
2.52E+19	4.26E+19	6.64E+19	9.75E+19	1.37E+20	1.18E+21
2.58E+19	4.34E+19	6.74E+19	9.86E+19	1.38E+20	1.18E+21
2.64E+19	4.41E+19	6.83E+19	9.97E+19	1.39E+20	1.19E+21
2.69E+19	4.47E+19	6.90E+19	1.01E+20	1.40E+20	1.19E+21
2.73E+19	4.53E+19	6.97E+19	1.01E+20	1.41E+20	1.19E+21
2.07E+19	3.66E+19	5.87E+19	8.81E+19	1.26E+20	1.15E+21
2.80E+19	4.63E+19	7.09E+19	1.03E+20	1.43E+20	1.20E+21
2.83E+19	4.67E+19	7.13E+19	1.03E+20	1.44E+20	1.20E+21
2.86E+19	4.70E+19	7.18E+19	1.04E+20	1.44E+20	1.20E+21
2.89E+19	4.73E+19	7.22E+19	1.04E+20	1.45E+20	1.20E+21
2.91E+19	4.76E+19	7.25E+19	1.05E+20	1.45E+20	1.20E+21
2.93E+19	4.79E+19	7.29E+19	1.05E+20	1.46E+20	1.20E+21
2.95E+19	4.82E+19	7.32E+19	1.05E+20	1.46E+20	1.20E+21
3.03E+19	4.91E+19	7.43E+19	1.07E+20	1.47E+20	1.21E+21
3.09E+19	4.98E+19	7.51E+19	1.08E+20	1.49E+20	1.21E+21
3.12E+19	5.02E+19	7.56E+19	1.08E+20	1.49E+20	1.21E+21
3.15E+19	5.06E+19	7.60E+19	1.09E+20	1.50E+20	1.21E+21
3.17E+19	5.08E+19	7.63E+19	1.09E+20	1.50E+20	1.21E+21
3.19E+19	5.10E+19	7.66E+19	1.09E+20	1.50E+20	1.21E+21
3.21E+19	5.13E+19	7.69E+19	1.10E+20	1.51E+20	1.21E+21
3.23E+19	5.15E+19	7.71E+19	1.10E+20	1.51E+20	1.21E+21
3.24E+19	5.17E+19	7.73E+19	1.10E+20	1.51E+20	1.22E+21
3.25E+19	5.18E+19	7.74E+19	1.10E+20	1.52E+20	1.22E+21
3.27E+19	5.19E+19	7.75E+19	1.10E+20	1.51E+20	1.21E+21

Table 3.5 Integrated Radiant Exitance (W/cm²)

Wavelength (microns)	Temperature (K)					
	200	273	293	300	400	500
1.00	3.10E-29	9.71E-21	3.82E-19	1.23E-18	2.71E-13	4.60E-10
1.50	6.05E-19	2.26E-13	2.50E-12	5.39E-12	1.70E-08	2.31E-06
2.00	2.42E-14	4.11E-10	2.60E-09	4.68E-09	2.42E-06	1.13E-04
2.50	1.29E-11	3.73E-08	1.69E-07	2.75E-07	4.63E-05	1.08E-03
3.00	8.56E-10	7.32E-07	2.64E-06	3.97E-06	3.07E-04	4.52E-03
3.50	1.66E-08	5.83E-06	1.77E-05	2.53E-05	1.12E-03	1.18E-02
4.00	1.48E-07	2.64E-05	7.09E-05	9.73E-05	2.84E-03	2.35E-02
4.50	7.84E-07	8.26E-05	2.01E-04	2.68E-04	5.67E-03	3.88E-02
5.00	2.89E-06	2.00E-04	4.51E-04	5.86E-04	9.61E-03	5.68E-02
5.50	8.20E-06	4.04E-04	8.56E-04	1.09E-03	1.45E-02	7.62E-02
6.00	1.92E-05	7.14E-04	1.43E-03	1.79E-03	2.02E-02	9.61E-02
6.50	3.89E-05	1.14E-03	2.19E-03	2.70E-03	2.64E-02	1.16E-01
7.00	7.01E-05	1.68E-03	3.10E-03	3.79E-03	3.28E-02	1.35E-01
7.50	1.16E-04	2.32E-03	4.17E-03	5.03E-03	3.94E-02	1.53E-01
8.00	1.78E-04	3.06E-03	5.34E-03	6.39E-03	4.58E-02	1.69E-01
8.50	2.57E-04	3.88E-03	6.61E-03	7.85E-03	5.21E-02	1.84E-01
9.00	3.54E-04	4.76E-03	7.93E-03	9.36E-03	5.81E-02	1.98E-01
9.50	4.69E-04	5.68E-03	9.30E-03	1.09E-02	6.39E-02	2.11E-01
10.00	6.01E-04	6.63E-03	1.07E-02	1.25E-02	6.93E-02	2.23E-01
10.50	7.48E-04	7.59E-03	1.20E-02	1.40E-02	7.43E-02	2.33E-01
11.00	9.08E-04	8.56E-03	1.34E-02	1.55E-02	7.91E-02	2.43E-01
11.50	1.08E-03	9.52E-03	1.47E-02	1.70E-02	8.35E-02	2.52E-01
12.00	1.26E-03	1.05E-02	1.60E-02	1.84E-02	8.75E-02	2.60E-01
12.50	1.45E-03	1.14E-02	1.73E-02	1.98E-02	9.13E-02	2.67E-01
13.00	1.65E-03	1.23E-02	1.85E-02	2.11E-02	9.48E-02	2.73E-01
13.50	1.85E-03	1.32E-02	1.96E-02	2.23E-02	9.80E-02	2.79E-01
14.00	2.05E-03	1.40E-02	2.07E-02	2.35E-02	1.01E-01	2.84E-01
15.00	2.46E-03	1.56E-02	2.27E-02	2.57E-02	1.06E-01	2.93E-01
16.00	2.86E-03	1.70E-02	2.46E-02	2.77E-02	1.11E-01	3.01E-01
17.00	3.26E-03	1.83E-02	2.62E-02	2.95E-02	1.15E-01	3.08E-01
18.00	3.63E-03	1.95E-02	2.76E-02	3.10E-02	1.18E-01	3.13E-01
19.00	3.99E-03	2.05E-02	2.89E-02	3.24E-02	1.21E-01	3.18E-01
20.00	4.33E-03	2.15E-02	3.01E-02	3.36E-02	1.23E-01	3.22E-01
21.00	4.65E-03	2.23E-02	3.11E-02	3.47E-02	1.26E-01	3.25E-01
22.00	4.94E-03	2.31E-02	3.20E-02	3.57E-02	1.27E-01	3.28E-01
13.00	1.66E-03	1.23E-02	1.85E-02	2.11E-02	9.48E-02	2.73E-01
24.00	5.46E-03	2.44E-02	3.36E-02	3.73E-02	1.31E-01	3.33E-01
25.00	5.70E-03	2.49E-02	3.42E-02	3.80E-02	1.32E-01	3.35E-01
26.00	5.91E-03	2.54E-02	3.48E-02	3.86E-02	1.33E-01	3.36E-01
27.00	6.12E-03	2.59E-02	3.53E-02	3.92E-02	1.34E-01	3.38E-01
28.00	6.30E-03	2.63E-02	3.58E-02	3.97E-02	1.35E-01	3.39E-01
29.00	6.47E-03	2.66E-02	3.62E-02	4.01E-02	1.36E-01	3.40E-01
30.00	6.63E-03	2.70E-02	3.66E-02	4.05E-02	1.36E-01	3.41E-01
35.00	7.27E-03	2.82E-02	3.81E-02	4.21E-02	1.39E-01	3.45E-01
40.00	7.70E-03	2.91E-02	3.90E-02	4.31E-02	1.40E-01	3.47E-01
45.00	8.00E-03	2.96E-02	3.97E-02	4.37E-02	1.41E-01	3.48E-01
50.00	8.22E-03	3.00E-02	4.01E-02	4.41E-02	1.42E-01	3.49E-01
55.00	8.38E-03	3.03E-02	4.04E-02	4.45E-02	1.43E-01	3.50E-01
60.00	8.50E-03	3.05E-02	4.06E-02	4.47E-02	1.43E-01	3.50E-01
70.00	8.66E-03	3.07E-02	4.09E-02	4.50E-02	1.43E-01	3.51E-01
80.00	8.76E-03	3.09E-02	4.11E-02	4.52E-02	1.44E-01	3.51E-01
90.00	8.82E-03	3.10E-02	4.12E-02	4.53E-02	1.44E-01	3.51E-01
100.00	8.86E-03	3.10E-02	4.12E-02	4.53E-02	1.44E-01	3.52E-01
all	9.07E-03	3.15E-02	4.18E-02	4.59E-02	1.45E-01	3.54E-01

600	700	800	900	1000	2000
6.83E-08	2.50E-06	3.82E-05	3.24E-04	1.82E-03	6.05E+00
6.56E-05	7.55E-04	2.16E-02	2.16E-02	7.21E-02	2.47E+01
1.56E-03	1.05E-02	4.54E-02	1.45E-01	3.75E-01	4.34E+01
9.37E-03	4.55E-02	1.54E-01	4.07E-01	9.08E-01	5.71E+01
2.87E-02	1.12E-01	3.23E-01	7.58E-01	1.54E+00	6.65E+01
6.06E-02	2.04E-01	5.25E-01	1.13E+00	2.15E+00	7.28E+01
1.02E-01	3.08E-01	7.33E-01	1.49E+00	2.71E+00	7.72E+01
1.50E-01	4.15E-01	9.30E-01	1.81E+00	3.18E+00	8.02E+01
1.99E-01	5.17E-01	1.11E+00	2.08E+00	3.57E+00	8.24E+01
2.48E-01	6.12E-01	1.27E+00	2.32E+00	3.89E+00	8.40E+01
2.94E-01	6.97E-01	1.40E+00	2.51E+00	4.15E+00	8.51E+01
3.37E-01	7.73E-01	1.52E+00	2.68E+00	4.37E+00	8.61E+01
3.76E-01	8.39E-01	1.62E+00	2.81E+00	4.55E+00	8.68E+01
4.12E-01	8.97E-01	1.70E+00	2.93E+00	4.70E+00	8.73E+01
4.43E-01	9.47E-01	1.77E+00	3.02E+00	4.82E+00	8.77E+01
4.71E-01	9.91E-01	1.84E+00	3.11E+00	4.92E+00	8.81E+01
4.96E-01	1.03E+00	1.89E+00	3.18E+00	5.01E+00	8.84E+01
5.18E-01	1.06E+00	1.93E+00	3.24E+00	5.08E+00	8.86E+01
5.38E-01	1.09E+00	1.97E+00	3.29E+00	5.14E+00	8.88E+01
5.56E-01	1.12E+00	2.01E+00	3.33E+00	5.20E+00	8.90E+01
5.71E-01	1.14E+00	2.04E+00	3.37E+00	5.24E+00	8.91E+01
5.85E-01	1.16E+00	2.06E+00	3.40E+00	5.28E+00	8.92E+01
5.97E-01	1.18E+00	2.09E+00	3.43E+00	5.32E+00	8.93E+01
6.09E-01	1.19E+00	2.11E+00	3.45E+00	5.35E+00	8.94E+01
6.18E-01	1.21E+00	2.13E+00	3.48E+00	5.38E+00	8.95E+01
6.27E-01	1.22E+00	2.14E+00	3.50E+00	5.40E+00	8.95E+01
6.35E-01	1.23E+00	2.15E+00	3.51E+00	5.42E+00	8.96E+01
6.49E-01	1.25E+00	2.18E+00	3.54E+00	5.45E+00	8.97E+01
6.60E-01	1.26E+00	2.20E+00	3.56E+00	5.48E+00	8.97E+01
6.70E-01	1.28E+00	2.21E+00	3.58E+00	5.50E+00	8.98E+01
6.77E-01	1.29E+00	2.23E+00	3.60E+00	5.52E+00	8.98E+01
6.84E-01	1.29E+00	2.24E+00	3.61E+00	5.53E+00	8.99E+01
6.89E-01	1.30E+00	2.24E+00	3.62E+00	5.55E+00	8.99E+01
6.94E-01	1.31E+00	2.25E+00	3.63E+00	5.56E+00	8.99E+01
6.98E-01	1.31E+00	2.26E+00	3.64E+00	5.57E+00	9.00E+01
6.18E-01	1.21E+00	2.12E+00	3.47E+00	5.37E+00	8.95E+01
7.05E-01	1.32E+00	2.27E+00	3.65E+00	5.58E+00	9.00E+01
7.07E-01	1.32E+00	2.27E+00	3.66E+00	5.59E+00	9.00E+01
7.09E-01	1.33E+00	2.28E+00	3.66E+00	5.59E+00	9.00E+01
7.11E-01	1.33E+00	2.28E+00	3.66E+00	5.59E+00	9.00E+01
7.13E-01	1.33E+00	2.28E+00	3.67E+00	5.60E+00	9.00E+01
7.15E-01	1.33E+00	2.28E+00	3.67E+00	5.60E+00	9.00E+01
7.16E-01	1.34E+00	2.29E+00	3.67E+00	5.60E+00	9.00E+01
7.21E-01	1.34E+00	2.29E+00	3.68E+00	5.61E+00	9.01E+01
7.24E-01	1.34E+00	2.30E+00	3.68E+00	5.62E+00	9.01E+01
7.25E-01	1.35E+00	2.30E+00	3.69E+00	5.62E+00	9.01E+01
7.26E-01	1.35E+00	2.30E+00	3.69E+00	5.62E+00	9.01E+01
7.27E-01	1.35E+00	2.30E+00	3.69E+00	5.63E+00	9.01E+01
7.28E-01	1.35E+00	2.30E+00	3.69E+00	5.63E+00	9.01E+01
7.29E-01	1.35E+00	2.30E+00	3.69E+00	5.63E+00	9.01E+01
7.29E-01	1.35E+00	2.31E+00	3.69E+00	5.63E+00	9.01E+01
7.29E-01	1.35E+00	2.31E+00	3.69E+00	5.63E+00	9.01E+01
7.29E-01	1.35E+00	2.31E+00	3.69E+00	5.63E+00	9.01E+01
7.35E-01	1.36E+00	2.32E+00	3.72E+00	5.67E+00	9.07E+01

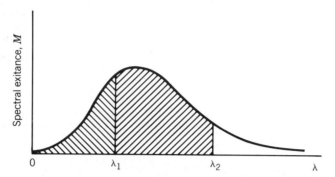

Figure 3.11. Wideband spectral integral. Area under curve between 0 and λ_1 = exitance with wavelength $\lambda < \lambda_1 = M_{0-\lambda_1}$. Areas under curve between 0 and λ_2 = exitance with wavelength $\lambda < \lambda_2 = M_{0-\lambda_2}$.

be used; this can be integrated directly:

$$M_q(0 \text{ to } \lambda) = \sigma_q T^3 e^{-x}(x^2 + 2x + 2)\frac{25.79436}{2\pi^3} \qquad (3.18a)$$

$$M_e(0 \text{ to } \lambda) = \sigma T^4 e^{-x}(x^3 + 3x^2 + 6x + 6)\frac{15}{\pi^4} \qquad (3.18b)$$

Except for this limiting case, manual calculations are generally so slow, difficult, and/or inaccurate as to not be worthwhile.

3.2.3. Accuracy of Radiometric Calculations

The uncertainty in radiometric predictions for our test setups is probably dominated by our inability to properly include reflections from nearby supporting structures and to measure the transmission of filters and window.

Our knowledge of the constants in the Planck's law equations, and the number of significant digits with which we carry along in the calculations, should not limit our accuracy. Let us estimate the accuracy with which we can perform the Planck's law calculation—it will give us insight into the limitations we face, and may suggest improvements.

To illustrate the propagation of errors in determining incidance at the detector plane, consider the case of total radiant incidance and the narrow-spectral-band case. The appropriate incidance formulas—slightly simplified—and the corresponding formulas for the relative error in the incidance due to the relative errors in the component variables are:

TOTAL

$$E_e = \epsilon\sigma T^4 \frac{d^2}{4D^2} \tag{3.19a}$$

$$\frac{\delta E_e}{E_e} = \frac{\delta\epsilon}{\epsilon} + \frac{\delta\sigma}{\sigma} + \frac{4\delta T}{T} + \frac{2\delta d}{d} + \frac{2\delta D}{D} \tag{3.19b}$$

NARROWBAND

$$E_e = \epsilon \frac{c_1 \,\Delta\lambda}{\lambda^4(e^x - 1)} \frac{d^2}{4D^2} \qquad x = \frac{c_2}{\lambda T} \tag{3.20a}$$

$$\frac{\delta E_e}{E_e} = \frac{\delta\epsilon}{\epsilon} + \frac{\delta c_1}{c_1} + \frac{\delta\Delta\lambda}{\Delta\lambda} + \left(\frac{xe^x}{e^x - 1} - 4\right)\frac{\delta\lambda}{\lambda}$$

$$+ \frac{xe^x}{e^x - 1}\frac{\delta T}{T} + \frac{xe^x}{e^x - 1}\frac{\delta c_2}{c_2} + \frac{2\delta d}{d} + \frac{2\delta D}{D} \tag{3.20b}$$

For small values of λT, the factor x can be very large, so errors in λ, T, and c_2 are amplified. For $\lambda = 3$ μm and $T = 300$ K, x is about 15. Tables 3.6 and 3.7 give typical values for the component variables, their absolute and relative uncertainties, and for the resulting effect on the uncertainty in the incidance. The uncertainties listed are representative of those in one laboratory and are probably not far different for many test setups, but all users must evaluate their own setups and corresponding errors.

Table 3.6 Contributions to Relative Uncertainty in Total Incidance Calculations (Typical Values)

Parameter	Value	Error Absolute	Error Relative	Contribution to Relative Error in Incidance
Emissivity	0.99	0.01	1%	1%
Temperature (K)	500	1.0	0.2%	0.8%
Aperture diameter (in.)	0.100	0.0002	0.2%	0.4%
Aperture to detector (in.)	10	0.050	0.5%	1%
Stefan–Boltzmann constant[a]	5.67051	0.00019	34 ppm[b]	34 ppm[b]
Total (RSS)				1.7%

[a] Value of Stefan–Boltzmann constant and its uncertainty are from Cohen and Taylor (1987); units are 10^{-12} W/(cm^2·K^4).

[b] Parts per million.

Table 3.7   Contributions to Relative Uncertainty in Incidance Calculations (Narrow Spectral Band, Typical Values)

Parameter	Value	Error Absolute	Error Relative	Contribution to Relative Error in Incidance
Emissivity	0.99	0.01	1%	1%
Temperature (K)	300	0.5	0.17%	2.5%
Wavelength (μm)	3.00	0.01	0.3%	3.3%
Spectral bandwidth (μm)	0.25	0.005	2.0%	2.0%
$c_1{}^a$	3.7417749	0.0000022	0.6 ppm[b]	0.6 ppm[b]
$c_2{}^a$	14387.69	0.12	8.4 ppm[b]	0.013%
Total (RSS)				4.7%

[a] Radiation constants and their uncertainties are from Cohen and Taylor (1987); units are 10^{+4} W·μm^4/cm^2 for c_1, and μm·K for c_2

[b] Parts per million.

Note the following points from those tables:

1. For the total incidance (all wavelengths) the uncertainties in emissivity, temperature, and detector to aperture distance are all roughly equal and limit our knowledge of incidance to about 2% (Table 3.6). Unless we can improve those measurements, there is little point in specifying the Stefan–Boltzmann constant to its available accuracy. The value 5.67 × 10^{-12} W/(cm^2·K^4) is convenient to remember; it is accurate to 0.01%— certainly good enough for our purposes.

2. At 3 μm, with a 300-K blackbody, the uncertainty in the sample spectral incidance prediction is about 5%, dominated by uncertainties in the temperature, wavelength, and the spectral bandpass (see Table 3.7). The uncertainties in wavelength and temperature are magnified because they appear in an exponent, and uncertainties at shorter wavelengths can be even worse.

3. Although not shown here, similar calculations can be done for longer wavelengths. The spectral bandpass and emissivity uncertainties limit the incidance uncertainty to about 2%.

4. For any wavelength, rounding the first radiation constant c_1 to three significant figures (3.74) introduces an error of only 0.05%; for four figures (3.742) the error is 0.005%.

5. The second radiation constant c_2 must be treated more carefully. The uncertainty in the best available estimate is 0.12 μm·K (Cohen and

Taylor, 1987); this yields an uncertainty of 0.01% for the short-wavelength example in Table 3.7. Rounding c_2 to 14,388 μm·K (a relative error of 0.002%) causes an error of 0.03% in the same example.

To recap:

- Incidance calculations are often limited in accuracy to 2 to 5% by our knowledge of the temperatures, wavelengths, and emissivities. (Actual incidance uncertainties are larger still, due to reflections.)
- Use of the following constants will not limit incidance accuracy by more than 0.1%:

$$\sigma = 5.67 \times 10^{-12} \text{ W/(cm}^2\text{·K}^4)$$

$$c_1 = 3.74 \times 10^4 \text{ W·}\mu\text{m}^4/\text{cm}^2$$

$$c_2 = 14388 \ \mu\text{m·K}$$

$$\sigma_q = 1.52 \times 10^{11} \text{ photons/(cm}^2\text{·s·K}^3)$$

$$c_1' = 1.884 \times 10^{23} \text{ photons } \mu\text{m}^3/(\text{cm}^2\text{·s})$$

These values are the "approximate constants" listed with equations (3.9a), (3.9b), (3.13a), and (3.13b).

We have not considered the transmittance of the spectral filters or windows in these calculations; their effect must be considered and a decision made about whether their uncertainties warrant efforts to improve the methods used to determine those values.

3.2.4. Signal from a Broad Spectral Source

Let's now turn to the problem of a detector with known spectral response viewing a broad spectral source. For this discussion we use the letter Q for the photon incidance and reserve the letter E for the radiant incidance, or for incidance in general.

In earlier discussions we derived an equation for the signal S in terms of the photon incidance Q, the detector photon responsivity $\mathcal{R}_q$, and the detector area A:

$$S = \mathcal{R}_q Q A \qquad (3.21)$$

This assumed that the responsivity was the same for all wavelengths included in Q. Except in very special situations, we must make allowance

for responsivity that varies with wavelength. To do that, we will multiply
the spectral responsivity (the responsivity at a given wavelength) times
the spectral photon incidance (the incidance per unit spectral interval) at
that wavelength, and then integrate over all wavelengths of interest:

$$S = \int_0^\infty \mathcal{R}_q(\lambda)Q(\lambda)A \, d\lambda \qquad (3.22)$$

If the "detector" we are considering is a detector–filter–window combination, the spectral responsivity includes the transmittance of the filter
and window.

Ideal Photon Detector. If the detector is an ideal photon detector with an
ideal spectral bandpass filter that transmits only from λ_1 to λ_2,

$$\mathcal{R}_q = \begin{cases} \text{constant} = \mathcal{R}_0 & \text{for } \lambda_1 < \lambda < \lambda_2 \\ \text{zero} & \text{otherwise} \end{cases} \qquad (3.23)$$

In this case, the signal is easy to predict:

$$\begin{aligned} S &= \mathcal{R}_q \int_{\lambda_1}^{\lambda_2} Q(\lambda) \, d\lambda \, A \\ &= \mathcal{R}_q Q(\lambda_1 \text{ to } \lambda_2)A \end{aligned} \qquad (3.24a)$$

Furthermore, we can calculate the detector responsivity directly from the
observed signal:

$$\mathcal{R}_q = \frac{S}{Q(\lambda_1 \text{ to } \lambda_2)A} \qquad (3.24b)$$

For a true photon detector, we need know only the in-band photon incidance, not its spectral distribution.

General Case. In the more general case, the spectral responsivity varies
with wavelength, so prediction of the signal and determination of the
responsivity at a given wavelength requires a little more work. To do the
necessary calculations, we need to know how the responsivity varies with
wavelength. Normally, we do not work with $\mathcal{R}(\lambda)$ directly. Instead, we
first define a *relative* responsivity $\mathcal{R}'(\lambda)$, the ratio of the responsivity at
wavelength λ to that at one special wavelength (we will call it λ_0):

$$\mathcal{R}'(\lambda) = \frac{\mathcal{R}(\lambda)}{\mathcal{R}(\lambda_0)} \qquad (3.25a)$$

Then whenever we need the spectral responsivity $\mathcal{R}(\lambda)$, we will write it as

$$\mathcal{R}(\lambda) = \mathcal{R}(\lambda_0)\mathcal{R}'(\lambda) \qquad (3.25b)$$

Normally—*but not necessarily*—λ_0 is the wavelength of peak response, so $\mathcal{R}'(\lambda)$ has a maximum value of unity.

Using this relative responsivity, equation (3.22) is

$$S = A\mathcal{R}_q(\lambda_0) \int_0^\infty \mathcal{R}_q'(\lambda)Q(\lambda)\, d\lambda \qquad (3.26)$$

As is often the case, assume that we know $Q(\lambda)$ and $\mathcal{R}_q'(\lambda)$. With these we can compute the integral either graphicallly, numerically, or (in rare cases) in closed form. Call the value of the integral Q_{eff}, the effective photon incidance at wavelength λ_0.

Once we know Q_{eff}, we can relate the expected broadband signal and the responsivity at λ_0, or the quantum efficiency η at λ_0, as seen in Equations 3.27a, 3.27b, and 3.28.

SIGNAL FROM A BROADBAND SOURCE: FROM RESPONSE PER PHOTON

$$S = \mathcal{R}_q(\lambda_0)Q_{\mathrm{eff}}A \qquad (3.27a)$$

$$S = \eta(\lambda_0)eGQ_{\mathrm{eff}}A \qquad (3.27b)$$

where

$$Q_{\mathrm{eff}} = \int_0^\infty \mathcal{R}_q'(\lambda)Q(\lambda)\, d\lambda \qquad (3.28)$$

These are just like our original monochromatic formulas, except that we have rigorously accounted for the variation of response with wavelength by using an ''effective'' incidance.

The same process allows us to relate the signal from a broad spectral source to the responsivity expressed in watts (Equations 3.29 and 3.30).

SIGNAL FROM A BROAD SPECTRAL SOURCE—FROM RESPONSE IN WATTS

$$S = \mathcal{R}_e(\lambda_0)E_{e,\text{eff}}A_d \tag{3.29}$$

where

$$E_{e,\text{eff}} = \int_0^\infty \mathcal{R}'_e(\lambda)E_e(\lambda)\,d\lambda \tag{3.30}$$

Note that it is not necessary to do two different integrations to obtain Q_{eff} and $E_{e,\text{eff}}$, since

$$E_{e,\text{eff}} = \frac{hc}{\lambda_0}\,Q_{\text{eff}} \tag{3.31}$$

This is true for any $\mathcal{R}(\lambda)$—we do not need to assume that the detector response is ideal, nor do we need to make any other assumptions. This can be seen in the following way:

$$E_{e,\text{eff}} = \int_0^\infty \mathcal{R}'_e(\lambda)E_e(\lambda)\,d\lambda$$

$$= \int_0^\infty \frac{\mathcal{R}_e(\lambda)}{\mathcal{R}_e(\lambda_0)}\,E_e(\lambda)\,d\lambda$$

$$= \int_0^\infty \frac{\mathcal{R}_q(\lambda)\,(\lambda/hc)}{\mathcal{R}_q(\lambda_0)\,(\lambda_0/hc)}\,E_e(\lambda)\,d\lambda$$

$$= \frac{hc}{\lambda_0}\int \mathcal{R}'_q(\lambda)\left[\frac{\lambda}{hc}\,E_e(\lambda)\right]d\lambda$$

$$= \frac{hc}{\lambda_0}\int \mathcal{R}'_q(\lambda)Q(\lambda)\,d\lambda$$

$$= \frac{hc}{\lambda_0}\,Q_{\text{eff}}$$

One interpretation of Q_{eff} or $E_{e,\text{eff}}$ is that they are the incidances which if concentrated at wavelength λ_0 would give the same signal as the original broad spectral source. All of the preceding has been rigorous and applies to any detector or detector filter combination; we have not made any assumptions about the spectral response of the detector.

Blackbody Values. For convenient test set use, blackbody test sets often include a table of incidances for different apertures and distances. Since there are many possible detector–filter–window combinations which may be tested on that set, it is not practical to provide a table of effective incidances for each of these possibilities. Instead, the total incidance ("blackbody incidance") is provided, and the user is left to convert that to an effective incidance using a *blackbody-to-peak conversion factor*, sometimes called a *cell response factor*:

$$Q_{\text{eff}} = \frac{Q_{\text{BB}}}{C_q} \tag{3.32}$$

where

$$C_q = \frac{\displaystyle\int_0^\infty Q(\lambda)\,d\lambda}{\displaystyle\int \mathcal{R}_q'(\lambda)Q(\lambda)\,d\lambda} \tag{3.33}$$

or

$$E_{\text{e,eff}} = \frac{E_{\text{BB}}}{C_e} \tag{3.34}$$

where

$$C_e = \frac{\displaystyle\int_0^\infty E_e(\lambda)\,d\lambda}{\displaystyle\int_0^\infty \mathcal{R}_e'(\lambda)E_e(\lambda)\,d\lambda} \tag{3.35}$$

Alternatively, one can compute a *blackbody responsivity* and convert that to the peak value: let

$$\mathcal{R}_{\text{BB}} = \frac{S}{E_{\text{BB}}A} \tag{3.36}$$

then

$$\mathscr{R}(\lambda_0) = \mathscr{R}_{BB}C \qquad (3.37)$$

These average or blackbody responsivities are weighted averages of the actual spectral responsivity. The weighting function is the spectral content of the assumed source, and it is important to specify the assumed source when we use these average or blackbody responsivities. If the incidance is modulated by a chopper at temperature T_{CH}, the derivation is similar but the result is a little different:

$$\mathscr{R}(\lambda_0) = \frac{S}{(E_{\text{eff,BB}} - E_{\text{eff,Ch}})A} \qquad (3.38)$$

$$\mathscr{R}(\lambda_0) = \frac{S}{(E_{BB}/C_{BB} - E_{Ch}/C_{Ch})A} \qquad (3.39)$$

If tables of $E_{net} = E_{BB} - E_{CH}$ are provided, it is more convenient to compute a conversion factor for the blackbody–chopper combination:

$$C_{BB,CH} = \frac{E_{BB} - E_{Ch}}{E_{\text{eff,BB}} - E_{\text{eff,Ch}}} \qquad (3.40)$$

so that

$$\mathscr{R}(\lambda_0) = \frac{S}{(E_{BB} - E_{Ch})A} C = \mathscr{R}_{BB}C_{BB,Ch} \qquad (3.41)$$

This entire derivation can be carried out in terms of either watts or photons, and different factors will result. Done consistently from beginning to end, the resulting peak values will be consistent, and

$$\mathscr{R}_e(\lambda_0) \frac{hc}{\lambda_0} = \mathscr{R}_q(\lambda_0) \qquad (3.42)$$

Note that the average responsivities will be less than the peak value, and that they will depend on the spectral content:

$$\mathscr{R}_{500} \neq \mathscr{R}_{499} \neq \mathscr{R}_{300}$$

One practical implication of this is that (unless the spectral response of all detectors in an array are identical) they will have different signals when viewing a uniform target *even if they were calibrated and balanced when viewing a uniform laboratory source.* This "spectral streaking" can be minimized by demanding that the detectors have matching spectral characteristics (within some tolerance) or by balancing them when viewing a source whose spectral content matches that of the target within an acceptable margin.

Example: The Human Eye. The spectral response of the eye is shown in Figure 3.12. The two curves show the *photopic* (daylight, due to the cones) and *scotopic* (low light levels, due to the rods) response curves. We often say that the visible portion of the spectrum extends from 0.4 to 0.7 μm, but note that the photopic spectral response of the eye exceeds 50% of its maximum value only between 0.5 and 0.6 μm. For radiometric estimates, a fair approximation to the spectral response of the eye would be a "spectral bandwidth" of 0.1 μm centered at 0.55 μm

Now consider the spectrum of light from the sun, as shown in Figure 3.13*a*. The total incidance is proportional to the area under the curve. Counting squares yields about 0.1 W/cm², in good agreement with the accepted value. The "effective" incidance, as described earlier, is the integrated value of the product of the sun's spectral distribution and the response of the eye. This is shown in Figure 3.13*b*; the integrated value is about 0.015 W/cm². If we were doing radiometric tests with the eye and sunlight, we could use an "effective incidance" value that was 15% of the total incidance, or use total incidance values and a blackbody to peak conversion factor of 6.7.

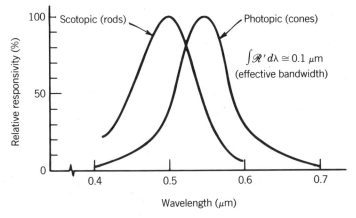

Figure 3.12. Spectral response of the eye.

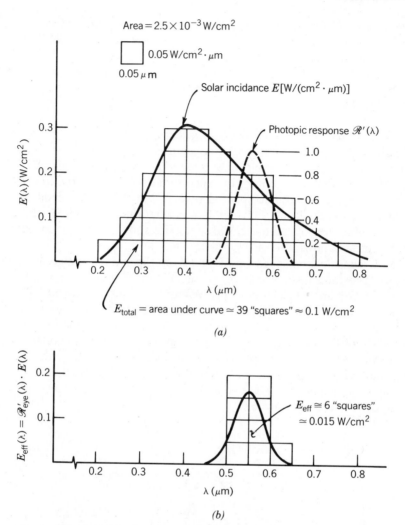

Figure 3.13. Incidance from the sun.

Photometry is the study of the radiometrics of visible light sensed by the human eye; people who specialize in photometry use a special unit—the *lumen*—for what we have called the effective power:

692 lumens = 1 watt of effective power

What we have called effective incidance is expressed in lumens per foot squared, which is termed a *footcandle*:

$$1 \text{ footcandle} = 1 \text{ lumen/ft}^2 \left(\frac{1 \text{ effective watt}}{692 \text{ lumens}} \right) \left(\frac{1 \text{ ft}}{12 \text{ in. } \times 2.54 \text{ cm/in.}} \right)^2$$

$$\approx 1.6 \times 10^{-6} \text{ effective watts/cm}^2$$

Example: Consider a 100-W incandescent bulb. If all of its power were made available as radiation at 0.55 μm, each watt would generate 692 lumens, for a total of 69,200 lumens. Bulb manufacturers now state the luminous output of their bulbs on the package; typical values are around 1700 lumens for a 100-W bulb. Thus the efficiency is about 1700 divided by 69,200, or 2.5%. The rest of the power goes into that portion of the spectrum to which the eye does not respond, and into heating the space around the bulb.   ∎

3.2.5.   Response per Degree Difference in Scene Temperature

We can predict the ac signal from a detector viewing a small temperature difference through a projected solid angle Ω^*:

$$S = S_2 - S_1 = A \left[\int \mathcal{R}(\lambda) M(\lambda, T_2) \, d\lambda - \int \mathcal{R}(\lambda) M(\lambda, T_1) \, d\lambda \right] \frac{\Omega}{\pi}$$

$$= \frac{A_d \Omega}{\pi} \int \mathcal{R}(\lambda) [M(\lambda, T_2) - M(\lambda, T_1)] \, d\lambda \tag{3.43}$$

It is sometimes desirable to express the responsivity in terms of ΔT: $\Delta T = T_2 - T_1$.

$$\mathcal{R}_{\Delta T} = \frac{S}{\Delta T} = \frac{S_2 - S_1}{T_2 - T_1} = \frac{A_d \Omega}{\pi} \int \mathcal{R}(\lambda) \frac{\Delta M}{\Delta T} \, d\lambda \tag{3.44}$$

As before, replace $\mathcal{R}(\lambda)$ with $\mathcal{R}(\lambda_0) \, \mathcal{R}'(\lambda)$:

$$\mathcal{R}_{\Delta T} = \frac{A_d \Omega}{\pi} \mathcal{R}(\lambda_0) \int \mathcal{R}' \frac{\Delta M}{\Delta T} \, d\lambda$$

If ΔT is small, the ratio $\Delta M/\Delta T$ can be replaced by dM/dT. In any case we can compute the integral. We will call it $\langle \Delta M/\Delta T \rangle$, the effective value

---

* Solid angles are discussed in Sections 3.3.1 and 3.3.2.

of $\Delta M/\Delta T$ for the system:

$$\left\langle \frac{\Delta M}{\Delta T} \right\rangle = \int_0^\infty \mathcal{R}'(\lambda) \frac{\Delta M}{\Delta T} \, d\lambda \qquad (3.45)$$

With that value calculated, we can relate $\mathcal{R}_{\Delta T}$ to $\mathcal{R}(\lambda_0)$ by Equation 3.46.

RESPONSE PER KELVIN FROM CONVENTIONAL RESPONSIVITY

$$\mathcal{R}_{\Delta T} = \mathcal{R}(\lambda_0) \frac{A_d \Omega}{\pi} \left\langle \frac{\Delta M}{\Delta T} \right\rangle \qquad (3.46)$$

This formula is true whether we use photons or radiant exitance, as long as the units for $\mathcal{R}$ and M are consistent (both in watts, or both in photons).

3.3.   SPATAL INTEGRAL

The incremental formula assumed a very small detector looking at a very small source of radiation. We want to be able to handle a detector looking at a larger source: for example, a detector looking out at the entire hemisphere in front of it, without any shielding.

To do this we can break up a hemisphere into a large number of small sources and apply our incremental formula to each of those sources. We then add up the effect of all those small sources. The process is known mathematically as integration, and the results for several geometries of interest are given in this section.

A detector viewing geometry is sometimes specified by the angle through which the detector can see a given object; this is called the *field of view* (FOV) for that object. Whenever possible, transmit information about the field of view by a sketch or drawing; a verbal or written statement of the field of view can be ambiguous or misleading. The field of view shown in Figure 3.14 might be described by different users as 30°, 45°, or 90°! Be alert to the possibility of confusion, and request a sketch whenever possible.

3.3.1.   Solid Angles

The energy falling on a detector is very dependent on the extent to which the source "surrounds" the detector. This is described by the *solid angle*

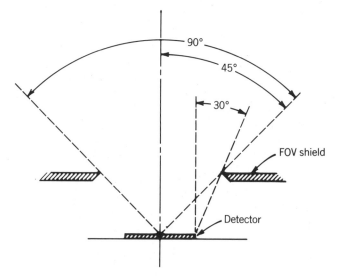

Figure 3.14. Field-of-view definitions.

through which the detector sees the source. Solid angles are a generalization of the common "plane" angles to three dimensions. Roughly speaking, we calculate the solid angle by dividing the surface area of the object viewed by the square of the distance to the object. For example:

- A 2 ft × 2 ft window, 10 ft away, subtends a solid angle of 0.04 sr: $(2 \times 2)/(10 \times 10)$.
- The sun subtends a solid angle of about 70×10^{-6} sr: the area is $(\pi/4)$ (864,000 miles)^2, and the distance from the earth to the sun is 93 million miles.
- The ice cream at the top of a cone subtends a solid angle, when viewed from the tip of the cone, of about 0.2 sr: $(\pi/4)$ (2 in.)^2/(4 in.)^2.
- From the center of a sphere, the solid angle subtended by the sphere is 4π steradians: The surface of a sphere is 4π times the radius squared, and when we divide by the distance to the sphere (squared), we are left with 4π. A hemisphere subtends 2π steradians.

A more exact definition of solid angle would note that the area we use should be the area on the surface of a sphere surrounding the point we are interested in (see Figure 3.15). For small surfaces nearly perpendicular to the viewer, the area divided by distance-squared formula is good enough for most work.

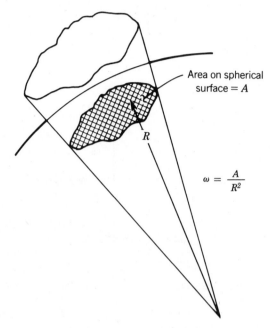

$$\omega = \frac{A}{R^2}$$

Figure 3.15. Solid angle.

3.3.2. Simple and Projected Solid Angles

All of the above applies to the *simple* or *true* solid angle. The incremental power transfer formula includes one more cosine factor than does the incremental form of the true solid angle. We call the resulting solid angle the *projected* or *effective* solid angle. For small angles subtended by surfaces that face each other, the two are nearly the same, and it is not necessary to distinguish between them. On the other hand, the projected solid angle is generally not much harder to compute than is the simple solid angle, so we might as well use the correct formula.

Bartell (1969a) provides a review of the projected solid angle concept. Nicodemus (1976) includes an even more complete and rigorous discussion of projected solid angles.

3.3.3. Solid Angles for Specific Geometries

Formulas for both the true solid angle and the projected solid angle are given in Figure 3.16 for the incremental limit, the general case, and for

Infinitesimal

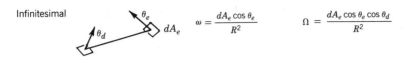

$$\omega = \frac{dA_e \cos \theta_e}{R^2} \qquad \Omega = \frac{dA_e \cos \theta_e \cos \theta_d}{R^2}$$

General case

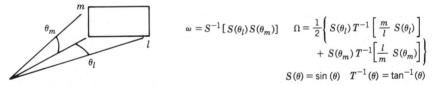

$$\omega = \int_{A_e} \frac{dA_e \cos \theta_e}{r^2} \qquad \Omega = \int \frac{dA_e \cos \theta_e \cos \theta_d}{r^2}$$

Hemisphere

$$\omega = 2\pi \qquad \Omega = \pi$$

Round aperture

$$\omega = 2\pi \left(1 - \cos \frac{\theta}{2} \right) \qquad \Omega = \pi \sin^2 \frac{\theta}{2}$$

$$= \pi \frac{r^2}{R^2 + r^2}$$

Rectangular aperture, one corner on axis

$$\omega = S^{-1}[S(\theta_l) S(\theta_m)] \qquad \Omega = \frac{1}{2} \left\{ S(\theta_l) T^{-1} \left[\frac{m}{l} S(\theta_l) \right] \right.$$

$$\left. + S(\theta_m) T^{-1} \left[\frac{l}{m} S(\theta_m) \right] \right\}$$

$$S(\theta) = \sin(\theta) \qquad T^{-1}(\theta) = \tan^{-1}(\theta)$$

Figure 3.16. Useful solid angle formulas. ω = true or simple solid angle. Ω = projected, or weighted solid angle.

three interesting geometries. The formula $\Omega = \pi \sin^2(\theta/2)$ will be encountered often.

Infinitesimal Limit. In the small source limit on axis, the projected solid angle reduces to the simple solid angle—that component of the area perpendicular to the detector–source line, divided by r^2. This approximation is valid to within 0.2% for total included angles of 10° or less, and within 2% for included angles of 30°.

General Case. To obtain the simple solid angle and projected solid angle in a general case, one must integrate the infinitesimal formula over the surface of the defining aperture. This integration is normally approximated by a computer-coded summation process. The desired solid angle is the sum of the solid angles for each of many small areas, using the appropriate angles and distance r for each small area. For a few simple cases closed-form solutions are possible. These include the hemisphere, a round aperture, and a rectangular aperture.

Hemisphere. Note that the projected solid angle for a hemisphere is only π, not 2π, steradians. Hudson (1969, p. 30) mentions confusion over this factor of 2 as one of the most common errors made by newcomers to radiometry. It is this value of π that relates the sterance L and the exitance M of a blackbody. The exitance is the integral of the sterance over the hemisphere into which the source emits, and the effective solid angle is π.

Round Aperture. The projected solid angle is $\pi \sin^2(\theta/2)$. Putting a round aperture directly above a detector reduces the projected solid angle from that of a hemisphere by a factor of $\sin^2(\theta/2)$. This reduces the background by the same factor, and the photon noise by the square root: $\sin(\theta/2)$. That factor is often used in D^* and NEP improvement calculations. A closed expression for an *off-axis* circular field of view is given a little later in this section.

Rectangular Aperture. Rectangular apertures are often used over linear detector arrays. Figure 3.16 gives the solid angles for such an aperture if the corner of the rectangle is directly over the detector. A more general situation (the off-axis case) can always be broken into four components, each of which has a corner over the detector so that the sum or difference of the solid angle of the four components is the desired solid angle (see Figure 3.17).

3.3.4. More Complicated Geometries: Angle Factors

The formulas described earlier handle most of the cases of interest in a detector test lab, but occasionaly more elaborate formulas are required. For example, we might have a large detector at a relatively small distance from a large source. Another example is the circular aperture off-axis from a small detector.

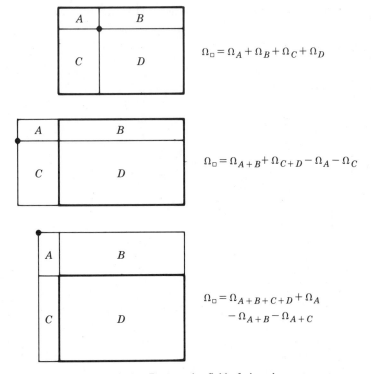

Figure 3.17.   Rectangular field of view, by parts.

Many of these geometries have been analyzed and the results reported in terms of *angle factors* (also referred to as *shape factors*). These may be found in books on heat transfer, in the sections on radiative transfer. Known angle factors can be manipulated to provide the angle factors for other geometries; this process is known as *angle factor algebra*.

In our notation, the angle factor $F_{dA_1-A_2}$ is just the projected solid angle subtended by area A_2 as seen from incremental area dA_1, divided by π:

$$F_{dA_1-A_2} = \frac{\Omega_{A2}}{\pi} \qquad (3.47)$$

If dA_1 is a Lambertian source, $F_{dA_1-A_2}$ is the fraction of the energy radiated by dA_1 that will reach A_2:

$$P \text{ to } A_2 = \epsilon \sigma T_1^4 \, dA_1 \, F_{dA_1-A_2}$$

Example: For a very small detector parallel to an off axis disk (Figure 3.18) the angle factor is given by Sparrow and Cess (1965) and Kreith (1962):

$$F_{dA_1-A_2} = \frac{1}{2}\left\{1 - \frac{(a^2 + b^2 + c^2) - 2b^2}{\sqrt{(a^2 + b^2 + c^2)^2 - 4a^2b^2}}\right\} \qquad (3.48)$$

In the limit as a goes to zero, this reduces to our familiar expression:

$$F_{dA_1-A_2} \rightarrow \frac{b^2}{b^2 + c^2} = \sin^2\frac{\theta}{2}$$

$$\Omega_{dA_1-A_2} \rightarrow \pi \sin^2\frac{\theta}{2}$$

$\blacksquare$

Effective Emissivity. One special case is important in dewar design. That is the case of two large surfaces so close together that each "sees" the other through a hemisphere. The ideal case is two concentric spheres, but two concentric cylinders or two parallel plane surfaces would behave similarly if their separation were small compared to their dimensions. The shape factor $F_{A_{inner}-A_{outer}}$ is unity if the inner surface can see only the outer one. But what about the emissivity?

If the emissivities are near unity, the problem is not hard: They each give off a power given by the Stefan–Boltzmann law, and each absorbs a large fraction of that power, so the net power transmitted is

$$P \approx \sigma \epsilon_1 \epsilon_2 (T_1^4 - T_2^4) \qquad (3.49)$$

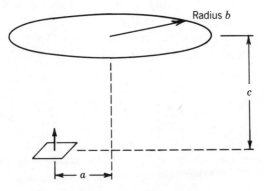

Figure 3.18. Off-axis disk.

If the emissivities are much less than unity (as they are if chosen to minimize the heat transfer), much of the incident power is reflected back to the source. Since the surfaces face each other, that power is *again* reflected, and "rattles around" until it is absorbed by either the original emitter or the facing surface. The derivation of a formula for the net power transfer ϕ requires that we follow a few such reflections and sum those that are absorbed on one side. This is done by Scott (1959); the result is

$$\Phi = \sigma \frac{\epsilon_1 \cdot \epsilon_2}{\epsilon_1 + \epsilon_2 - \epsilon_1 \epsilon_2} (T_1^4 - T_2^4) \qquad (3.50)$$

If $\epsilon_1 = \epsilon_2 \ll 1$, then

$$\Phi \simeq \sigma \frac{\epsilon}{2} (T_1^4 - T_2^4) \qquad (3.51)$$

The addition of another shield between these two which can "float" to some intermediate temperature will cut the power transmitted by another factor of 2. Further shields will further decrease the power. One way to introduce additional shields conveniently is in the form of multilayer insulation (MLI); it consists of many layers of highly reflective material separated by good thermal insulators.

3.4. TIME DEPENDENCE (MODULATION)

It is often convenient to work with an alternating-current (ac) signal instead of a dc signal: amplifier stability is easier to maintain, and both amplifier and detector noise are lower. To use an ac amplifier, we must convert our constant incidance into an alternating source. This is called modulating ("chopping") the source, and several methods are available. One common chopper is a toothed wheel that rotates. The shape of the teeth and the shape of the blackbody aperture determine the resulting waveform.

3.4.1. Fourier Analysis of Waveforms

It can be shown that any periodic waveforms—such as the chopped incidance from our blackbody, or the resulting signal from the detector—can be written mathematically (and treated experimentally) as the *sum* of a number of separate *sinusoidal* waves. An infinite number of such component sinusoidal waves is required to represent the initial waveform exactly, but for our purposes we will need only the first term, and for almost every case we can get away with considering the first few and neglecting the rest.

The series of infinite but simple terms that replaces our complicated waveform is called the Fourier series, after Baron Jean Baptiste Joseph Fourier (a French mathematician, physicist, and Egyptologist, 1768–1830). The first term in the Fourier series is the *fundamental* component, and its frequency is the *fundamental frequency*. The fundamental frequency is the same frequency as that of the arbitrary waveform.

3.4.2. Modulation Factors

With very few exceptions we want to work with a detector signal that is sinusoidal at a particular frequency. For example, we might work with an 800-Hz sinusoid, as opposed to a square wave or some other shape. This is true for at least two reasons:

1. The responsivity of a detector is dependent on frequency, so to specify the responsivity uniquely, we must also specify the frequency.
2. Noise can be reduced if we narrow the bandpass.

Thus we restrict the electrical signal to one frequency (by using electrical filters which reject all but the fundamental component of the signal waveform), and then we measure the rms value of the resulting sinusoid. The responsivity was defined as electrical output divided by IR input. Since we measure the rms of the fundamental output, we want to use the rms of the fundamental component of the IR input. Direct application of Planck's law generally yields the *peak-to-peak value* of our modulated incidance pattern. To get the rms of the fundamental from the peak-to-peak value, we will multiply by the *modulation factor* (MF; sometimes called the *waveform factor*) as in Equation 3.52.

APPLICATION OF THE MODULATION FACTOR

rms of fundamental

$$= \text{MF} \times \text{peak-to-peak value of original waveform} \qquad (3.52a)$$

where we define

$$\text{MF} = \frac{\text{rms of fundamental component}}{\text{peak-to-peak value of total waveform}} \qquad (3.52b)$$

For a square wave, the modulation factor is $\sqrt{2}/\pi$, about 0.45. Modulation factors are provided in Table 3.8 for a rotating chopper blade with radial teeth. Calculation of modulation factors for arbitrary chopping methods are discussed next, and formulas for common geometries are provided.

General Case. To calculate the modulation factor for a general waveform $E(t)$, we determine the amplitude of the fundamental Fourier component, then use the relation that the rms of the fundamental is the square root of 2 over two times the amplitude (Equation 3.53).

Table 3.8 Modulation factors

DAP/TW	CASE II	CASE III: NUMBER OF TOOTH-SLOT PAIRS				
		50	20	10	5	2
0.00	0.4502	0.4502	0.4502	0.4502	0.4502	0.4502
0.10	0.4488	0.4488	0.4488	0.4488	0.4488	0.4488
0.20	0.4446	0.4446	0.4446	0.4446	0.4446	0.4446
0.30	0.4378	0.4378	0.4378	0.4378	0.4378	0.4377
0.40	0.4283	0.4283	0.4283	0.4283	0.4282	0.4279
0.50	0.4163	0.4163	0.4163	0.4163	0.4162	0.4154
0.60	0.4020	0.4020	0.4020	0.4019	0.4017	0.4002
0.70	0.3855	0.3855	0.3854	0.3853	0.3850	0.3821
0.80	0.3670	0.3670	0.3669	0.3668	0.3662	0.3613
0.90	0.3467	0.3467	0.3466	0.3464	0.3454	0.3377
1.00	0.3249	0.3249	0.3248	0.3244	0.3231	0.3113
1.10	0.3018	0.3018	0.3017	0.3012	0.2993	0.2822
1.20	0.2777	0.2777	0.2775	0.2769	0.2744	0.2502
1.30	0.2529	0.2529	0.2527	0.2519	0.2487	
1.40	0.2277	0.2276	0.2273	0.2264	0.2224	
1.50	0.2022	0.2022	0.2018	0.2007	0.1959	
1.60	0.1769	0.1768	0.1764	0.1751	0.1694	
1.70	0.1519	0.1518	0.1514	0.1499	0.1435	
1.80	0.1275	0.1274	0.1270	0.1253	0.1183	
1.90	0.1040	0.1039	0.1034	0.1016	0.0942	
2.00	0.0816	0.0815	0.0809	0.0791	0.0715	
2.10	0.0604	0.0603	0.0598	0.0579	0.0507	
2.20	0.0406	0.0405	0.0400	0.0383	0.0319	
2.30	0.0225	0.0224	0.0219	0.0204	0.0155	
2.40	0.0060	0.0059	0.0055	0.0043	0.0017	
2.50	-0.0087	-0.0087	-0.0090	-0.0098	-0.0093	
2.60	-0.0215	-0.0215	-0.0216	-0.0218	-0.0173	
2.70	-0.0324	-0.0324	-0.0323	-0.0318	-0.0223	
2.80	-0.0414	-0.0414	-0.0411	-0.0397	-0.0243	
2.90	-0.0485	-0.0484	-0.0479	-0.0455	-0.0235	
3.00	-0.0538	-0.0537	-0.0528	-0.0493	-0.0202	
3.10	-0.0573	-0.0571	-0.0560	-0.0512	-0.0152	
3.20	-0.0592	-0.0589	-0.0574	-0.0513		
3.30	-0.0595	-0.0591	-0.0573	-0.0499		

DAP = aperture diameter
TW = tooth width (along arc for case III)

Program AMODF - revised 9 Sept 1988 - JDV

MODULATION FACTOR: GENERAL CASE

$$\text{MF} = \frac{\sqrt{2}}{2} \frac{C}{E_{max} - E_{min}} \qquad (3.53a)$$

where

$$C = \sqrt{A^2 + B^2} \qquad (3.53b)$$

$$A = \frac{2}{T} \int_0^T E(t) \cos \frac{2\pi t}{T} \, dt \qquad (3.53c)$$

$$B = \frac{2}{T} \int_0^T E(t) \sin \frac{2\pi t}{T} \, dt \qquad (3.53d)$$

$$T = \text{period of the waveform}$$

If one can select the time origin so that the wave is made symmetric in time $[E(t) = E(-t)]$, then $B =$ zero
and $C = A = \frac{4}{T} \int_0^{T/2} E(t) \cos \left(\frac{2\pi t}{T} \right) dt.$

The formulas given are for the fundamental component, but equivalent amplitudes A_j, B_j, C_j and modulation factors MF_j can be obtained for higher-order components $j = 2, 3, \ldots$ by replacing the term $2\pi t/T$ in the integrals with $j \cdot 2\pi t/T$.

Special Cases. Five chopper configurations of interest are shown in Figure 3.19. The first is a very small aperture modulated by a large tooth–slot combination. It will yield a square wave. In case II the aperture is no longer infinitely small, but we retain the simplifying assumption of linear motion of the chopper. Case III is an accurate model of most blackbody modulation: a round aperture chopped by radial teeth. Cases IV and V are round blackbody apertures chopped by round chopper holes.

Case I: Square Wave. A large toothed chopper moving past a very small hole yields very nearly a square wave. This ideal shape is never exactly achieved, but it represents a limiting case that is approximated very well in many setups. It is useful to consider for several reasons:

- The resulting modulation factor (0.45) can be used as a rough value for "quick-and-dirty" calculations.

Case I: square-wave chopping

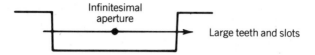

Case II: rectangular teeth, linear motion, circular aperture

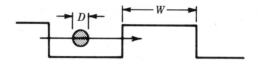

Case III: radial teeth, circular motion, circular aperture

Case IV: circular chopper holes, linear motion, circular apertures

Case V: circular chopper holes, circular motion, circular aperture

Figure 3.19. Chopper configurations.

- The value can be used as a check on formulas or computer calculations for the more realistic cases.
- The derivation and calculation provides some insight and practice for the more realistic cases.

Application of the general equations yields the square wave modulation factor (Equation 3.54).

MODULATION FACTOR: SQUARE WAVE

$$\text{MF} = \frac{\sqrt{2}}{\pi} \simeq 0.4502 \qquad (3.54)$$

Case II: Round Aperture, Square Teeth. This waveform results when rectangular teeth move linearly across a circular aperture. Test sets do not have such a chopping configuration, but the results approximate those for common configurations (case III) within 0.1% for typical chopper dimensions, and the formula is easy to evaluate using available tables. Thus it is a convenient and often acceptable approximation for the more realistic case III.

The waveform analysis was done by McQuistan (1958); it leads to the modulation factor shown in Equation 3.55*a*.

MODULATION FACTOR: ROUND APERTURE, SQUARE TEETH

$$\text{MF}_2 = 2\,\frac{\sqrt{2}}{\pi}\,\frac{J_1(z)}{z} \qquad (3.55a)$$

where $\quad z = \dfrac{\pi \times \text{aperture diameter}}{2 \times \text{tooth width}}$

$J_1(z)$ = first-order Bessel function of the first kind (available in most books of tabulated mathematical functions)

An alternative but entirely equivalent expression is given by Guenzer (1976):

$$\text{MF}_2 = \frac{\sqrt{2}}{\pi}\,[J_0(z) + J_2(z)] \qquad (3.55b)$$

Table 3.8 provides modulation factors for case II and case III as a function of the aperture diameter-to-tooth width ratio. Note that the case II values approximate those of case III very closely for most chopper geometries.

Case III: Round Aperture, Radial Teeth. This case is that encountered in many test stations: A rotating chopper with radial teeth moves across the round blackbody aperture. The modulation factor can be derived easily (Eq. 3.56) from the analysis of the waveform done by McQuistan (1959).

MODULATION FACTOR: ROUND APERTURE, RADIAL TEETH

$$\text{MF}_3 = \frac{\sqrt{2}}{\pi} F\left[\frac{-N}{2}, \frac{+N}{2}, 2, \left(\frac{d}{D}\right)^2\right] \tag{3.56}$$

where $F[\cdot]$ = hypergeometric function

N = number of tooth–slot pairs

d = aperture diameter

D = pitch diameter of the chopper

Although the hypergeometric function sounds rather intimidating, it is easily programmed for computer calculation. The listing for a Fortran function (AMODF3) that calculates the modulation factor in this way is supplied later in this chapter. Arguments called for in the subroutine are the aperture diameter, the chopper pitch diameter, the number of tooth–slot pairs, and the maximum allowable error.

Table 3.8 lists case III modulation factors for several configurations. To use the table, first calculate the equivalent tooth width, defined as the arc at the pitch diameter:

$$W = \frac{\pi \times D}{2 \times N} \tag{3.57}$$

Then calculate the ratio of the aperture diameter d to the tooth width W. Enter the table with that ratio and the number N of tooth–slot pairs, interpolating as necessary. Note that for a given ratio the modulation factor is not sensitive to N. The formula of case II will work within about 0.1% for reasonable chopper dimensions if the tooth width described above is used.

Cases IV and V: Round Aperture, Round Chopper Holes. I know of no published closed form for the modulation factor for these cases. If needed, the formulas for the intersection of the two circles as a function of time could be used with the general method described earlier in the section on modulation to calculate the modulation factor for the configuration of interest. The numerical values for cases IV and V are probably very similar.

3.4.3. Sine-Wave Modulation

It is sometimes convenient to have the blackbody incidence modulated as a pure sine wave. This might be done to measure the frequency response of the detector by varying the chopping frequency and measuring the output voltage with a simple ac meter. (The fundamental of the output could be isolated using a wave analyzer, but that requires adjusting the wave analyzer frequency every time the modulator frequency was moved. If the waveform were a pure sine wave, a simple ac meter could be used.) The modulation factor of a sine wave is $\sqrt{2}/4 \simeq 0.3536$.

Two chopper schemes that generate sine waves are described next. To verify that a sine-wave modulator in working properly, look at the resulting signal with a wave analyzer. These should be no signal at the harmonics ($2f_0$, $3f_0$, etc.); the amount observed is a measure of the deviation from the pure sine wave.

Modulation with a Tuning Fork Chopper. Choppers are available that consist of a pair of vanes attached to the ends of electrically driven "tuning forks"; these oscillate at a fixed frequency. The dimension of the illuminated area is defined in one dimensions by a mechanical slit that is independent of the chopper, and in the other dimension by the opening between the vanes.

The travel of the vanes can be adjusted so that at one end of their "swing" they either overlap, just meet, or fail to meet. The resulting waveforms are shown in Figure 3.20. Note that if the vanes overlap, the waveform that results is *not* sinusoidal.

If the resulting incidence must be known, the total swing $2L$ determines the maximum "open aperture" area. To determine $2L$, adjust the amplitude until the pattern in Figure 3.20a is seen, and then measure the "rest width" L_0. The total swing will be $2L_0$ if the chopper obeys simple harmonic motion.

Modulation with a Sine-Wave Aperture. Another way of generating a sinusoidal incidence is with a conventional chopper and an aperture that just fills the space between the chopper teeth, and has the "eye" shape shown in Figure 3.21. The radial height is proportional to the cosine of the chopper turning angle. The area exposed is proportional to the integral of the height y and the travel $d\theta$; that is a sine function.

Problems with this scheme are that:

1. The aperture must be tailored to match the chopper.
2. Cutting such an aperture accurately requires precise layout and photoetch work, both of which are time consuming.

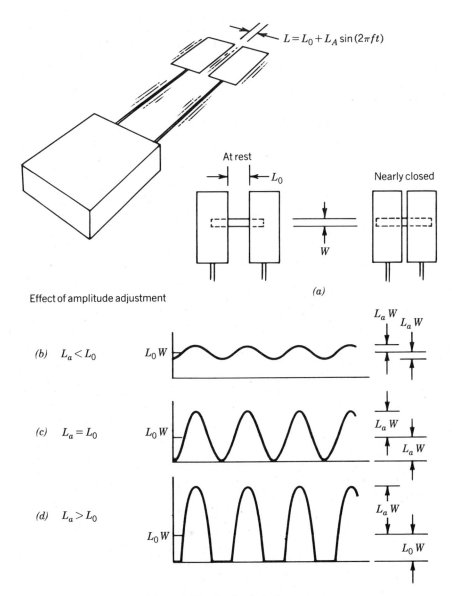

$$L = L_0 + L_A \sin(2\pi ft)$$

At rest

L_0

Nearly closed

W

(a)

Effect of amplitude adjustment

(b) $L_a < L_0$

$L_0 W$

$L_a W$ $L_a W$

(c) $L_a = L_0$

$L_0 W$

$L_a W$ $L_a W$

(d) $L_a > L_0$

$L_0 W$

$L_a W$

$L_0 W$

Figure 3.20. Tuning fork chopper.

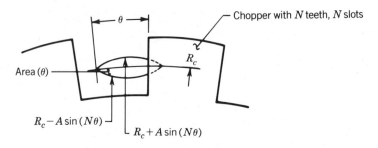

$d\,\text{Area} = r\,d\theta\,dr$

$$\text{Area} = \int_0^\theta \int_{R_c - A\sin(N\theta)}^{R_c + A\sin(N\theta)} r\,dr\,d\theta \quad = \frac{1}{2}\int_0^\theta r^2 \Big|_{R - A\sin(N\theta)}^{R + A\sin(N\theta)} d\theta$$

$$= \frac{1}{2}\int_0^\theta \Big\{[R + A\sin(N\theta)]^2 - [R - A\sin(N\theta)]^2\Big\}d\theta$$

$$= \frac{1}{2}\cdot 4\,AR_c \int_0^\theta \sin(N\theta)\,d\theta$$

$$= \frac{2AR_c}{N}[1 - \cos(N\theta)]$$

Potential misalignment (exaggerated)

Figure 3.21.   Sine-wave aperture.

3. The aperture must be installed so that it just fills the tooth–slot. (Otherwise, the resulting wave will not be an accurate sinusoid.)

3.5.   EXAMPLES AND CASES OF SPECIAL INTEREST

Let us now apply the methods that we have discussed to a few cases of special interest: the incidance in a dewar due to the background and the blackbody source, and the effective incidance from the sun and moonlight.

We begin with the incremental power transfer equation (3.1), but extend it using the expressions that we have developed for the spectral integral, the spatial integral, and the modulation factor. Because we often use a specially designed blackbody, the emissivity of the source is near unity. We want to calculate all the energy that falls on the detector, so we take the absorptivity to be unity.

We often use a filter whose transmittance is either constant (in the spectral range where it is not zero) or varies so little that we can use some average value without too much error. The spectral integral in that case is just the in-band exitance times the average transmittance.

For the normal test setup, the detector is small, and the source and detector face each other directly. We often use circular apertures to define the blackbody area and the background. If the aperture is circular with diameter d at a distance D, the projected solid angle Ω is

$$\Omega = \pi \frac{(d/2)^2}{D^2 + (d//2)^2} \tag{3.58}$$

The exitance M in the spectral band of interest will be in either watts or photons, for one of three spectral possibilities: narrowband, wideband, or all wavelengths.

The background is not modulated, but the blackbody incidance may be modulated. If modulated, we will assume a modulation factor MF of 0.45 for now. (If necessary, we can substitute the more exact methods later.) If not modulated, we can use the same formula with a modulation factor of 1.0.

The different combinations for most special cases can be combined symbolically:

$$E = \qquad M \qquad\qquad \Omega/\pi \qquad\qquad MF$$

$$E = \begin{cases} \sigma T^4 \text{ or } \sigma_q T^3 & & 1.0 \\ \\ M(0 \text{ to } \lambda_2) - M(0 \text{ to } \lambda_1) & \dfrac{(d/2)^2}{D^2 + (d/2)^2} & 0.45 \\ \text{(see Tables 3.4 and 3.5)} & & \\ \\ M(\lambda_c, T)\, \Delta\lambda \text{ [see Tables 3.2} & & MF \\ \text{and 3.3 or formulas (3.9}a) & & \\ \text{and (3.9}b)] \end{cases}$$

We will illustrate the use of these formulas by calculating the background and blackbody incidances at the detector plane of the test setup of Figure

3.22 with using the narrow bandpass filter of Figure 3.23 and the wide bandpass filter of Figure 3.24.

3.5.1. Background in a Test Dewar—Narrow Bandpass Filter

The background in the test dewar of Figure 3.22 is limited geometrically by the 0.4-in.-diameter field of view aperture 2.5 in. from the detector. The projected solid angle is about 0.020 sr:

$$\Omega = \pi \frac{(d/2)^2}{D^2 + (d/2)^2}$$

$$= \pi \frac{(0.2)^2}{(2.5)^2 + (0.2)^2}$$

$$= 0.01998 \text{ sr}$$

We will approximate the spectral transmittance of the spectral filter (Figure 3.23a) by the ideal case of Figure 3.23b. The in-band exitance is given by the spectral exitance [from Planck's law—formula (3.9a) or Table 3.2]

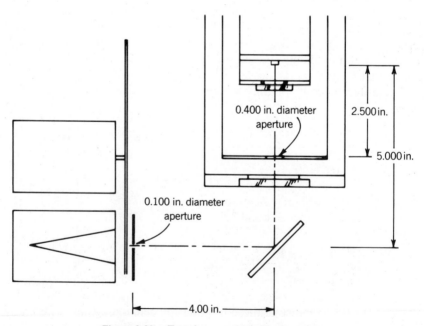

Figure 3.22.   Test dewar and blackbody setup.

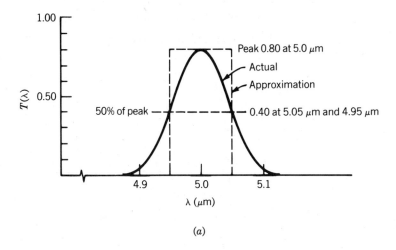

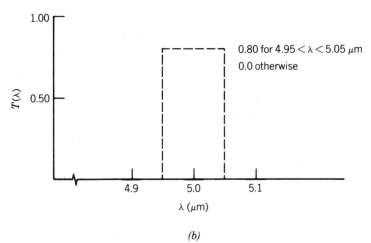

Figure 3.23. Transmittance curves for narrowband spectral filter: (*a*) actual; (*b*) model.

at the center wavelength, times the spectral bandwidth at the half-peak wavelengths, times the peak transmittance:

$$M = M(\lambda_c) \, \Delta\lambda \, T_{\text{pk}}$$
$$= [1.64 \times 10^{15} \text{ photons/(cm}^2\text{·s·}\mu\text{m})] \, (0.1 \ \mu\text{m}) \, (0.80)$$
$$= 1.312 \times 10^{15} \text{ photons/(cm}^2\text{·s)}$$

The background is not modulated, so we use a modulation factor of 1.0,

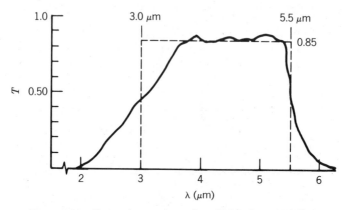

Figure 3.24. Transmittance curve for wideband spectral filter.

and arrive at a background incidance of about 7.8×10^{12} photons/(cm²·s):

$$E = M \frac{\Omega}{\pi} \tau_{\text{window}} \text{ (MF)}$$

$$= [1.31 \times 10^{15} \text{ photons/(cm}^2\text{·s)}] \left(\frac{0.02 \text{ sr}}{\pi \text{ sr}} \right) (0.94) (1.0)$$

$$= 7.84 \times 10^{12} \text{ photons/(cm}^2\text{·s)}$$

3.5.2. Background in a Test Dewar—Wide Bandpass Filter

Now calculate the background if we use the wide bandpass filter of Figure 3.24. The geometrical part of the calculation is done exactly as in the preceding example. It is only the spectral calculation that must be done differently.

The *in-band* background incidance includes the incidance from 3.0 to 5.5 μm. Using Table 3.4 for a 293-K source, this is

0 to 5.5 μm:	2.10×10^{16} photons/(cm²·s)
0 to 3.0 μm:	3.76×10^{13} photons/(cm²·s)
3.0 to 5.5 μm:	2.14×10^{16} photons/(cm²·s)

(The exitance from zero to 3.0 μm is negligible compared to that from

3.0 to 5.5 μm: The in-band exitance is only slightly sensitive to the exact filter cut-on wavelength.)

Using the same projected solid angle and a modulation factor of 1.0, we calculate an in-band incidence of 1.1×10^{14} photons/(cm²·s):

$$E(3.0 \text{ to } 5.5 \text{ μm}) = M(3.0 \text{ to } 5.5 \text{ μm}) \frac{\Omega}{\pi} \tau_{\text{filter}} \tau_{\text{window}}$$

$$= (2.10 \times 10^{16} \text{ photons/(cm}^2\text{·s)}) \left(\frac{0.02 \text{ sr}}{\pi} \right) (0.85 \, (0.94)$$

$$\cong 1.1 \times 10^{14} \text{ photons/(cm}^2\text{·s)}$$

3.5.3. Incidence from a Blackbody Test Set (All Wavelengths)

The preceding examples dealt with the background in the test dewar. Now let's calculate the signal flux density, the rms of the fundamental of the incidence from the blackbody. For the setup of Figure 3.22, the source is a 500-K blackbody, with a solid angle defined by the 0.100-in.-diameter aperture at a distance of 9 in. The projected solid angle is 9.7×10^{-5} sr:

$$\Omega = \pi \frac{(d/2)^2}{D^2 + (d/2)^2}$$

$$= \pi \frac{(0.100/2)^2}{(9.0)^2 + (0.100/2)^2}$$

$$= 9.70 \times 10^{-5} \text{ sr}$$

The aperture diameter is much smaller than the width of the chopper teeth, so the chopper–aperture combination yields nearly a square wave, and the modulation factor will be very close to 0.45. This is quite often the case. [We could refer to the modulation factor table (Table 3.8) if we wanted to be more rigorous.]

The chopper is mounted between the aperture and the blackbody, so as the chopper rotates, the detector sees no changes except thru the blackbody aperture. Thus the difference between the maximum incidence and the minimum incidence is just the incidence from the blackbody less that from the room-temperature chopper, both viewed through the 9.7×10^{-5} sr solid angle calculated above. The modulated waveform is shown in Figure 3.25.

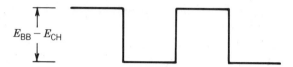

Figure 3.25.   Modulated incidance from a blackbody.

The rms of the fundamental component is about 4.28 μW/cm:

$$E = (M_{BB} - M_{CH}) \frac{\Omega_{BB}}{\pi} (MF)$$

$$= \sigma(T_{BB}^4 - T_{CH}^4) \frac{\Omega_{BB}}{\pi} (MF)$$

$$= [5.67 \times 10^{-12} \text{ W/(cm}^2 \cdot \text{K}^4)] [(500 \text{ K})^4 - (300 \text{ K})^4]$$

$$\times \left(\frac{9.7 \times 10^{-5} \text{ sr}}{\pi} \right) (0.45)$$

$$\simeq 4.28 \text{ μW/cm}$$

Note that we did not include the transmission of the dewar window or the spectral effects of any filters in the dewar. That was deliberate: These "blackbody incidance" values are prepared for the convenience of test set users, and each user will have a different window and spectral filter. In addition, the spectral characteristics of the detector will determine the fraction of the incidance that is useful. Users can provide their own "blackbody-to-peak" conversion factor, or "usefulness fraction," to convert the blackbody incidance to a useful or effective incidance.

Other differences between this example and the earlier two (background incidance) are:

- The source temperature is a blackbody at some high temperature instead of room temperature (293 K). This does not affect the formulas used, merely the value we enter into the formula.
- We have chosen to express the incidance in watts instead of photons used for most of the previous examples. This means that we use different (but entirely analogous) formulas.
- The incidance is modulated, so we must consider the *difference* in incidance from the source and the chopper, and the modulation factor associated with the chopper. This makes the problem a bit more difficult.

- The geometry is defined by a blackbody aperture, while in the previous cases it was limited by the field of view shield in the dewar. The formulas are the same, but the values used differ.

3.5.4. Sunlight on the Earth

For our purposes the sun can be treated as a 5900-K blackbody with a diameter of 864,000 miles, at a distance of 92.9 million miles. Neglecting any atmospheric absorption, the total incidance at the earth's surface is about 0.15 W/cm^2:

$$E = \sigma T^4 \frac{(d/2)^2}{D^2 + (d/2)^2}$$

$$\simeq [5.67 \times 10^{-12} \text{ W/(cm}^2 \cdot \text{K}^4)] (5900 \text{ K})^4 \left[\frac{(0.864/2)^2}{(92.9)^2 + (0.864/2)^2} \right]$$

$$\simeq 0.15 \text{ W/cm}^2$$

This agrees well with the commonly accepted value cited by Hudson (1969) of 0.14 W/cm^2. Only about two-thirds of the energy reaches the earth's surface; 0.1 W/cm^2 is a convenient figure to remember.

3.5.5. Sunlight Reflected from a Window

Occasionally, the geometry is just right for our eyes to see the reflection of the sun from a distant window or piece of metal. Will sunlight reflected from a 1-ft^2 glass window 5 miles away be visible? As indicated in Figure 3.26, the mirror defines the field of view through which the sun can be seen—it acts like an aperture over a 5900-K blackbody.

The solid angle is very small, so we can use the area-over-distance-squared approximation:

$$\Omega \simeq \frac{A}{D^2}$$

$$= \frac{1 \text{ ft}^2}{(5 \times 5280 \text{ ft})^2}$$

$$\simeq 1.43 \times 10^{-9} \text{ sr}$$

If the reflectance were perfect, the resulting incidance at our eye would

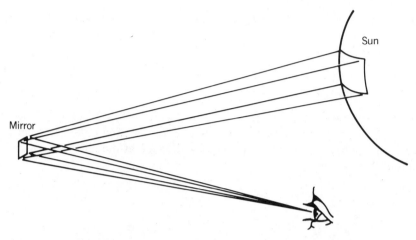

Figure 3.26. Sunlight reflected from a window.

be about 0.94 μW/cm^2:

$$E = \sigma T^4 \frac{\Omega}{\pi}$$

$$= [5.67 \times 10^{-12} \text{ W/(cm}^2 \cdot \text{K}^4)] \, (5900 \text{ K})^4 \left(\frac{1.43 \times 10^{-9} \text{ sr}}{\pi \text{ sr}} \right)$$

$$= 0.94 \times 10^{-6} \text{ W/cm}^2$$

Of this about 15% or 0.14 μW/cm^2 is useful to the eye. If the reflector is glass, we must include a reflectance of about 0.10 (uses the reflectance formula of Figure 9.4 and a refractive index of 1.6), so the effective incidance is about 14×10^{-9} W/cm^2:

$$(0.94 \times 10^{-6} \text{ W/cm}^2) \, (0.15) \, (0.10) \simeq 14. \times 10^{-9} \text{ W/cm}^2$$

That is about the effective incidance from the full moon, so sunlight reflected from such a window should be clearly visible.

Figure 3.27 shows the effective incidances from this and a number of other sources; it allows a quick estimate of visibility of sources of practical interest.

3.5.6. Moonlight on the Earth

The moon provides an instructive example of radiometric calculations. Most of the "moonlight" we see is sunlight reflected from the moon's

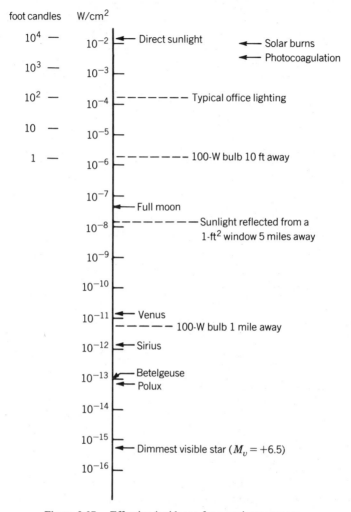

Figure 3.27.   Effective incidence from various sources.

surface. To calculate the incidence due to moonlight, first determine the incidence of sunlight on the moon, then see how much of that reaches the earth. The moon reflects about 10% of its incident radiation. Unlike a mirror, this reflected radiation is spread out in all directions: the moon acts like a source of $1/r^2$ radiation. The geometry is shown in Figure 3.28 and the calculations in the following list:

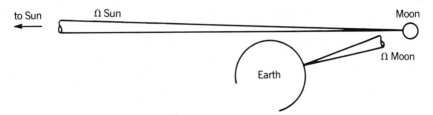

Figure 3.28. Moonlight on the earth.

Calculations for Moonlight on the Earth

Sun's temperature = 5900 K
Sun's diameter = 886.4 × 10³ miles
Sun-to-earth distance = 92.9 × 10⁶ miles
Moon-to-earth distance = 238.5 × 10³ miles
Earth's diameter = 8000 miles
Moon's diameter = 2160 milles
Moon's albedo ≃ 0.1

$$E_1 = E_{\text{at moon, from sun}} = \epsilon\sigma T_s^4 \frac{\Omega_{\text{sun}}}{\pi}$$

$$\simeq 0.16 \text{ W/cm}^2$$

$$E_2 = E_{\text{at earth, from moon}} = \rho E_1 \frac{\Omega_{\text{moon}}}{\pi} \tau_{\text{atmosphere}}$$

$$\simeq 0.22 \text{ W/cm}^2$$

$$E_{\text{effective}} = E_2 \times 0.15\%$$

$$\simeq 3.3 \times 10^{-8} \text{ W/cm}^2$$

The resulting effective (for the eye) incidance is about $3.3 \cdot 10^{-8}$ W/cm². Published visual magnitude and spectral incidance values are about 4 × 10^{-8} W/cm².

Even when the moon is not illuminated by the sun, it emits *some* radiation purely because of its own temperature. Can we see the self-emission of the moon? Probably not: We have heard that the unlit side of the moon is very cold. Even if the unlit side were at room temperature, the incidance would be like that from a room-temperature object when viewed in a dark closet—our eyes just do not see the radiation from objects at room temperature or lower. We can see sometimes see the unlighted crescent because of the sunlight reflected from the *earth* to the moon, and then back to the earth.

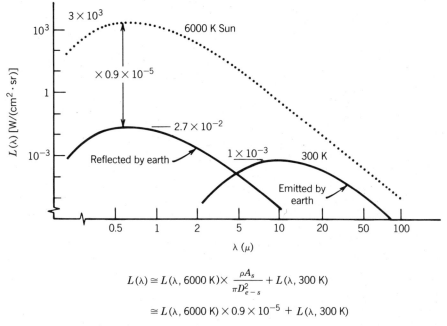

$$L(\lambda) \cong L(\lambda, 6000 \text{ K}) \times \frac{\rho A_s}{\pi D^2_{e-s}} + L(\lambda, 300 \text{ K})$$

$$\cong L(\lambda, 6000 \text{ K}) \times 0.9 \times 10^{-5} + L(\lambda, 300 \text{ K})$$

Figure 3.29.   Sterance from the earth.

3.5.7.   Sterance from the Earth

What incidence levels should we expect for a sensor in space looking down on the earth? To calculate the incidence, we would need to know the solid angle through which the sensor could "see" the earth. Instead, calculate the *sterance*—for which we do not need the sensor information. The calculated reflected solar component and the self-emitted components are shown in Figure 3.29.

3.6.   FORTRAN COMPUTER CODE FOR RADIOMETRIC CALCULATIONS

3.6.1.   Fortran Program AMODF: Prints Table of Modulation Factors

```
      PROGRAM AMODF
C     creates a table of modulation factors
C     NOTATION: DA      = Diameter of blackbody aperture
C               TW      = tooth width
C               DW      = dia. of chopper wheel (pitch dia.)
C               N       = number of tooth-slot pairs
C               ERR     = max. allowable error (1.E-03 = 0.1 %)
C
```

```
      DIMENSION NN(6)
      DIMENSION VALUE(7)
C
      NN(1) =    50
      NN(2) =    20
      NN(3) =    10
      NN(4) =     5
      NN(5) =     2
      NN(6) =     1
C
      OPEN(91, FILE = 'AMODF.DAT', STATUS = 'NEW')
C
      WRITE (91, 3)
3     FORMAT (1X, ' Table          Modulation Factors  ' / )
      WRITE (91, 4)
4     FORMAT (1X, ' _____  ',
     1 ' _____ ')
      WRITE (91, 5) NN
5     FORMAT (  1X, /
     2 29X, '  CASE III: NUMBER OF TOOTH-SLOT PAIRS ' /
     3 23X,
     3 ' _____ ' /
     4 3X,  ' DAP/TW   CASE II ' /
     5 1X, 17X, 6I9 )
C
      WRITE (91,4)
      WRITE (91,7)
7     FORMAT (1X)
C
      ERR  = 0.00001
      PI   = 3.141593
C
      DO 950 I = 1, 34
      TW      = 1.0
      DAP     = 0.1 * (I - 1)
      RATIO   = DAP / TW
C
20    CONTINUE
      VALUE(1) =       AMODF2(DAP, TW, ERR)
      JPRINT   = 1
C
      DO 750 J = 2, 7
      N       =       NN(J-1)
      DW      =       2. * N * TW / PI    ! circumferance / pi
      IF (DW.LE.DAP)  VALUE(J) = 9.        ! unphysical
      IF (DW.LE.DAP)  GO TO 750
      VALUE(J) =      AMODF3(DAP, DW, N, ERR)
      JPRINT   = JPRINT + 1
750   CONTINUE
C
950   WRITE(91, 987) RATIO, (VALUE(J), J = 1, JPRINT)
987   FORMAT (1X, F9.2, 1X, 7F9.4)
      WRITE (91, 4)
C
      WRITE (91, 999)
999   FORMAT (1X /
     1    1X, 'DAP = aperture diameter ' /
     2    1X, 'TW  = tooth width ' //
     3    1X, 'Program JDV-003, revised 19 August 1988' )
      CLOSE (91)
      END
```

3.6.2. Fortran Subroutine AMODF2: Generates Case II Modulation Factors

```
          FUNCTION AMODF2(DA, TW, ERR)
C
C         Modulation factor for case II - linear tooth motion.
C         Recursion relation for Bessel Function from CRC tables.
C         The series alternates; summation stops when last term
C         calculated is less than the allowable error.
C
C         REFERENCE: McQuistan, J. Opt. Soc. Am. 48, 63 (1958)
C                    CRC tables
C
C         NOTATION:
C                    DA  = Diameter of blackbody aperture
C                    TW  = tooth width
C                    ERR = maximum allowable error (1.E-03 = 0.1 %)
C
          PI        = 3.141593
          Z         = PI * DA / (2. * TW)
C
          TERM      =   0.5            ! First term in J1(Z)/Z
          L         =   1
          SUM       =   TERM
C
20        L         = L + 1
          RECUR     = - Z*Z / (4. * L*(L-1) )    ! recursion
          TERM      = TERM * RECUR
          SUM       = SUM + TERM
          ERROR     = ABS(TERM)
          IF (ERROR - ERR) 95, 20, 20
C
95        AMODF2 = 2 * SQRT(2.) / PI * SUM
C
          RETURN
          END
```

3.6.3. Fortran Subroutine AMODF3: Generates Case III Modulation Factors

```
          FUNCTION AMODF3(DA, DW, N, ERR)
C
C         Mod. factor for rotating wheel with radial teeth.
C         Uses recursion relation for hypergeometric series.
C         Series alternates; summation stops when last term
C         calculated is less than the allowable error.
C
C         REFERENCES:
C                 McQuistan (1959)
C                     J. Opt. Soc. Am. 49, 70
C                 Margenau and Murphy (1956)
C                     The Mathematics of Physics and Chem., p72
C                     Van Nostrand, Princeton
C         NOTATION:  N   = Number of teeth in chopper
C                    DA  = Diameter of blackbody aperture
C                    DW  = Pitch diameter of chopper wheel
C                    ERR = maximum allowable error (1.E-03 = 0.1 %)
C                    Z   = DA/DW
```

```
        REAL L
C
        A         = -N/2.
        B         = +N/2.
        L         =   0.
        G         =   2.
        Z         = (DA/DW)
        PI        = 3.141593
        TERM      =   1.
        SUM       = TERM
C
20      RECUR     = Z*Z * ( (L+A) * (L+B) ) / ( (L+1) *( L+G) )
        TERM      = TERM * RECUR
        SUM       = SUM + TERM
        L         = L + 1
        ERROR     = ABS(TERM)
        IF (ERROR - ERR) 99, 20, 20
C
99      AMODF3    = SQRT(2.) / PI * SUM
C
        RETURN
        END
```

3.6.4.   Fortran Subroutine AFLUX: Generates rms Incidances

```
        FUNCTION AFLUX(TBB, TCH, DIST, DAP, DCH, N)
C
C       BB Incidance
C             (RMS value of fundamental component)
C       Assumes perfect emissivity for BB and chopper
C
C       Notation
C          TBB = Blackbody Temperature, deg K
C          TCH = Chopper Temperature,   deg K
C          N   = Number of tooth-slot pairs
C          DCH = Chopper "pitch" diameter
C          DAP = Blackbody aperture diameter
C          DIST= BB aperture to detector distance
C
C       (Use same units for distance, diameters)
C
        PI    = 3.1415926
        SIGMA = 5.67051 E-12
C        Stefan-Boltzmann Constant
C        from   Cohen and Taylor (1987)
C               "Fundamental Physical Constants ..."
C               J. Res. Natl. Bur. Stand. 92, 85 (1987)
C               or NBS Special Pub. 731, July 1987
C
        EBB   = 1.
        ECH   = 1.
C
        WF        = AMODF3(DAP, DCH, N, 1.0E-04)
        S_ANGLE
1         = PI * (DAP/2)**2 / (DIST**2 + (DAP/2.)**2)
C
        AFLUX = SIGMA / PI
1           * S_ANGLE * (EBB*TBB**4 - ECH*TCH**4) * WF
C
        RETURN
        END
```

3.6.5.   Fortran Program FLUX TABLES: Prints Tables of Blackbody Incidance

```
       PROGRAM FLUX TABLES
C
C      Computes, prints flux densities
C              for several apertures,distances
C      Uses subroutine AMODF3, AFLUX
       DIMENSION DAP(5), HEAD(30), MODF(5), FLUX(5)
1      READ(2,201) HEAD, DCH, NT, DAP
201    FORMAT (30A1, F5.2, I5, 5F6.4)
11     READ (2,202) TBB, TCH, D1, D2, DELTA
202    FORMAT(5F10.2)
       WRITE(3,301) HEAD, DCH, TBB, NT, TCH
301    FORMAT(1H1,35X,'BLACKBODY INCIDANCE'/31X,30A1//
      1 26X,'Chopper-',F6.2,' in.dia.', F6.0,' deg K BB'/
      2 34X,I6,' teeth',2X,          F6.0,' deg K Ch.' )
C
       LINES = 0
       DO 50 J=1, 5
       APD   = DAP(J)
       IF(APD.GE..25) APD = APD/25.4
C                allows input in mm or inches
50     MODF(J) = AMODF3(ADP,DCH, NT, 1.E-04)
C
       WRITE(3,302) DAP, MODF
302    FORMAT(1H0,21X,'AP. DIA.-INCH', 3X, 5(F6.4,2X) /
      1 22X, 'MOD. FACT.', 6X, 5(F6.4,2X) / )
       WRITE(3,303)
303    FORMAT (22X, 'DISTANCE - CM')
       DIST  = 10.
       DISTA = DIST + .005
60     DO 70 J = 1, 5
       APD   = DAP(J)
       D     = DIST/2.54
C      Read in cm, convert D to inches
C           (APD, CHD are in inches)
70     FLUX(J) = AFLUX (TBB, TCH, D, APD, DCH, NT)
      1                 * 1.0E+06        ! microwatt/cm2
C
       IF(LINES.EQ.40) WRITE(3,305) DISTA,FLUX
       IF(LINES.NE.40) WRITE(3,304) DISTA,FLUX
304    FORMAT(   24X,F6.2,5X,5F8.3)
305    FORMAT(1H1,23X,F6.2,5X,5F8.3)
       DIST  = DIST + DELTA
       LINES = LINES+1
       IF (DIST.GT.D2) GO TO 99
       GO TO 60
99     WRITE (3, 100)
100    FORMAT(1H0,20X,
      1'FLUX       = MODF * SIGMA * (TBB**4-TCH**4) * OMEGA'//
      2'OMEGA / PI = (APD/2)**2/(DIST**2+(DAP/2)**2'         //
      3'SIGMA      = 5.6697E-12      W/cm**2/K**4'            //
      3'Program FLUX_TABLE,  Revised 12 Sept 88 - JDV ')
C
       GO TO 11
       END
```

REFERENCES

American National Standards Institute (1986). "Nomenclature and Definitions for Illuminating Engineering," ANSI/IES RP 16-1986, Illuminating Engineering Society, New York.

Bartell, F. O. (1989a). "Projected Solid Angle and Blackbody Simulators," *Appl. Opt.* 28, 1055.

Bartell, F. O. (1989b). "Blackbody, blackbody simulator, blackbody simulator cavity, blackbody simulator cavity aperture, and blackbody simulator aperture are each different from one another" *Proc SPIE* 1110, 183.

Bartell, F. O., and W. L. Wolfe (1976). "Cavity Radiators: An Ecumenical Theory," *Appl. Opt.* 15, 48.

Bell, Ely E. (1959). "Radiometric Quantities, Symbols and Units," *Proc. IRE* 47, 1432.

Chandos, Raymond J., and Robert E. Chandos (1974). "Radiometric Properties of Diffuse Wall Cavity Sources," *Appl. Opt.* 13, 2142.

Cohen, E. Richard, and Barry M. Taylor (1987). "The 1986 CODATA Recommended Values of the Fundamental Physical Constants," *J. Res. Natl. Bur. Stand.* 92, 85. The recommended values are available on a pocket size card— National Bureau of Standards Special Publication 731, NBS, Washington, D.C.

Commission Internationale de l'Éclairage (1970) *International Lighting Vocabulary*, 3rd Ed., Publ. CIE no. 17 (E-1.1.). Bureau Central de la CIE, Paris. The CIE is the International Commission on Illumination.

Guenzer, C. S. (1976). "Chopping Factors for Circular and Square Apertures," *Appl. Opt.* 15, 80.

Hudson, Richard D., Jr. (1969). *Infrared System Engineering*, Wiley, New York.

Jones, R. (1963). "Terminology in Photometry and Radiometry," *J. Opt. Soc. Am.* 53, 1314.

Kreith, Frank (1962). *Radiation Heat Transfer*, International Textbook, Scranton, Pa.

Margenau, Henry, and George Mosley Murphy (1956). *The Mathematics of Physics and Chemistry*, Van Nostrand, Princeton, N.J.

McAdam, David (1967). "Nomenclature and Symbols for Radiometry and Photometry," *J. Opt. Soc. Am.* 57, 854.

McQuistan, R. B. (1958). "On an Approximation to Sinusoidal Modulation," *J. Opt. Soc. Am.* 48, 63.

McQuistan, R. B. (1959). "On Radiation Modulation," *J. Opt. Soc. Am.* 49, 70.

Muray, J. J., Fred E. Nicodemus, and I. Wunderman (1971). "Proposed Supplement to the SI Nomenclature for Radiometry and Photometry," *Appl. Opt.* 10, 1465–1468.

Nicodemus, F. E. (1976) ed., National Bureau of Standards Self Study Manual on Optical Radiation Measurements. NBS Technical Notes 910-1 through 910-

8. Out-of-print Government Printing Office document, now available from the National Technical Information Service (NTIS), U.S. Department of Commerce, Springfield, VA. This is a comprehensive, rigorous treatise on radiometry. Because of its completeness and rigor it is not short nor easily read. Chapters 1 through 3 of part 1 (Concepts) are in TN 910-1; they cover the material discussed in this book.

Pivovonsky, M., and M. Nagle (1961). *Tables of Blackbody Radiation Functions,* Macmillan, New York.

Scott, R. B. (1959). *Cryogenic Engineering,* D. Van Nostrand, Princeton, N.J.

Sparrow, E. M., and R. D. Cess (1965). *Radiation Heat Transfer,* Brooks/Cole, Monterey, Calif.

Spiro, Irving J. (1974) "Radiometry and Photometry," *Optical Engineering* 13, G183. "Radiometry and Photometry" is a regular column in *Optical Engineering*; nomenclature is discussed in most of the 1974 and early 1975 issues.

Wolfe, William L. (1980) "Radiometry," Chapter 5 in *Applied Optics and Optical Engineering,* volume VIII; ed. by Robert R. Shannon and James C. Wyant, Academic Press, New York.

Wolfe, William L., and George J. Zissis, eds. (1978). *The Infrared Handbook,* Environmental Research Institute of Michigan, Ann Arbor, Michigan.

SUGGESTED READINGS

Boyd, Robert W. (1983) *Radiometry and the Detection of Optical Radiation,* Wiley, New York.

Dereniak, Eustace L., and Devon G. Crowe (1984). *Optical Radiation Detectors,* Wiley, New York. Excellent and up-to-date references; covers radiometry, detectors, and charge transfer devices. Recommended.

Geist, J. (1976). "Trends in the Development of Radiometry," Opt. Eng 15, 537.

Geist, Jon, and Edward Zalewski (1973). "Chinese Restaurant Nomenclature for Radiometry," *Appl. Opt.* 12, 435.

Holman, J. P. (1981). *Heat Transfer,* McGraw-Hill, New York.

Holter, Marvin R., et al. (1962). *Fundamentals of Infrared Technology,* Macmillan, New York.

Hudson, Richard D., Jr. (1969). *Infrared System Engineering,* Wiley, New York.

Jameson, John, Raymond McFee, Gilbert Plass, Robert Grube, and Robert Richards (1963). *Infrared Physics and Engineering,* Inter-University Electronics Series, Mc-Graw-Hill, New York.

Jones, R. C., D. Goodwin, and G. Pullan (1960). *Standard Procedure for Testing Infrared Detectors and Describing Their Performance,* Office of Director of Defense Research and Engineering, Washington, D.C.

Kreith, Frank (1962). *Radiation Heat Transfer,* International Textbook, Scranton, Pa.

Kruse, P. W., L. D. McGlaughlin, and R. B. McQuistan (1962). *Elements of Infrared Technology*, Wiley, New York.

Nicodemus, F. E. (1970). "Optical Resource Letter on Radiometry," *Am. J. Phys.* 38, 43–49.

Nicodemus, F. E. (1971). *Proposed Military Standard—Infrared Terms and Definitions* part 1 of 2 parts. Philco-Ford Corp., Newport Beach, California. Available from National Techical Information Service (NTIS) as AD 758,341. This document contains a thorough list of terms and definitions. It has not been adopted for official use.

Nicodemus, F. E., ed. (1971). *Radiometry—Selected Reprints,* American Institute of Physics, New York.

Nicodemus, F. E. (1974). *Reference Book on Radiometric Nomenclature* U.S. Naval Weapons Center, China Lake, California.

Nicodemus, F. E. (1976) ed., National Bureau of Standards Self Study Manual on Optical Radiation Measurements. NBS Technical Notes 910-1 through 910-8. Out of print Government Printing Office document, now available from the National Technical Information Service (NTIS), U.S. Department of Commerce, Springfield, VA. This is a comprehensive, rigorous treatise on radiometry. Unfortunately, because of its completeness and rigor, it is not short nor easily read.

Pivovonsky, M., and M. Nagle (1961). *Tables of Blackbody Radiation Functions*, Macmillan, New York.

Smith, R. A., F. E. Jones, and R. P. Chasmer (1957). *The Detection and Measurement of Infrared Radiation,* Oxford University Press, Oxford.

Sparrow, E. M., and R. D. Cass (1966). *Radiation Heat Transfer*, Brooks/Cole, Monterey, Calif.

Wolf, Helmut (1983). *Heat Transfer*, Harper & Row, New York.

Wolfe, William L. (1980). "Radiometry," Chapter 5 in *Applied Optics and Optical Engineering*, Vol. 8, Robert R. Shannon and James C. Wyant, eds. Academic Press, New York.

Wyatt, Clair L. (1978). *Radiometric Calibration: Theory and Method*, Academic Press, New York.

Wyatt, Clair L. (1987). *Radiometric System Design*, McMillan, New York.

PROBLEMS
General Radiometry

1. The total exitance (W/cm^2) from a blackbody is given by

$$M = \sigma T^4$$

where ϵ = Stefan–Boltzmann constant

$$\simeq 5.67 \times 10^{12} \text{ W/(cm}^2 \cdot \text{K)}$$

Verify that the units are correct.

2. Estimate the total power emitted by a heated swimming pool. (You pick the size of the pool, its temperature, and its emisivity. Remember that water absorbs most IR very well.)

3. Calculate the background in the test dewar shown in Figure 3.30.

4. Calculate the blackbody incidance on the detector plane for the blackbody setup shown in Figure 3.29.

Spectral Content

1. Planck's law for the spectral radiant exitance is

$$M_e(\lambda_1 T) = \frac{2\pi h c^2}{\lambda^5(e^x - 1)} = \frac{c_1}{\lambda^5(e^x - 1)}$$

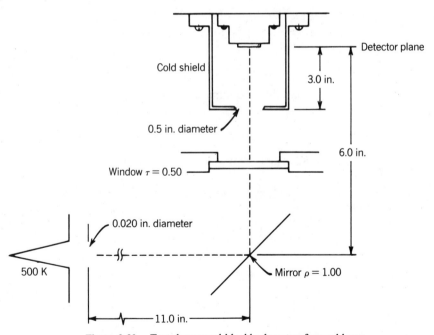

Figure 3.30. Test dewar and blackbody setup for problems.

where $\quad x = \dfrac{hc}{\lambda kT} = \dfrac{c_2}{\lambda T}$

c_1 = first radiation constant for radiant exitance

$\quad = 2\pi hc^2$

$\quad \approx 3.74 \times 10^4 \ \text{W·µm}^4/\text{cm}^2$

c_2 = second radiation constant

$\quad = hc/k$

$\quad \approx 14{,}388 \ \text{µm·K}$

Using the given formulas and values for h, k, and c, confirm the values and units of c_1 and c_2.

2. Given that the radiant spectral sterance and exitance peak at $\lambda T = 2898$ µm·K, confirm that

$$M_{\text{pk}}(T) = 1.28 \times 10^{-15} \ \text{W}/(\text{µm·cm}^2\text{·K}^5)T^5$$

3. The power into an $180°$ field of view from a source of emissivity ϵ, area A, and temperature T is given by

$$P = \sigma\epsilon AT^4$$

where $\quad \sigma$ = the Stefan–Boltzmann constant

$$= \dfrac{2\pi^5 k^4}{15h^3 c^2}$$

$$\approx 5.67 \times 10^{-12} \ \text{W}/(\text{cm}^2\text{·s·K}^4)$$

Using the formula and values given for h, k, and c, confirm the value and units given for σ.

4. Planck's law for the spectral photon exitance is

$$M_q = \dfrac{2\pi c}{\lambda^4(e^x - 1)}$$

where $\quad x = \dfrac{hc}{\lambda kT} = \dfrac{c_2}{\lambda T}$

$c_1' = $ first radiation constant for the photon exitance

$= 2\pi c$

$\simeq 1.884 \times 10^{23}$ photons$\cdot\mu m^3/(cm^2\cdot s)$

$c_2 = $ second radiation constant $= hc/k$

$\simeq 14{,}388\ \mu m\cdot k$

Using the given formulas and values for h, k, and c, confirm the value of c_1' and c_2.

5. Given that the photon sterance and photon exitance peak at $\lambda T = 3669.7\ \mu m\cdot K$, confirm that

$$M_{pk} = [2.1 \times 10^7\ \text{photons}/(cm^2\cdot s\cdot\mu m\cdot K^4)]T^4$$

6. The rate ϕ at which, photons are emitted into an 180° field of view from a source of emissivity ϵ, area A, and temperature T is given by

$$\phi = \sigma_q \epsilon A T^3$$

where $\sigma_q = $ photon equivalent of the Stefan–Boltzmann constant

$$= \frac{4\pi^4 k^3}{25.79436 h^3 c^2}$$

$\simeq 1.52 \times 10^{11}$ photons$/(cm^2\cdot s\cdot K^3)$

Using the formula and values given for h, k, and c, confirm the value and units given for σ_q.

7. For a 1500-K globar with a 1-cm^2 emitter area, find:
   (a) The peak spectral radiant exitance
   (b) The wavelength at which it occurs
   (c) The (total) radiant exitance
   (d) The (total) photon exitance
   (e) The fraction of the exitance between 3.0 and 5.0 μm.

8. A spike filter (narrow bandpass filter) transmits 80% of the incident radiation with wavelengths between 2.3 and 2.4 μm and rejects all other wavelengths. What fraction of the energy from an 800-K blackbody is transmitted by this filter?

9. A Ge:Hg detector is sensitive only to wavelengths of less than 10.4 μm. What fraction of the power from a 500-K blackbody does this detector "see"? From a 300-K blackbody?

10. An InSb detector is sensitive only to wavelengths of less than 5.3 μm. What fraction of the photons from a 500-K blackbody does this detector "see"? From a 300-K blackbody?

The Spatial Integral

1. Calculate the solid angle and projected solid angle for given conditions. Unless otherwise stated, assume that the aperture faces the detector directly.

    (a) Circular aperture: 3 in. diameter, 12 in. from detector

    (b) Unknown shape, area = 2 cm², distance = 2 m

    (c) Circular aperture, 3-ft diameter, 12 feet from the detector, and tilted at a 45° angle.

2. If an unshielded InSb detector "sees" 1×10^{16} photons/(cm²·s), what would its background be if it were shielded with the aperture of Problem 1(a)?

3. A linear array of detectors 0.5 in. long is shielded by a 1.0 in. × 0.5 in. slot shield that is 2.0 in. from the array (Figure 3.31). What is the effective solid angle for:

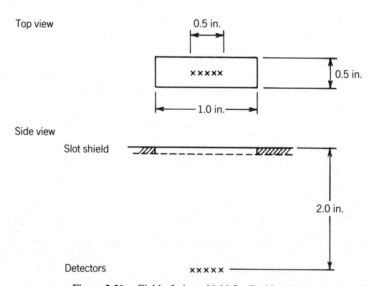

Figure 3.31. Field-of-view shield for Problem 3.

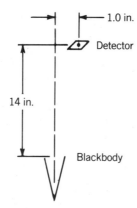

Figure 3.32. Detectors and blackbody position for Problem 4.

   (a) A detector at the center of the array?

   (b) A detector at the end of the array?

4. Consider the effect on the incidance if a detector dewar is slid 1.0 in. off the blackbody axis, as shown in Figure 3.32. Assume that no vignetting occurs; that is, nothing obscures the path from the detector to the aperture.

   (a) What is the cosine of the angles between the "line of sight" and the surface normals?

   (b) By how much does the detector to aperture distance increase?

   (c) By how much does the incidance at the detector decrease as a result of this offset?

Modulation

1. *Estimate* the modulation factor for an 8-in. chopper with two teeth and two slots, chopping a 0.050-in.-diameter aperture.

2. Consider a 5-in.-diameter blade with 32 teeth chopping a 0.20-in.-diameter aperture. Without recourse to Bessel functions or computers, can you tell whether the modulation factor will be close to 0.45? Now use the Bessel function formula to calculate the modulation factor.

3. What is the modulation factor for a pure sine wave? *Hint:* The rms of a sine wave is $\sqrt{2}/2$ times the amplitude, and the amplitude is by definition half of the peak to peak for a sine wave.

4

The Test Set

In this chapter we focus on the features and techniques that should be considered when specifying, designing, modifying, or troubleshooting blackbody test setups. ("Test setups" here includes the source, its mounting and modulation, and the test dewar or chamber.) Once put in writing, these guidelines may seem so obvious as to be trivial, and it may be difficult to believe that one could overlook such obvious problems when designing and building a test set or in debugging it when things go wrong. Yet it does happen. Each of the features listed here represents a lesson learned—at least once—the hard way. The time and energy spent debugging the problems could be better spent making sure that they don't happen in the first place.

Test engineers' data are continually challenged—are the data really correct? Detector builders may want very much to believe that poor results are really due to test set problems, and the customer will want to be sure that good results are real. These concerns have—unfortunately—often been justified. It is quite common to find errors in test setups and data, particularly with new or recently modified equipment. It is incumbent on test engineers to monitor their setups to discover and eliminate potential errors.

The chapter begins with a brief description of the types of radiometric test sets normally encountered. A list of things to check when troubleshooting follows. The remainder of the chapter covers the design issues that must be considered to avoid test set problems.

4.1. TYPES OF TEST SETS

4.1.1. Blackbody and Dewar

For many applications, the detectors to be tested can be mounted in a simple dewar, shielded with cold metal caps and cooled spectral filters. The IR radiation is supplied by a blackbody mounted under a bench, and the electronics are mounted in a nearby rack or panel. This kind of a setup is versatile and will undoubtedly be found in test labs for many years. For production testing, methods that will provide faster turnaround times are desirable, and if complex optics are required, it may be more convenient to house them in a permanent chamber.

4.1.2. Chambers

Test chambers can vary in size from 1 ft or so in diameter up to entire walk-in rooms 30 ft in diameter. The design of these chambers can require the contributions of architects and structural engineers as well as the optical, cryogenics, electronics, and detector expertise normally associated with testing.

4.1.3. Focused Optics: Spot and Line Scanners

To characterize the spatial behavior of detectors requires an illuminated spot or slit, preferably much smaller than the size of the detectors of interest. Test setups which provide that small spot or slit are generally diffraction limited, so the spot is seldom as small as one would desire. In addition, the distance from the optics mount to the detector plane (the working distance) is often short, and the design of a workable system requires ingenuity and compromise.

   A special challenge of these spot or line scanners is in the focusing operation. Whether done manually or automated, considerable effort and experience are required to get the small spot on the small detector, to minimize the spot size, and to demonstrate that it was indeed at its best focus during a given test. One procedure is described in Section 5.10.1.

   Another complication is the characterization of spot or slit profile. Because most detectors are about the same size as the smallest spots available, it is necessary to "back out" the spot profile from the observed scan data. This requires a fairly accurate knowledge of the spot profile.

   The minimum resolvable spot diameter is given by the diffraction equation

$$x = 2.44 \lambda f/\text{number}$$

In the simplest testing, one confirms approximate detector size or locations by moving a small spot across the detector and noting the points at which the signal drops to 50% of its maximum value. If the spot is small compared to the detector, *and* if the detector response is uniform, *and* if the spot is symmetric, the 50% points occur when the spot is centered on the edges of the detector.

For more detailed information, one measures signal versus position, then extracts from that information the modulation transfer function of the detector–optics system. In any case, some characterization of the setup is essential if one is to be able to interpet the data. To describe a spot scanner adequately for testing decisions, one needs the following information as a minimum:

- *Working Distance*. The distance from the focal point to the closest mechanical element. If your detector is set back 4 in. from a window, you cannot use a spot scanner with a 3-in. working distance.
- *Spectral Limitations*. If refractive optics are used, what wavelengths will the lens material transmit? Are there spectral filters in place? Can they be replaced?
- *Effective f/Number*. With this and the wavelength of interest, you can calculate the diffraction diameter.
- *Source Diameter and System Magnification*. From this you can calculate the geometrical spot size (the image size ignoring diffraction effects).

Using the information above, you can calculate the probable spot (or slit) size and decide how you will need to interpret the data. Table 4.1 is an example of a convenient reference sheet for spot scanners. It lists information of interest for three typical spot scanning systems; Figure 4.1 shows the resulting spot diameter as a function of wavelength for those systems.

4.1.4. Spectrometers

Measurement of the spectral response of an IR detector is normally done by comparing the signal from the detector under test to that from a standard detector while both are illuminated with a known wavelength. The essential equipment include a source of IR, a monochrometer, and a reference detector.

The monochrometer selects and passes only the radiation in a narrow bandpass of known wavelength. This can be done with prisms, grating, or by setting up interference patterns with parallel reflectors. Commercial instruments are listed in a table (Table 20-7) by Zissis and LaRocca (1978). The same reference discusses the various instrument types in detail. Moore et al. (1983, sec. 4.7) discuss optical dispersing instruments and provide a table comparing the various types. Shannon and Wyant (1979) also discuss spectral dispersing instruments.

Table 4.1 Reference Sheet for Spot Scanner

Optics Name	3.5 in.	Beck	PRESS
Type	Reflective	Reflective	Refractive
Working distance (in.)	3.5	1.1	1.4
Clear aperture diameter, d (in.)	0.960	0.750	0.625
Stop to focus, d (in.)	4.08	1.5	1.5
D/d (f/number)	4.25	2.0	2.4
Diffraction diameter[a] at 5.0 μm	51.9 μm (2.1 mils)	24.4 μm (1.0 mils)	29.3 μm (1.1 mils)
Magnification, M[b]	5 × 36.5	15 × 36.5	4.5 × 36.5
Ideal image diameter[a] (0.100 in./M)	13.9 μm (0.55 mils)	4.6 μm (0.18 mils)	15.4 μm (0.61 mils)

[a] Diffraction diameter and ideal image diameter are calculated from simple theory. The number of significant digits shown is *not* an indicator of accuracy or confidence.

[b] A magnification factor of 36.5 is built into the spot scanner body; the other factors (5, 15, 4.5) are characteristic of the removable optics.

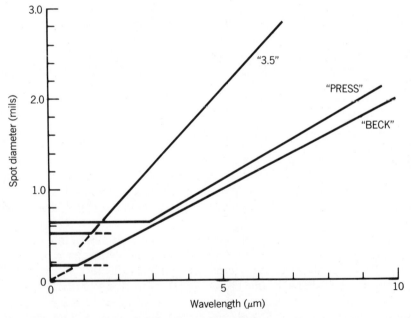

Figure 4.1. Spot size versus wavelength for three spot scanners. Assumes: 0.100-in. source and simple diffraction limit. See Table 4.1 for optics characteristics.

The grating spectrometer takes advantage of constructive interference from a series of stepped reflecting surfaces. Spectral calibration is predictable from the grating and optical geometry. An additional filtering scheme is necessary to achieve complete separation of wavelengths.

Prism instruments are just improved versions of the glass prism spreading a rainbow of visible light. Various prism materials are available to cover different portions of the visible and IR spectrum. The prisms may be damaged if not kept dry and or if dropped or mishandled. Calibration depends on the index of refraction with wavelength, so it is nonlinear and requires an empirical formula to interpolate between calibration wavelengths.

Fourier transform instruments use the interference from two reflecting surfaces to enhance the transmitted energy of some wavelengths and diminish that from others. The wavelengths transmitted vary as the separation between the reflectors is changed. Signal from the detector is recorded as one reflector is moved. A Fourier transform of the signal versus position scan data yields signal versus wavelength. This system yields relatively high signals because a great deal of energy is transmitted at one time. The raw data do not provide any insight into the real spectral response, and the data reduction requires involved computations. The system requires reliable measurement of the mirror positions.

Rough *spectral calibration* can be done using the atmospheric absorption at 1.4, 1.9, 2.7, 4.3, and 15 μm as reference points. Accurate wavelengths of absorption bands useful for spectral calibration are provided by Billings (1972), Rao et al. (1963), and IUPAC (1961).

To determine the absolute response of a detector at a given wavelength would require that we know the incidance from the monochromoter. This is seldom available with the required accuracy. Instead, the energy is switched between the unit under test (UUT) and a standard detector whose spectral response is already known. At each wavelength of interest the incidance is determined from the signal of the standard detector and used to compute the spectral response of the detector under test. Since the two detectors do not see the source through the same geometry, the spectral response calculated in this way is not correct, but is incorrect by some factor, which (we hope) is the same for all wavelengths:

$$\mathscr{R}_{UUT}(\lambda) = \mathscr{R}_{std}(\lambda) \frac{S_{UUT}}{S_{standard}} \times \text{constant}$$

Thus this procedure yields only the shape of the spectral response curve, not the absolute spectral response. This information can be combined with the (broad spectral band) "blackbody responsivity" to determine absolute response at all wavelengths—see Section 3.2.4.

The accuracy of the result depends on the accuracy with which the spectral response of a standard detector is known, the accuracy with which the signal from the standard and the test detectors is measured, and the assumption that the constant is truly independent of wavelength. Accurate spectral measurements are not easy to make.

4.1.5. Special Applications

Special setups are needed to obtain the low backgrounds required of some space applications and to deal with the resulting high detector impedances and low signal levels. Evaluation of performance at high frequencies requires the ability to modulate incidance at those frequencies, and to process the resulting signals with minimal distortion. Testing at gigahertz frequencies is possible.

Future technology will demand larger and larger arrays of detector elements. Testing such arrays demands acquiring and processing ever-increasing amounts of data, at increasing rates. Such testing relies on multiplexing data from many elements to one data stream, and complex signal processing to extract and condense desired information to a manageable size. The processing may be done "on-chip", by a warm, dedicated electronics board, and/or in the test set. Much of the design of the test equipment requires the assistance of specialists in digital electronics, software architecture, and programming. Part of the challenge (and fun) for the detector test engineer in this environment is to predict how the test set will be used, to specify what is needed, and to work with other specialists to make it come together.

4.2. DESIGN AND DOCUMENTATION

4.2.1. Requirement Definition

The single largest source of problems in test set acquisition is the failure of users to define adequately what is wanted. This can occur because we often cannot predict accurately (at the beginning of a contract) what the needs will be. This can also occur because of time pressures ("We have to get the fabrication started now, or it won't be ready in time"), because the user assumes that the test set people know and understand all that the user knows, and because it is tedious to write out clearly what one wants. In any case it is well worthwhile to prepare a complete written

statement of what the test set is expected to do. Checklist 4.1 may assist
in that preparation.

4.2.2. Design Reviews

Early in the conceptual stage of a project, and again when the proposed
design is clearly defined, the plans should be reviewed by three or four
people, including the user of the test station and people with experience
on similar test equipment. Scale assembly drawings should be available,
and should be checked for reflections and vignetting. Unless the setup is
very similar to units already working in a satisfactory way, it may be very
helpful to build a mockup.

Ease of operation should be considered: Can the blackbody aperture
plate be moved readily? How difficult will it be to gain access to parts
that need maintenance? Can the blackbody cavity be probed conveniently? It may be that various design goals cannot be met, but it is worthwhile for everyone who has an interest in the set to recognize and acknowledge the trade-offs at this stage, rather than being disappointed
later.

Checklist 4.1: Test Set Specification*

General description
   Units to be tested, and interfaces: mechanical, thermal, and
      electrical connections
   Tests to be performed
   Operating concept
      Extent of operator control desired
Detailed description of tests to be performed
   Nature of tests
   Environment: with range and accuracy of variables
      Optical
      Thermal
      Vacuum
      Electrical
   Output (mode, format, contents, archiving)
   Accuracy of results

* Includes requirements for the entire test set: hardware, firmware, software.

Checklist 4.1 (*Continued*)

Throughput
  Time per test, or tests per given time period
  Allowable downtime for calibration and maintenance
Software
  Acquisition
  Reduction algorithms
  Displays and reporting
  Data archiving
  Re-reduction capability
Quality assurance requirements
Safety of unit under test, and reliability
  Failure modes analysis
  Safety interlocks
  Allowable downtime for repair
Documentation required
  Mechanical drawings
  Optical schematics
  Setup
    Floor plan
    Utilities needed
    Procedure for setup, alignment, and checkout
  Specifications for components
    Mechanical
    Optical
    Electronic
  Operating manuals
    Critical spares list
  Diagnostic procedures
  Calibration information
    Procedures
    Schedule
Acceptance test plan and procedure   (How will vendor demonstrate
  that the machine meets the stated specifications?)
Validation plan and procedure   (How will the user demonstrate that
  the machine does what is needed?)

4.2.3. Scale Drawings

Separate scale drawings or sketches of the blackbody setup and dewar are invaluable in locating problems. Checking for potential vignetting, reflections, and determining the correct distances for calculations are all easier and less susceptible to error if a good scale drawing is available.

This drawing should show the blackbody, apertures, filters, mirrors and other optical elements, baffles, and any supporting structures that might vignette or cause reflections. Although the scale used is theoretically unimportant, it turns out to be very convenient if a 1 : 1 scale is used. Errors are reduced because of the ease with which comparisons with "the real thing" can be made, and the standardization allows convenient overlaying of dewar and blackbody drawings.

Everyone involved should discipline themselves to provide and maintain such a sketch. This requires that time and money be budgeted for this part of the design or debugging task. It will probably occur to someone that time and money could be saved by deleting this effort, but I submit that such a decision is shortsighted and does not pay off.

4.2.4. Emissivity and Reflectivity in the Infrared

The complete design of a setup includes the specification of the surface finish. It is often desirable to make some surfaces nonreflecting, or "black." However our everyday understanding of "black" and "shiny" is not necessarily a good guide to what is black in the infrared. The reflectance of most materials depends on the wavelength and on how the surface is prepared. If the surface is coated, the method of priming and application can alter the reflectance dramatically.

One method of obtaining black anodized aluminum, for example, yields a specular reflectance of nearly 80% at 2 μm, while other treatments for aluminum cut this to 10%.

Infrared Black Coatings. 3M Black Velvet paints are easily applied and convenient for laboratory work, but not recommended for space or high vacuum applications because of their outgassing rate and loose particle count. Three varieties are common: 3M 101–C10, 3M 401–C10, and 3M 3101; they are available from Minnesota Mining and Manufacturing Company, Minneapolis, Minnesota.

Cat-a-lac Black and CTL-15 are epoxy systems with low reflectance and low outgassing rates. Cat-a-lac Black is manufactured as Sikkens 463-3-8 Flat Black Epoxy by Sikkens Aerospace Finishes Division of Akzo Coatings America, Inc. 434 W. Meats Avenue, Orange, California 92665; CTL–15 is made by Chemical Technology Labs, Inc., 16401 S. Avalon,

Gardena, California 90248. Black chrome is described by Driver (1977); anodized aluminum is also used, but has high reflectance at grazing angles.

The reflectance of all black coatings is dependent on the surface finish of the base material and the method of application. Control of these is essential if reproducible results are to be achieved.

Both specular and diffuse reflectance are often of interest, and both are very angle dependent. Thus the choice and preparation of black surfaces requires and generally deserves specialized research, and measured reflectances in the wavelengths of interest should be used when making design decisions.

Complilation of infrared emissivity and/or reflectivity data is difficult because of the volume of potential data: One really needs the reflectance as a function of surface material, preparation, wavelength, and incident and exit angles. (Refer to our discussion of bidirectional reflectance distribution function in Chapter 9.) In some cases there is so much data available that it is difficult to work with.

The summaries by Wolfe in Chapter 7 of *The Infrared Handbook* (1978a) and Chapter 7 of the *Handbook of Optics* (1978b) provide good starting points. A list of recent references to optical blacks is included at the end of this chapter. Siegel and Howell (1975) devote one chapter to radiative properties of real materials.

Infrared Mirrors and Low-Emissivity Surfaces. Data for materials of high reflectivity do not present as severe a problem. Chapter 7 of *The Infrared Handbook* (Wolfe, 1978a) and Section 8 of the *Handbook of Optics* (Dobrowolski, 1978) provide good overviews.

As an example of the subtle problems that can cause errors, consider the reflectance of aluminum mirrors coated with silicon oxides to protect them. Cox et al. (1975) report that such mirrors have very high reflectance—about 97%—for near-normal incidence, but at angles beyond 40° from the normal the reflectance decreases to 54% in the 8- to 12-μm band. Other wavelengths are reflected well even at large angles. A similar problem is encountered with Al_2O_3 (Cox and Hass, 1978a), but Y_2O_3 and HfO_2 provide good reflectance even in the 8- to 12-μm region and large angles.

One question that reoccurs is: How thick must metal coatings be to provide nearly optimum reflectance in the IR? Bennett et al. (1962) show that to be within about 1% of the maximum reflectance requires about 300 Å of aluminum; to be within 0.1% requires a thickness of 600 Å.

4.3. BLACKBODY TEST SET

Many sources of infrared radiation are available for use in the laboratory. The choice of a source for a particular application depends on the spectral region of interest, the exitance needed, and whether the exitance must be accurately known. The blackbody is commonly used because its radiant output can be accurately calculated from its temperature.

Other sources include filament lamps, arc lamps, and globars. Auxiliary equipment used with these sources include integrating spheres, which remove the spatial structure from source, providing a uniform output, and collimators, which provide a source that appears to be a long distance away. Blackbodies and other sources are discussed by Hudson (1969, chap. 3), in *The Infrared Handbook* (chap. 2, by LaRocca, 1978), in the *Handbook of Optics* (Sec. 3 by Zissus and LaRocca, 1978), and by Moore et al. (1983).

4.3.1. Blackbodies

The most common source for detector testing in the 3- to 20-μm region is a cavity type blackbody simulator—a heated cavity with diffuse walls and a large wall area-to-opening ratio. To the extent that such a cavity simulates a true blackbody, its exitance can be predicted using Planck's law; this is discussed in Chapter 3. Hereafter we will follow convention and refer to such a cavity simply as "a blackbody."

A few of the numerous sources of information on blackbody design and use are listed here; they, in turn, will lead the reader to other sources. Cussen (1982) describes the construction and calibration of blackbodies intended for different purposes. Moore et al. (1983) describe the construction of an "NBS-type blackbody source." Articles by Chandos and Chandos (1974) and Bartell and Wolfe (1976) analyze the effective emissivity of a blackbody in terms of the emissivity of the walls and the cavity configuration; they also cite the older articles. Fu et al. (1988) describe the analysis of a conventional blackbody system. A list of articles on blackbody design and usage, including some new or novel systems or designs, is included at the end of this chapter.

4.3.2. Mounting of the Blackbody

When deciding how to mount a blackbody, consider the need to probe the cavity for temperature calibration, and make sure that a probe can be inserted conveniently. Placing the cavity facing upward is traditional, but it makes the accumulation of dust and other debris in the cavity almost certain. A horizontal position avoids that accumulation, but it often necesitates the use of a folding mirror, and that adds uncertainty and complexity to the radiometrics. If the cavity is mounted facing upward, a cover should be provided and in place whenever the blackbody is not in use.

4.3.3. Measurement of Blackbody Temperature

Application of Planck's law requires knowledge of the temperature of the emitting surface of the blackbody. Commercial blackbodies generally contain a temperature sensor (generally either a platinum resistance thermometer or a thermocouple) embedded in the block from which the cavity is machined. The controller monitors that temperature sensor and maintains it at a selected temperature. By design the block and the cavity walls are generally nearly isothermal, and the cavity temperature is thus very close to that selected.

To determine how close the selected and actual temperatures agree, or to verify the proper operation of the controller system, one must measure the temperature of the emitting surface. This is very difficult to do. Two general methods are available: probing the cavity with a temperature sensor, or using an optical pyrometer to measure the surface temperature. Before discussing the details of these methods, note that temperatures for blackbodies are sometimes reported in degrees Celsius (centigrade) and sometimes in kelvin. Be alert to the units used!

Pyrometer Determination of Cavity Temperature. Since this is a noncontact method, it eliminates possible mechanical damage and thermal contact problems. It has some problems: the resolution with which the temperatures can be read, the calibration of the pyrometer itself, and the emissivity of the cavity. The latter should not be a serious problem since blackbody cavities have emissivities near unity. This method is not often used for blackbody temperature measurement.

Probing the Cavity. Probing the cavity directly with a temperature sensor is the simplest method and is the one generally used. It has problems: (1) the danger of damaging or changing the surface finish, (2) getting the probe to reach the wall temperature, and (3) the usual traceability problems for high-temperature measurements. To minimize damage to the cavity walls, use fine probes and place them carefully.

Getting the probe to reach the wall temperature has two components: obtaining intimate thermal contact with the cavity wall, and avoiding cooling the point of contact by the probe (one end of which is by necessity at room temperature). The cooling effect can be minimized by the use of long, small-diameter probes.

I know of no literature that claims to estimate accurately the uncertainties involved in determining the temperature of the emitting surface by this method. The traceability/accuracy issue is better understood, although it is sometimes difficult to get calibration personnel to commit themselves to specific uncertainties. Thermocouples are generally used for this probing. The common materials are given in Table 4.2; the "type" designations are those of the Instrumentation Society of America (ISA). The use of thermocouples is summarized in a manual from the American Society for Testing and Materials (ASTM, 1981) and in a commercial catalog/handbook from the Omega Engineering Company (Omega, 1988). It includes references to other surveys and original work, as well as instrumentation, noise rejection, and diagnostic tips.

General calibration tables are available showing the thermocouple electromotive force (EMF) as a function of temperature for all of the standard materials (Powell et al., 1979; Omega, 1988); however, many applications warrant a special calibration for each individual device. Thermocouple wire can be calibrated periodically along its length, and certified to comply with NBS calibration tables within specified accuracies.

The useful temperature range depends on both the thermocouple material and the availability of equipment to calibrate it. The calibration equipment, in turn, may be limited by contractual issues as well as technical problems, so that the user must consult calibration and quality assurance personnel for specific requirements. Representative values are given in Figure 4.2.

The voltage–temperature characteristics of thermocouple wire will change if the wire is strained. At least one source (Cussen, 1982) comments that to preserve the thermocouple calibration it should not be inserted into a hot cavity, then withdrawn. Instead, the blackbody should be turned off, the thermocouple inserted, then the thermocouple and cavity heated together, then cooled together before removing the thermocouple. It is important to recognize that it is not the junction that generates the EMF, but the length of wire in which the temperature gradient exists. Cutting off the junction and rewelding the wire will not restore the calibration if the wire was stressed.

One rule of thumb used by calibration personnel is to claim an uncertainty in the measured value (the blackbody temperature in our case) no

TABLE 4.2 Thermocouple Types and Characteristics

Type	Composition
E	Chromel/constantan
K	Chromel/alumel
J	Iron/iron-constantan
R	Platinum/platinum–13% rhodium
S	Platinum/platinum–10% rhodium
T	Copper/constantan

better than four times the uncertainty in their instrumentation. For example, if they adhere to that guideline, and believe their TC calibration to within ±0.5 K, they would not be willing to certify the blackbody temperature to better than ±2 K. A 10:1 ratio is sometimes used. Your customer or quality assurance practices may impose still other requirements. The accuracies shown in Figure 4.2 are for better than typical thermocouple setups: These are what one could obtain by taking special pains to select and maintain the TC and measuring equipment. It does not include a 4:1 or 10:1 ratio.

4.3.4. Measurement and Documentation of Blackbody Aperture Diameters

Because the blackbody apertures are generally small, it is necessary to know their diameters quite accurately. If we allow a 2% contribution to

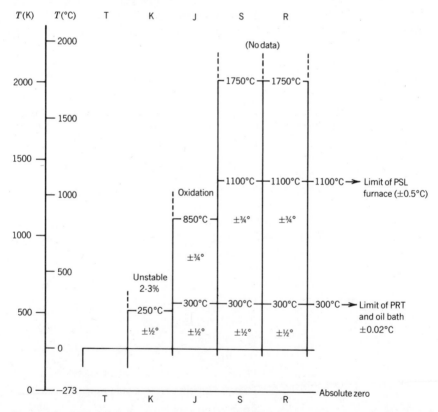

Figure 4.2. Thermocouple ranges and accuracy. Estimated accuracy for measuring setup assumes specially selected and maintained TCs, used only for selected BB calibration.

the error in incidance due to uncertainties in the diameter of a 0.020-in. aperture, we must know its diameter within 0.0002 in. This requires careful work.

Measuring the Apertures. To make these measurements, it may be necessary to remove the apertures from their mountings. If you do so, *first* note the position of each aperture in its mounting. (You will probably want to return them to the same holes later. Even if you do not, someone may later want to know which aperture came from which hole.) Then remove the apertures *very carefully*.

Examine each aperture under a microscope and note any apparent flaws. The effect on the incidance may be estimated, and depending on the precision desired, the apertures may be replaced or compensations made in the incidance calculations. Using a traveling microscope, measure the diameter twice in each of two perpendicular directions. If the results are not consistent within the repeatability expected from the microscope, repeat the measurements and discard the nonrepeatable reading. Compute and record the average of the diameters.

If you have data taken as indicated here, you will know how much confidence to place in your aperture diameters. A good set of aperture diameter data is shown in Figure 4.3. A record of the observations, raw data, and calculated average diameters should be preserved and made

	Aperture #		
	1	2	3
d1	0.2497	0.1249	0.0751
	0.2498	0.1248	0.0738*
			0.0748
d2	0.2541	0.1274*	0.0749
	0.2545	0.1251	0.0748
		0.1249	
average	0.2520	0.1249	0.0749
uncertainty (+_)	0.0025	0.0002	0.0002
	(1.0%)	(0.2%)	(0.3%)

* indicates values which were not used in the
  averages.

d1 and d2 are diameters (inches) measure at
90 degrees to each other. Measurements were
made on travelling microscope # 34789
by A. Stirling Workman, 25 December 1986

Figure 4.3. Sample aperture diameter data.

available for convenient reference. This allows subsequent users to assess the reliability and accuracy of the average values.

Method of Averaging. For a well-made aperture, the diameters measured will all be very similar and the method of averaging used will not make a significant difference. If the hole is elongated, treating it as an ellipse should yield a slightly more accurate estimate of the area than will use of a simple arithmetic average:

$$A = \frac{d_1 d_2 \pi}{4}$$

Since a circle is a special case of an ellipse, there is no loss of generality by using the formula for an ellipse.

4.3.5. Shutters and Choppers

Test set design should generally allow the introduction of both a room-temperature shutter and a chopper blade between the blackbody and the detector. This allows the test set to provide significantly more data with little additional complexity: One can measure signal from a hot source (the blackbody), from the room-temperature background, or from the difference between the two.

Shutter Material. Shutters built into test stations are generally of blackened metal. If a shutter is not built in but is needed for diagnostic work, materials convenient at hand are sometimes used. In either case, take care to use a material known to be opaque. In one case a black, velvet-like cloth was used, and only after several hours of work was it discovered that it had a transmittance of about 15%.

Location of the Shutter and Chopper. Put the shutter on the blackbody side of the area defining apertures, not on the detector side. Refer to Figure 4.4. In configuration (a) the change in incidance ("shutter out minus shutter in") is due only to the area defined by the blackbody aperture; nothing else in the detector field of view changes at all. In configuration (b), the incidance differences are difficult to compute unless one assumes that the temperature and emissivity of the shutter and aperture plate holder are identical. That assumption is not generally valid, although for large blackbody apertures or temperatures, one may be able to get away with it.

Position the chopper in the same place—between the aperture and the

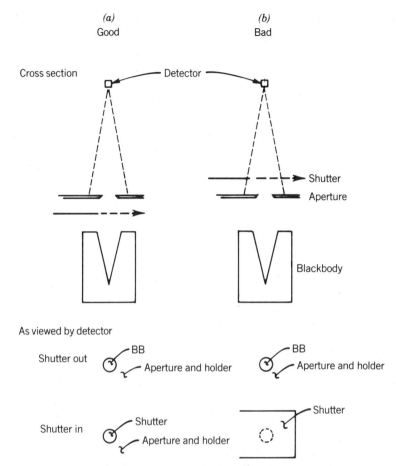

Figure 4.4. Shutter should be mounted between the blackbody and the blackbody area defining aperture.

blackbody. A chopper is just a moving shutter, so the reasons given above for shutter placement apply also to the chopper.

Temperature and Emissivity of the Shutter and Chopper. The temperature and emissivity of the shutter and chopper enter our incidance calculations. Assuming that the shutter or chopper is at room temperature is often adequate, but the validity of this assumption and its effect on the incidance should be reviewed periodically. The blackbody will heat the chopper and shutter, and the increased temperatures can cause significant errors in some cases. For high blackbody temperatures it may be essential to place a shield between the blackbody and the chopper or shutter, to use a low emissivity coating on the blackbody side, or to cool the shutter and chopper.

4.4. BETWEEN THE SOURCE AND DETECTOR

4.4.1. Folding Mirrors

Although folding mirrors can allow a compact and/or convenient setup, the mirror reflectance introduces radiometric complications. The reflectance is not exactly unity, so its value must be measured, or estimated from literature values. The reflectance may degrade with time, due to oxidation or the accumulation of dust. The extent of degradation can be minimized by the selection of coating materials, but when problems occur a measurement may be necessary. The dust problem can be minimized by mounting the mirror so that the reflecting surface does not face upward, placing the mirror where it can be inspected and cleaned easily, and providing and using a dust cover.

4.4.2. Orientation of Knife-Edged Apertures and Baffles

To minimize reflections from aperture edges, area-defining apertures are generally made very thin and selectively etched from one side: One side has a very accurate diameter, the other side is relieved, as shown in Figure 4.5. (This is done by using bimetallic stock or plated stock.) For the purposes of our discussion these apertures can be considered to be knife edged, as shown in part (*b*) of the figure.

The issue that then arises is: Which side of the aperture should face the detector, or does it matter? If the apertures are thin, it probably does not matter. If the geometry is complicated, a detailed ray trace is required to provide the optimum design. In general, however, there is less chance of excess radiation reaching the detector if the small side of the aperture is toward the device (blackbody or detector) from or to which the rays are most nearly parallel. This is motivated by Figure 4.5. Configuration (*a*) keeps reflected energy out of the detector cavity and prevents the detector from ''seeing'' the blackbody via reflections better than does configuration (*b*).

4.4.3. Unwanted Reflecting Surfaces

Reflecting surfaces are often inadvertently built into the blackbody setups; these can cause spurious signals that are 2 to 5 times greater than the designed signal. They must be eliminated or at least dramatically atten-

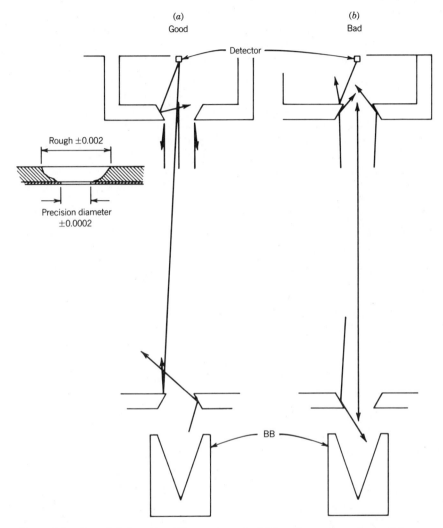

Figure 4.5. Area defining side of the aperture should face the most nearly parallel rays.

uated. One way to reduce reflections is to blacken everything possible. Two precautions come to mind here:

1. Do not assume that all is well just because the surfaces appear black to your eye. The degree to which things are black in the infrared is discussed in section 4.2.4. It depends on the material and on the angles of incidence and reflection.

2. Some areas cannot be blackened without compromising their intended function, so we cannot just adopt the rule "If in doubt, paint it black." Dewar walls (which are generally shiny to reduce heat transfer) fall into this category.

Geometrically, surfaces can be classified by the angle they make with respect to rays between the blackbody and the detector. Those surfaces that are nearly perpendicular to rays from the blackbody to the detector are not generally a problem because reflections from them will tend to bounce away from the detector, striking the detector only after many reflections, if at all. Even so, they should be blackened. Those nearly parallel to the rays are a serious problem and are discussed in the following section.

These "problem" reflecting surfaces are often (but not necessarily) cylindrical: support tubes for mounting detectors, shutter housings, and holes in support plates. Cylindrical surfaces are particularly bad because they can cause a light pipe or focusing effect, reflecting light into the dewar and detector. However, even flat surfaces can cause unacceptable reflections. These reflections are enhanced because the specular reflectance of most materials is very nearly unity at the grazing angles that often occur. Figure 4.6 illustrates two extremes in their potential for reflections. Part (d) shows a setup with a high probability of spurious signal due to reflections. In part (c) the probability of reflections is much less.

Methods of reducing reflections are illustrated in Figure 4.7; these include:

a. Make the holes larger, so that the cylindrical surface is out of the "line of sight" between blackbody and detector.
b. Minimize the potential reflecting surface area by using plates as thin as possible.
c. Eliminate the specularity by roughening or serrating the surface.
d. Insert one or more baffles on the blackbody side to stop radiation from hitting the cylindrical surfaces and/or on the detector side, to keep the reflected light from getting to the detector.
e. Line the cylindrical surface with a low-reflectance material. This is recommended only as a last resort, because the grazing angle reflectance of any material chosen will probably be high.

4.4.4. Accuracy Required in Measuring Detector-to-Blackbody Distances

The distance from the detector plane to the blackbody area defining aperture affects the blackbody incidence calculation, so the necessary com-

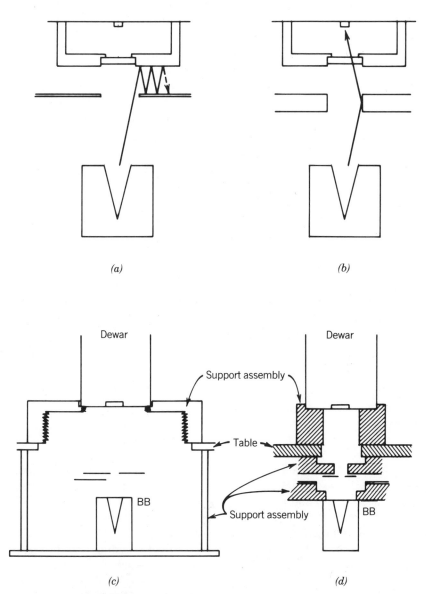

(a)   *(b)*

(c)   *(d)*

Figure 4.6.   Problem and nonproblem reflections: (*a*) nonproblem reflections; (*b*) problem reflections; (*c*) few problem reflections; (*d*) many problem reflections.

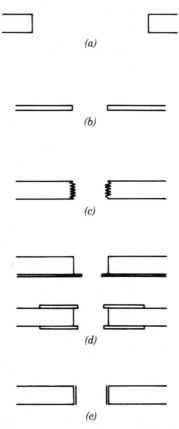

Figure 4.7. Methods of reducing reflections: (*a*) surface removed from field of view; (*b*) thinner material; (*c*) roughened surface; (*d*) baffles added; (*e*) surface lined.

ponents should be readily available. As an indication of the accuracy required, consider a typical situation. Assume that the desired accuracy in incidance is 5% and that the error budget for distance effects is 3%. Since the incidance varies as 1 over r^2, this means that we must know the distance within 1.5%.

Assume that the total distance is calculated by adding two distances, and that the errors in measuring these distances are not correlated, so that we can allow each distance to be in error by about 1% of the total distance. For a blackbody-to-detector distance of 10 in., this means that we need to know the individual components within 0.10 in. This is generally easy to achieve. Distance measurements to 0.020 in. are probably not worthwhile.

4.5. DEWARS

4.5.1. Design Decisions for Background Reduction

Often we are required to take blackbody data on a detector at a given background level. Someone needs to determine a combination of spectral filters, field-of-view shields, blackbody temperature, apertures and distances that will provide the required background, and adequate signal-to-noise ratio, and allow prediction of the performance at the wavelengths of interest.

There is no direct method that will lead to a solution in all cases. Instead, one usually explores a few options, selects the most promising, and modifies it until a satisfactory design is reached. We will attempt to show by example how this can be done. Consider the requirement to test a 0.25-in.-long array of PV InSb detectors at a background of 5×10^{12} photons/(cm^2·s). To determine if we will have adequate signal and signal-to-noise ratios, we will need the expected responsivity and noise equivalent incidance (NEI) values for the detector. It is convenient to have the responsivity per flux density: V/[photon/(cm^2·s)], A/[photon/(cm^2·s)], V/(W/cm^2)] or A/(W/cm^2). We can determine these from previous tests or by using the methods of Chapter 2. For our example, assume that the responsivity is 15×10^{-15} V/[photon/(cm^2·s)] and the NEI is 2×10^{11} photons/(cm^2·s).

Begin by assuming a 180° field-of-view shield, a CaF$_2$ window, and no spectral shielding. The InSb detector is blind to wavelengths beyond 5.5 μm. Using the methods of Chapter 3 we calculate that the room-temperature background is about 2×10^{16} photons/(cm^2·s). This is about 4000 times greater than we desire, so we must reduce the field of view, or the spectral bandpass, or add a neutral density filter, or some combination of these. First let's try to reduce the field of view. The reduction in projected solid angle is $\sin^2(\theta/2)$; to make this equal 1/4000 would require a half-angle of about 0.9°. If the cold shield is placed 3 in. from the detector, we will need a 0.095-in.-diameter hole in the field-of-view shield. This will provide the right background, but will obscure most of the array—we will not be able to illuminate the entire array with a distant blackbody.

Evidently we cannot get all of the required background reduction with a cold shield alone, so we will need a spectral filter in conjunction with the cold shield. We pay a penalty for the use of the spectral filter: the signal from the blackbody will be reduced. Assume that (like many labs) we have 500- and 800-K blackbodies with apertures that range from 0.050 to 0.500 in. in diameter, and that total distances from the blackbody aperture to the detector can be as little as 8 in. We calculate the incidance we can expect with three filters, each with a bandpass of 0.2 μm, centered at 3, 4, and 5 μm, using two blackbody temperatures (500 K, 800 K) and the two blackbody aperture extremes. Table 4.3 shows the incidances that we can expect with the 12 different combinations. It also lists the signal-to-noise ratios that we will see, based on the expected NEI. From this table we can pick out the acceptable combinations.

Table 4.3 shows that we can achieve the desired background with all three of the filters proposed. With the 5-μm filter, the field-of-view limiting aperture is only 0.23 in. in diameter; this is shorter than the length of our array, so it will probably vignette the signal from the blackbody, or at least make careful alignment necessary. If a better solution is not evident, we may try a narrower spectral filter at 5 μm, but set this option aside for the moment.

The 4-μm filter provides the desired background when used with a 0.51-in.-diameter cold aperture; this is twice the length of the array and should not cause any vignetting or alignment problems. With a 500-K blackbody, the signal-to-noise ratios range from 15 (smallest aperture) to almost 250 (largest aperture). This is marginal—we would like to have several apertures available, all yielding signal to noise ratios greater than 100. With the 800-K blackbody the signal-to-noise ratios range from 232 to 3705. This is a feasible option. Use of a wider spectral bandpass—say twice as wide, 0.4 μm—and cold aperture with 0.7 times the diameter would yield the same background but twice the signal-to-noise ratio; it would provide more options by allowing at least some testing to be done on the 500-K blackbody, but that is not essential and may be a luxury we must do without.

The 3.0-μm filter provides the required background with a 2.4-in.-diameter cold aperture; the signal-to-noise ratios with the 500 K-blackbody are too small, but those with the 800-K blackbody are acceptable. We could certainly get by with a much smaller cold aperture and a much wider spectral filter, but the 4-μm filter is probably a better choice because it is nearer the peak wavelength of the detector. Most interest in detector performance will probably concern its behavior near the wavelength of peak response, and the closer to that wavelength we can test, the better.

With the hindsight we have gained with the first three proposals, consider one more: a 0.3-in.-diameter cold shield and a spectral filter cutting off all wavelengths beyond 4 μm. The cold shield reduces the background by a factor of 400, so we can tolerate an in-band exitance of (5×10^{12}) $(400) = 2 \times 10^{15}$ photons/(cm^2·s). Inspection of Table 3.4 shows that we can limit the 293-K exitance to about 2×10^{15} with a filter that cuts off at about 4 μm, or a little beyond that. A 4-μm cutoff filter limits the 500-K exitance to 4×10^{17}, more than three times the in-band exitance from

Table 4.3 Background Reduction Worksheet

Spectral Band	$M(\lambda_c)$	$\Delta\lambda$	$M(\lambda_1\text{ to }\lambda_2)^a$	Reduction to Reach 5×10^{12}	0.050 in.-diam. aperture at 8 in. E^a	0.050 in.-diam. aperture at 8 in. S/N	Half Angle (deg)	0.200 in.-diam. aperture at 8 in. E^a	0.200 in.-diam. aperture at 8 in. S/N	Diameter (in.) (for $D = 3$ in.)
$T = 293\ K$										
0 to 5.5 μm			2.14×10^{16}	4280			0.9			0.092
0.2 μm at 5.0 μm	1.64×10^{16}	0.2	3.28×10^{15}	656			2.2			0.234
0.2 μm at 4.0 μm	3.43×10^{15}	0.2	6.86×10^{14}	137			4.9			0.514
0.2 μm at 3.0 μm	1.81×10^{14}	0.2	3.62×10^{13}	7			21.8			2.402
$T = 500\ K$										
0.2 μm at 5.0 μm	9.57×10^{17}	0.2	1.91×10^{17}		1.9×10^{12}	27		3.0×10^{13}	427	
0.2 μm at 4.0 μm	5.53×10^{17}	0.2	1.11×10^{17}		1.1×10^{12}	15		1.7×10^{13}	247	
0.2 μm at 3.0 μm	1.59×10^{17}	0.2	3.18×10^{16}		3.1×10^{11}	4		5.0×10^{12}	71	
$T = 800\ K$										
0.2 μm at 5.0 μm	8.49×10^{18}	0.2	1.70×10^{18}		1.7×10^{13}	237		2.7×10^{14}	3790	
0.2 μm at 4.0 μm	8.30×10^{18}	0.2	1.66×10^{18}		1.6×10^{13}	232		2.6×10^{14}	3705	
0.2 μm at 3.0 μm	5.80×10^{18}	0.2	1.16×10^{18}		1.1×10^{13}	162		1.8×10^{14}	2589	

[a] Units are photons/(cm²·s).

the 0.2-μm filter at 4 μm that we considered earlier. This would yield signal-to-noise ratios from 50 to 750 with the 500-K blackbody. This combination of field of view shield and filter seems like a pretty good setup—the field of view aperture is larger than the array, and we can obtain adequate signal-to-noise ratios without excessive blackbody temperatures.

Finally, check the geometry of our dewar and blackbody setup to make sure that the detector can be completely illuminated by the blackbody apertures selected. It is useful to do this with a scale drawing or sketch, and then, if there is any question about the accuracy with which the necessary details can be seen on the sketch, to double check it with a few calculations. Figure 4.8 shows the array, the cold aperture, and the blackbody aperture. The x and y dimensions were not drawn to the same scale: The diameters and the array length are shown 10 times larger than the corresponding distances between the apertures and the array. This is a useful trick to provide more accuracy in plotting the small array and aperture dimensions. It prohibits the use of protractors to measure angles, but does not invalidate any conclusions about vignetting.

In real life, the availability of spectral filters would be an important consideration. Costs and delivery times of the ideal filter might convince you to settle for something a little less perfect if you had it on the shelf.

4.5.2. Effect of Ambient Temperature on Background

The background "seen" by a detector viewing a room-temperature source can be very sensitive to the cutoff wavelength and the temperature of the room. It is common to use 300 K for room temperature—it is a nice round number and easy to remember—but it may lead to unacceptable errors. Note that 300 K = 80°F! A better "standard" value to use is 293 K (20°C, 68°F).

Table 4.4 compares the backgrounds from a 293-K source and one at 300 K for detectors with different cutoff wavelengths. For a 3-μm cutoff detector, the background changes by about 7% per degree celsius and day-to-day variations in room temperature can cause observable changes in detector current and noise.

4.5.3. Field of View Baffles

Reflection of background incidance within the dewar tends to increase the actual background flux density above that predicted from Planck's law by a factor that is difficult to predict analytically, but that seems to be between 25% and a factor of 2, depending on the dewar geometry.

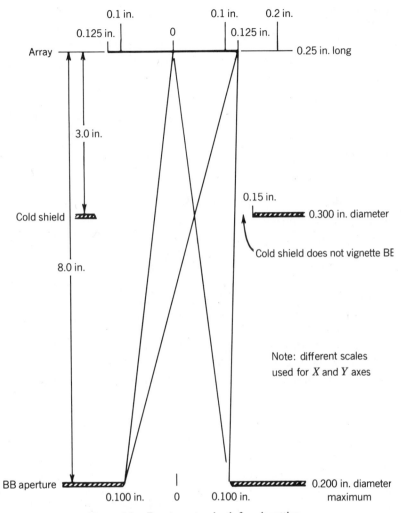

Figure 4.8. Ray trace to check for vignetting.

Design features that reduce these reflections are shown in Figure 4.9; these include:

1. Blackening those dewar surfaces that either receive direct background radiation, or face surfaces that receive direct background radiation, or are close to the detectors
2. Insertion of blackened baffles around the detector

Table 4.4   Effective Background[a]

Wavelength (μm)	Background Temperature		
	293 K (68°F)	300 K (80°F)	ΔE Due to $\Delta T = 7$ K
0 to 3.0	3.75×10^{13}	5.64×10^{13}	1.89×10^{13} (50%)
0 to 5.5	2.10×10^{16}	2.66×10^{16}	0.56×10^{16} (27%)
0 to 10.0	4.23×10^{17}	4.90×10^{17}	0.67×10^{17} (16%)

[a] Units are photons/cm²·s.

To reduce the backgrounds to the levels at which some of our products will be used, we often use a combination of spectral filters, neutral density filters, and small fields of view.

For low background testing, *elaborate precautions must* be taken to avoid background leaks. The metal-to-metal joint in part (*b*) of Figure 4.10 will leak some radiant energy, so interlocking grooves are used [see part (*c*)]. That scheme traps air in the shield, so vents are required to release air when pumping the dewar. Several types of vents are listed here [see Figure 4.10(*d*)] along with a very rough estimate of the photon incidance that they allow to pass through to the detector (assuming a Ge:Hg detector: $Q_{max} = 1 \times 10^{18}$ photons/(cm²·s). The purpose of this list is not so much to quantify the methods as to remind us of the difficulty in reducing leaks.

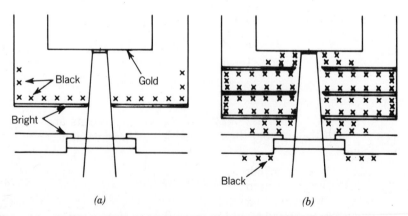

Figure 4.9. Baffling in FOV shields improves shielding effectivity; (*a*) minimum shielding: $Q_{BK} \sim 3 \times$ simple calculation; (*b*) better: $Q \simeq 1.25 \times$ simple calculation.

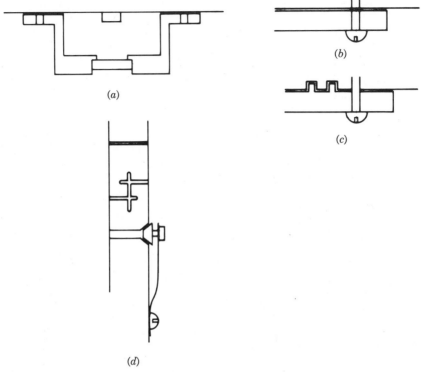

Figure 4.10. Baffles and labyrinths for venting shields.

A straight hole or slot, 0.030 in. in diameter and 0.10 in. long: 1×10^{13}

A simple labyrinth slot: 1×10^{12}

Valve with lapped seat, operated with a bimetallic strip (open when warm, closed when cold): 1×10^9

4.5.4. Windows and Filters

Problems commonly encountered with optical components are summarized here. Refer to Chapter 9 for a more detailed discussion.

Transmittance Data for Windows. The transmittance of window materials is graphed in many handbooks, texts, and wall charts. These are very handy for comparing the characteristics of various materials and for selecting suitable materials for a given application. For consistency and accuracy, the transmittance of window materials should *not* be read from those graphs unless absolutely necessary: the graphs are often copies of copies, sometimes copied by hand, and therefore are susceptible to error in reproduction.

If possible, measure the transmittance of the window directly. However, even if a measured value is available, it should be compared to the theoretical or expected transmittance from published data. Tabulated values are available in several handbooks, generally as a function of wavelength. If the transmittance is not listed, it can be calculated from the index of refraction, as described in Section 9.1.5. To assist others in auditing your calculations, indicate in your notes the method used to arrive at the transmittance you used; include the index of refraction, the source of that index, and the formula used.

Reflections from Windows. Windows with a high index of refraction reflect a significant amount of incident radiation. (For example, Ge reflects 53%, KRS-5 reflects 28%.) We can consider the effect of the transmittance loss on the first-pass incidence easily enough (see Figure 4.11), but it is more difficult to predict how much of the energy that is reflected from the detector surface will strike the window and be returned to the detector.

This kind of reflection is especially difficult because interference effects can cause nonuniformities in the incidence along the length of an array. Rather than attempt to correct for these effects analytically, it is easier to select a window material with a low index of refraction.

Spectral Filters. It is common to use cooled filters, sometimes at a non-

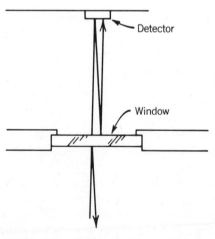

Figure 4.11. Reflectance from a window.

normal angle of incidence, but most of the readily available spectral curves are taken at room temperature and normal incidence. The correction of filter data for temperature and incidence angle is discussed briefly in Chapter 9, but actual measurements should be used whenever possible. In any case, see that these effects have been considered in all radiometric calculations.

Spectral filters often are assumed to have zero transmittance outside their passband. This assumption may introduce a substantial error. The need for more careful measurements may be checked easily by calculating the effect of a 0.2% out-of-band transmittance. If this effect is large, you had better verify the out-of-band transmittance with a measurement accuracy of 0.2% or better. Review of articles describing problems encountered with cold spectral filters (Stierwalt, 1974; Stierwalt and Eisenman, 1978) will give the experimenter a healthy respect for potential difficulties!

Reflecting Filters in Series. The composite transmittance of two reflecting filters placed in series is often assumed to be the product of their individual transmittance. This is approximately true only if their transmittances are relatively high (i.e., near unity) or if the energy reflected from one filter cannot reach the detector after reflection from the other filter. Whenever possible, avoid the use of two filters in series. If this cannot be avoided, mount them at an angle to each other (Figure 4.12) and do a ray trace to confirm that reflected rays cannot reach the detector.

Transmittance of the Dewar Window Reduces the Background. One factor in the background calculation is the transmittance of the window. The background in a dewar with an uncoated Ge window ($\tau = 0.45$) is roughly half of the background in a dewar with a near perfectly transmitting window.

This is occasionally questioned: "Since the window is at room temperature, doesn't it emit radiation that compensates exactly for the radiation that it keeps from reaching the detector?" The answer is generally no: Since our windows generally have a low emissivity, their emission is negligible.

Said slightly differently: The total background flux density has two components:

1. The energy generated outside the dewar, attenuated by the transmittance of the window
2. The incidance from the window itself: Treat the window as a graybody whose emissivity is that of the coated window

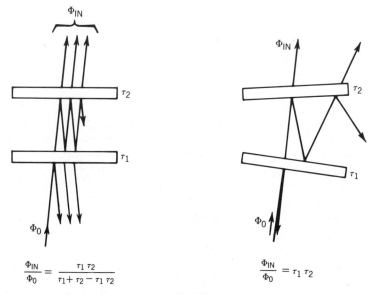

$$\frac{\Phi_{IN}}{\Phi_0} = \frac{\tau_1\,\tau_2}{\tau_1 + \tau_2 - \tau_1\,\tau_2} \qquad\qquad \frac{\Phi_{IN}}{\Phi_0} = \tau_1\,\tau_2$$

Figure 4.12. Two filters in series.

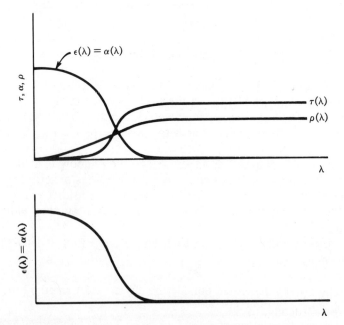

Figure 4.13. Emissivity of windows and filters is low in the passband, but can be high outside the passband.

The imperfect transmittance of the windows we use is generally due to reflective losses; it is not due to absorption. Thus the emissivity of the window is low—a few percent at most—and component 2 above is very small. Note, however, that this is not always true: Some window materials are absorbing, at least in some spectral regions. To the extent that the window does absorb, its emissivity is nonzero and will contribute to the detector incidance. This is illustrated in Figure 4.13.

4.6. TEST SET EVALUATION AND TROUBLESHOOTING

Clearly, there are an infinite number of possible test setups, and no one can address all the possible problems that can occur. On the other hand, it is surprising how many times the same problems will reoccur, and how experience will assist in detecting and eliminating those problems as rapidly as possible.

If test results are suspect and the problem cannot be readily identified, it is helpful to have a methodical approach to help identify the problem. Although each test setup is unique and thus subject to unique problems, Checklist 4.2 is offered as a "starter kit." Consideration of the items on the list could also be used as a form of periodic preventive maintenance. If a detector is available, use the detector and Checklist 5.2, on "confidence building."

Checklist 4.2: Test Set Evaluation and Troubleshooting

Assumptions about the dewar
  Window and spectral filters
    Check the assumed transmittance versus wavelength, including out-of-band leakage.
    Could scratches or uncoated areas allow leakage?
    Was the transmittance measured at (or correctly extrapolated to) the operating temperature and incidence angle?
  Are the surfaces around the detector and optical path blackened or baffled to prevent unwanted reflections?
  Is the detector at the assumed temperature?
  Could icing, condensation, or other contamination on the windows, filters, or detector be reducing the apparent responsivity?
Electronics: Verify the gain and frequency response of the amplifier and filter. Ideally, this is done with a circuit that closely simulates the detector circuit.

Checklist 4.2 (*Continued*)

Test set

    Visually check the apertures and optical path

        Is there dust in apertures?

        Are the folding mirrors dirty?

        Is there an obstruction in the optical path?

        Are there supports or baffles that might reflect?

        Is the dewar window clean and smooth?

    Examine documentation, and evaluate for completeness and carefulness of preparation:

        Aperture diameters, distance from aperture to tabletop, and from tabletop to detector.

        Waveform factor, chopper dimensions, number of teeth.

        Blackbody temperature calibration records.

        Is a table of incidance values available?

    Does the incidance calculation correctly include the effect of chopper temperature? Are assumptions about the chopper temperature apt to cause significant errors?

    Is the modulation factor correct? (It should be about 0.45 for a small aperture.)

    Verify one or more of the incidance values using Planck's law.

    Ask some questions:

        Has the test set been modified or disassembled recently?

        Are other users having problems?

        Have other users verified the test set is working well?

    Check measurements—depending on the magnitude of the suspected error:

        Rough measurement to find a large error, or

        Careful measurement: distances with vernier calipers, aperture dimensions with traveling microscope.

        Have calibration personnel verify blackbody temperature.

4.7. THE RADIOMETRIC CALIBRATION PROBLEM

Calibration of IR test setups is a problem because there are no accepted radiometeric standards for the infrared. Without a calibrated detector we cannot calibrate the test set, and vice versa. I recommend the comments by Hudson (1969, pp. 79, 80) on this problem to anyone interested in radiometry.

4.7.1. Radiometric Standards

Blackbodies can be built that satisfy stringent requirements on emissivity, temperature uniformity and accuracy, and accuracy of the aperture area. These use the melting point of various pure metals for temperature control (Cussen, 1982). Unfortunately, these do not provide much help with the practical radiometric problems in a day-to-day test operation because most radiometric problems are due to dewar-related effects (windows, filters, reflections) or lack of engineering discipline (distances or temperatures changed without the accompanying calculational changes).

Eisenman and Bates (1964) describe an improved thermal detector used for calibration work. A process for calibrating silicon photovoltaic detectors for use as absolute radiometric standards has been proposed and implemented (Geist, 1979; Geist et al., 1980), but nothing similar has yet been done beyond 2 μm. Note that a calibrated detector would not necessarily solve all the IR calibration problems: Each user has specific mechanical configuration, spectral range, and operating temperature requirements.

4.7.2. Radiometric Verification

Until standards are available, we cannot do a true calibration. Instead, we are limited to a process of verification or "confidence building" done in two parts. The first part is a prediction of the expected incidance using measured parameters and Planck's law. That is done routinely. The second part of the radiometric verification is an independent check of the incidance. It is desirable—perhaps essential—to perform that second step, but it is not easy, and it is often not done.

Step 1: Calculation of Expected Incidance. To predict the expected incidance we measure the blackbody temperature, the transmittance of the optical components, and the distances and aperture diameters, then calculate the incidance using Planck's law. If the test station design is free of spurious reflections, and if the measurements and calculations are done accurately, the resulting incidance is accurate. The test setups should be "audited" using checklists 4.2 and 5.2 as guidelines, to ensure that these conditions are satisfied.

The process described above yields the best available prediction of test station incidance. It is the fundamental step in characterizing the setup.

Step 2: Independent Verification of Test Station Performance. The problem now is that we want a way to verify independently and directly that all is well—to demonstrate that the incidance calculated is actually present, or to measure directly the extent to which the actual incidance differs from the value calculated. The techniques used to obtain and demonstrate the accuracy of the test sets vary from one laboratory to another and are not documented in the literature or in government standards. The method described here is a series of confidence-building measurements. The procedure invvolves obtaining several detectors and testing them on a variety of test stations. A systematic comparison of the test results for the detector–test station combinations and with theoretically predicted performance provide a consensus of the "correct" responsivity for each detector and therefore of the incidance of each test set. Although not a calibration in the usual sense of the word, this process establishes the radiometeric parameters with high confidence. The cooperation of groups responsible for the more conventional calibration efforts (mechanical, electronic, temperature) should be solicited, and their expertise used wherever possible.

Obtain at least one detector for use as a transfer standard. Three or more would be better. They should be fairly rugged, easy to handle, and representative of the spectral range and temperature of the detectors with which the test set will ultimately operate. (Using this work as a base and a model, the calibration effort can later be extended to other portions of the IR spectrum.)

Test stations that may be useful for the program should be audited to ensure that the necessary information (configuration, temperatures) is available and correct, and that the configuration complies with the guidelines established. Where necessary, modifications should be made to bring the stations into conformance with the guidelines. The transfer standards are then tested on the selected test stations. This is broken into two general steps:

A. Check for agreement among the blackbodies available. If they are in agreement, the chances are very good that they also conform to Planck's law, since deviations from Planck's law are not likely to be consistent from test set to test set. This step does *not* verify that the dewars are accurately characterized.

Signals from standard detector–dewar units are measured on the blackbody test sets, and the resulting data reduced. For each standard detector–dewar unit, the responsivities obtained on each of the blackbody sets should be compared to each other. Widely discrepant values can be set aside and the remainder averaged to provide the best estimate of the responsivity of the detector–dewar unit. This process should be repeated for each of the available standard detector–dewar units. For each test set, the ratio of the measured to best-estimate responsivity for each standard detector–dewar unit can be calculated and the results used to characterize the accuracy of that test set.

B. The best-estimate responsivity of each detector–dewar combination can be compared to the value expected from theoretical predictions of detector performance. Good agreement here is an indication that the dewar (with its window and filters) has been characterized accurately. This is limited to the confidence that one has in the theoretical prediction of detector performance.

Once the test sets are characterized, the unmounted standard detectors can be mounted and tested in each of several dewars used for day-to-day testing. For each standard detector the results from the different dewars should be compared and the results averaged after setting aside widely discrepant values. The accuracy of each dewar is then assessed by comparing the responsivity it "measures" for each detector with the best-estimate responsivity of the detector and the theoretical performance of the detector. For those dewars or blackbodies that yield discrepant results, the construction and documentation should be reviewed and recommendations made for test or modifications. When these have been made, the test with the transfer standards should be repeated.

Data on the tests should be stored in a controlled "standards library" and sorted in two ways: by transfer standard and by test station. The standards and their test history become an invaluable resource in checking out new stations or in auditing the continued accuracy of old ones. The results of this program include a set of transfer standards, designed for that purpose, and well characterized, as well as a data base for each "participating" test station, providing an estimated uncertainty and evidence that the stations radiometric characteristics are known within that uncertainty.

REFERENCES

American Society for Testing and Materials (1981). *Manual on the Use of Thermocouples in Temperature Measurement,* ASTM, Philadelphia, Pa.

Bartell, F. O., and W. L. Wolfe (1976). "Cavity Radiators: An Ecuminical Theory," *Appl. Opt.* 15, 84.

Bennett, H. E., Jean M. Bennett, and E. J. Ashley (1962). "Infrared Reflectance of Evaporated Aluminum Films," *J. Opt. Soc. Am.* 52, 1245–1250.

Billings, Bruce H., ed. (1972). "Wavelengths for Spectrographic Calibration," in *American Institute of Physics Handbook,* 3rd ed., McGraw-Hill, New York.

Chandos, Raymond J., and Robert E. Chandos (1974). "Radiometric Properties of Diffuse Wall Cavity Sources," *Appl. Opt.* 13, 2142.

Cox, J. Thomas, and Georg Hass (1978a). "Aluminum Mirrors Al_2O_3 Protected, with High Reflectance at Normal but Greatly Decreased Reflectance at Higher Angles of Incidence in the 8–12 Micron Region," *Appl. Opt.* 17, 333.

Cox, J. Thomas, and George Hass (1978b). "Protected Al Mirrors with High Reflectance in the 8–12 μm Region from Normal to High Angles of Incidence," *Appl. Opt.* 17, 2125.

Cox, J. Thomas, Georg Hass, and W. R. Hunter (1975). "Infrared Reflectance of Silicon Oxide and Magnesium Flouride Protected Aluminum Mirrors at Various Angles of Incidence from 8 to 12 μm," *Appl. Opt.* 14, 1247.

Cussen, Arthur J. (1982). "Overview of Blackbody Radiation Sources," in *Infrared Sensor Technology, Proc. SPIE,* 344, 2.

Dobrowolski, J. A. (1978). "Properties of Optical Materials," Section 8 in *Handbook of Optics,* Walter G. Driscoll, ed., William Vaughan, assoc. ed., McGraw-Hill, New York.

Eisenman, W. L., and R. L. Bates (1964). "Improved Black Radiation Detector," *J. Opt. Soc. Am.* 54, 1280.

Fu, C., N. H. Anger, R. Kaehms, and K. B. Jaeger (1988). "Characteristics of a Simple Blackbody Measurement System," *Proc. SPIE* 940, 102–116.

Geist, Jon (1979). "Quantum Efficiency of the p-n Junction in Silicon as an Absolute Radiometric Standard," *Appl. Opt.* 18, 760.

Geist, J., et al. (1980). "Spectral Response Self-Calibration and Interpolation of Silicon Photodiodes," *Appl. Opt.* 19, 3795.

Hudson, Richard D., Jr. (1969). *Infrared System Engineering,* Wiley, New York.

International Union of Pure and Applied Chemistry (1961). *Calibrating Wave Numbers for the Calibration of Infrared Spectrometers,* IUPAC, Butterworth, London.

LaRocca, Anthony J. (1978). "Artificial Sources," Chapter 2 in *The Infrared Handbook,* William L. Wolfe and George J. Zissis, eds., Environmental Research Institute of Michigan, Ann Arbor, Michigan.

Moore, John H., Christopher C. Davis, and Michael A. Coplan (1983). *Building Scientific Apparatus,* Addison-Wesley, Reading, Mass.

Omega Engineering Co. (1988). *Temperature Measurement Handbook and Encyclopedia,* Omega Engineering Co., Stamford, Conn.

Powell, Robert L., William J. Hall, Clyde H. Hyink, Jr., Larry L. Sparks, George W. Burns, Margaret G. Scroger, and Harmon H. Plumb (1974). *Thermocouple Reference Tables Based on the IPTS-68.* Monograph 125, NBS National Bureau of Standards, Washington, DC. (Also available from Omega Press, Stamford, Conn., as OP–3.)

Rao, K. Narahari, R. V. de Vore, and Earle K. Plyler (1963). "Wavelength Calibrations in the Far Infrared (30 to 1000 Microns)," *J. Res. Natl. Bur. Stand.* 67A, 351.

Shannon, R. R., and J. C. Wyant, eds. (1979). *Applied Optics and Optical Engineering, Volume 5.* Academic Press, New York.

Siegel, Robert, and John R. Howell (1975). *Thermal Radiation Heat Transfer,* McGraw-Hill, New York.

Stierwalt, Donald L. (1974). "Geometrical Effects on the Out-of-band Transmittance of Interference Filters", reported by R. Barry Johnson in the column "Infrared", *Opt. Eng.* 13, G115–117.

Stierwalt, D. L., and W. L. Eisenman (1978). "Problems in Using Cold Spectral Filters with LWIR Detectors," *Proc. SPIE* 132, 134–140.

Wolfe, William L. (1978a). "Optical Materials," Chapter 7 in *The Infrared Handbook,* William L. Wolfe and George J. Zissis, eds., Environmental Research Institute of Michigan, Ann Arbor, Michigan.

Wolfe, William L. (1978b). "Properties of Optical Materials," Chapter 7 in *Handbook of Optics,* Walter G. Driscolll, ed., William Vaughan, assoc. ed., McGraw-Hill, New York.

Zissis, George, and Anthony J. LaRocca (1978). "Optical Radiators and Sources," Section 3 in *Handbook of Optics,* Walter G. Driscoll, ed., William Vaughan, assoc. Ed., McGraw-Hill, New York.

SUGGESTED READING

IR Reflectance of Optical Blacks

Betts, D. B., F. J. J. Clarke, L. J. Cox, and J. A. Larkin (1985). "Infrared Reflection Properties of Five Types of Black Coating for Radiometric Detectors," *J. Phys. E (GB)* 18, 689–696.

Clarke, F. J. J., and J. A. Larkin (1985). "Measurement of Total Reflectance, Transmittance, and Emissivity over the Thermal IR Spectrum," *Infrared Phys.* 25, 359–367. Includes data for a number of black coatings.

Driver, P. M., R. W. Jones, C. L. Riddiford, and R. J. Simpson (1977). "A New Chrome Black Selective Absorbing Surface," *Solar Energy* 19, 301–306.

Houck, J. R. (1982). "New Black Paint for Cryogenic Infrared Applications," *Proc. SPIE* 362, 54–56.

Isard, J. O. (1980). "Surface Reflectivity of Strongly Absorbing Media and Calculation of the Infrared Emissivity of Glasses," *Infrared Phys.* 20, 249–256.

Pompea, S. M., D. W. Bergener, D. F. Shepard, S. Russak, and W. L. Wolfe (1983). "Preliminary Performance Data on an Improved Optical Black for Infrared Use," *Proc. SPIE* 400, 128–133. Martin black, and an improved Infrablack.

Siegel, Robert, and John R. Howell (1975). *Thermal Radiation Heat Transfer,* McGraw-Hill, New York. Chapter 5 is "Radiative Properties of Real Materials."

Smith, S. M. (1981). "The Far Infrared Reflectance of Optical Black Coatings," in *Proceedings of the 6th International Conference on Infrared and Millimeter Waves,* Institute of Electrical and Electronics Engineers, New York.

Smith, S. M. (1984). "Specular Reflectance of Optical-Black Coatings in the Far Infrared," *Appl. Opt.* 23, 2311–2326. Seven black coatings, 12 to 500 μm.

Stierwalt, D. L. (1979). "Infrared Absorption of Optical Blacks," *Opt. Eng.* 18, 147–151. Anodized aluminum with different surface finishes, anodized beryllium, 3M Black Velvet, Cat-a-Lac black paint, and black chrome.

Strimer, P., X. Gerbaux, A. Hadni, and T. Souel (1981). "Black Coatings for Infrared and Visible, with High Electrical Resistivity," *Infrared Phys.* 21, 37–39.

Wolfe, William L. (1978). "Optical Materials," Chapter 7 in *the Infrared Handbook,* William L. Wolfe and George J. Zissis, eds., Environmental Research Institute of Michigan.

Wolfe, William L. (1978). "Properties of Optical Materials," Chapter 7 in *Handbook of Optics,* Walter G. Driscoll, ed., William Vaughan, assoc. ed., McGraw-Hill, New York.

Zissis, George (1978). "Radiometry," Chapter 20 in *The Infrared Handbook,* William L. Wolfe and George J. Zissis, eds., Environmental Research Institute of Michigan, Ann Arbor, Michigan.

IR Reflectance of "Mirrors" (High-Reflectivity Coatings)

Bennett, H. E., Jean M. Bennett, and E. J. Ashley (1962). "Infrared Reflectance of Evaporated Aluminum Films," *J. Opt. Soc. Am.* 52, 1245–1250.

Cox, J. Thomas, and Georg Hass (1978a). "Aluminum Mirrors Al_2O_3 Protected, with High Reflectance at Normal but Greatly Decreased Reflectance at Higher Angles of Incidence in the 8–12 Micron Region," *Appl. Opt.* 17, 333. Al_2O_3 does not eliminate the 8- to 12-μm problem.

Cox, J. Thomas, and Georg Hass (1978b). "Protected Al Mirrors with High Reflectance in the 8–12 μm Region from Normal to High Angles of Incidence," *Appl. Opt.* 17, 2125. Use of Y_2O_3 and HfO_2 eliminates the 8- to 12-μm problem.

Cox, J. Thomas, Georg Hass, and W. R. Hunter (1975). "Infrared Reflectance of Silicon Oxide and Magnesium Fluoride Protected Aluminum Mirrors at Various Angles of Incidence from 8 to 12 μm," *Appl. Opt.* 14, 1247. Poor reflectance for large incidence angles for 8 to 12 μm; otherwise, high reflectance at normal incidence and for other wavelengths.

Dobrowolski, J. A. (1978). "Properties of Optical Materials," Section 8 in *Handbook of Optics,* Walter G. Driscoll, ed., William Vaughan, assoc. ed., McGraw-Hill, New York.

Hass, Georg (1955). "Filmed surface for reflecting optics" *J. Opt. Soc. Amer* 45, 945.

Wolfe, William L. (1978). "Optical Materials," Chapter 7 in *The Infrared Handbook*, William L. Wolfe and George J. Zissis, eds., Environmental Research Institute of Michigan, Ann Arbor, Michigan.

Blackbody Theory and Design

Bartell, F. O. (1981). "New Designs for Blackbody Simulator Cavities," *Proc. SPIE* 308, 22–27.

Bartell, F. O., and W. L. Wolfe (1976). "Cavity Radiators: An Ecuminical Theory," *Appl. Opt.* 15, 84.

Chandos, Raymond J., and Robert E. Chandos (1974). "Radiometric Properties of Diffuse Wall Cavity Sources," *Appl. Opt.* 13, 2142. Overview of cavity theory; cites earlier work.

Cussen, Arthur J. (1982). "Overview of Blackbody Radiation Sources," in Infrared Sensor Technology, *Proc SPIE* 344, 2.

Duncan, W. D., D. J. Robertson, and J. M. D. Strachan (1986). "Novel Variable Temperature Blackbody for the Far-Infrared," *Infrared Phys.* 26, 403–411. Blackbody temperatures from 77 to 310 K.

Fu, C., N. H. Anger, R. Kaehms, and K. B. Jaeger (1988). "Characteristics of a Simple Blackbody Measurement System," *Proc. SPIE* 940, 102–116. Analysis of a conventional blackbody system.

Hudson, Richard D., Jr. (1969). *Infrared System Engineering,* Wiley, New York. Overview of blackbody design and theory.

Jones, J. A., and A. G. Beswick (1987). "Miniature High Stability High Temperature Space Rated Blackbody Radiance Source," *Proc. SPIE* 750, 10–17. Operates at 1300 K, with a 49°C case temperature; dissipates 75 W. Size is 2.5 in. × 2.5 in. × 3.0 in. Fabrication is difficult.

Kuiming, Gao, Zhang Wei, Xie Zhi, and Shang Guanghui (1988). "Development of an Extended Blackbody Source for the Calibration of Infrared Systems. II," *Proc. SPIE* 940, 131–136. Companion to the paper by Wei.

LaRocca, Anthony J. (1978). "Artificial Sources," Chapter 2 in *The Infrared Handbook*, William L. Wolfe and George J. Zissis, eds., Environmental Research Institute of Michigan, Ann Arbor, Michigan.

Wei, Zhang, Gao Kuiming, and Xie Zhi (1988). "Development of an Extended Blackbody Source for the Calibration of Infrared Systems. I," *Proc. SPIE* 940, 125–130. 40 to 50°C, uniform to 1.4°C. Uses heat pipes. Companion paper is by Kuiming.

Zissis, George (1978). "Radiometry," Chapter 20 in *The Infrared Handbook*, William L. Wolfe and George J. Zissis, eds., Environmental Research Institute of Michigan.

Zissis, George J. and Anthony J. LaRocca (1978). "Optical Radiators and

Sources," Section 3 in *Handbook of Optics*, Walter G. Driscoll, ed., William Vaughan, assoc. ed., McGraw-Hill, New York.

PROBLEMS

1. List the equipment you would like in a completely equipped detector test lab.
2. Sketch the layout and mounting of a blackbody that would allow convenient access but would minimize the chance of reflections.
3. Indicate on the sketch for Problem 2 the blackbody temperature and the mechanical distances, aperture sizes, and their required tolerances if you are to predict the incidence within 5 percent.
4. If a blackbody test set is available to you, examine it for potential problems. Use Checklist 4.2 as a starting point.
5. Discuss with peronnel at your facility the radiometric "calibration" procedures they use. To what extent are they traceable to NBS or other facilities?

5

Detector Testing

Testing here includes the characterization (determining the characteristics of the detectors) and the judgment process of comparing the observed characteristics with one or more specifications or requirements. The testing of infrared detectors is an important and interesting part of the IR business. It often allows interaction with the customer (or at least his specification) and the fabrication personnel. It is a challenging job because the work is always being scrutinized and critiqued: The detector fabrication people will mistrust reports of poor detector performance, and the customer will want to be convinced that good performance reports are correct. It is also challenging because the test group is often under schedule and budget pressures: Since the testing operation comes last, the product delivery schedule may have been missed even before the unit gets to the test operation.

Many tools are available for characterization, and the ones selected depend on the type of detector, the specification, and the resources of the testing group. This chapter is intended to survey the entire test process. We begin with the considerations and operations done before the test starts, and continue through each of the tests normally encountered, describing how they are performed and what can be learned from them.

5.1. PRELIMINARIES

Test situations can range from the most straightforward (the test group knows exactly what is expected of them and how they are to proceed) to complex, uncertain, developing programs with many technical, budgetary, and schedule constraints. The material that follows assumes a middle-of-the-road situation. Some situations will not need the detail discussed here, while others will need a good deal more. In any case, this can serve as a checklist to help prevent misunderstandings.

213

5.1.1. Information Needed

Before you can perform a test, it is advisable (and in some cases essential) to obtain the information in Checklist 5.1. The amount of information needed depends on the role the test personnel are expected to play. In some cases they will be expected only to report results, but in general they have a responsibility to verify that the results are reasonable, so that they can catch errors in their own work and point out problems or trends to the people who use the data. The test work will be more meaningful, more interesting, and more effective if the test charter includes knowing why tests are being done, and at least some interpretion of the results.

Most of the material listed is straightforward or has been discussed earlier. Serialization is a seemingly minor point, but it is important. If several similar units are tested, confusion and errors can be avoided if each unit is identified with a unique, permanent number. Everyone involved should want such a serial number, and the test group has a right to require it on any unit they accept for test.

Checklist 5.1: Checklist of Information to be Obtained before Testing

Detector information
    Type and material
    Size (sensitive area)
    Operating temperature
Dewar or test chamber installation information
    Mounting configuration
    Valve type and pumping required
    Cooling method to be used
    Field of view
    Distance from detector to an accesible reference surface
    Window material, and spectral curves if coated
    Filter spectral curves (at operating temperature)
    Dewar labeling and serialization
    Wiring, cable, or pin identification
    Temperature sensors and their operation method and calibration data
    Special handling, pumping, cooling, and electrical precautions needed
Test requirements
    Documentation expected or needed

Test plans
Reports, including content and format
Tests desired; parameters to be reported
Accuracy expected or needed
Budget and schedule constraints
Witnessing or other participation by customer representatives

5.1.2. Test Planning

Once the information about the detector and test requirements are available, the test personnel can combine that with their knowledge of their resources and limitations to lay out a test plan that can then be reviewed with the requestor, customer, and management as appropriate. The test plan normally provides more detail and insight than the test request, but does not include all of the detail needed for operators to actually perform the test. The test plan should define any issues that may be controversial in enough detail that—once approved by the requestor—it will serve as a technical contract to which the test group will perform.

Report Definition. It is helpful to draft a final report during the planning stage. The concreteness of the report will clarify for both the tester and the customer the requirements and definitions. The type of information, the amount of data, and its format need to be considered. For large arrays or programs with many technical requirements, the report can be so voluminous that it needs one or two levels of summaries to allow readers to find what they want without wading through many pages that they do not want.

Test Accuracy. The test plan should address the accuracy that is expected in the data. Equipment, personnel, time, budget, and detector features all affect the accuracy, so expected or possible accuracies are unique to the laboratory and the job. Despite this variability, it is natural to want to know what accuracies "should" be expected. As a starting point for discussions, Table 5.1 attempts to provide "typical" accuracies for a few common tests.

5.1.3. Test Procedure

The test procedure includes the specific steps that the test personnel will follow to perform the tests called out in the test plan. As in any work we

Table 5.1 Typical Accuracy for Detector Testing

Parameter	Probable Accuracy: 1σ (%)		
	Routine	With Care	Exceptional
Incidance			
Blackbody	10	7	4
Background	20	10	8
Noise bandwidth	5	3	2
Signal	4	2	1
Noise	10	7	4
Blackbody responsivity	12	8	5
Blackbody D^*	15	10	7
Spectral response			
Wavelength (μm)	0.1	0.05	0.02
Spectral responsivity (%)	15	10	7

do, it is important that the person doing the work have a clear agreement with his supervisor and with the customer on what he or she is to do, and how it is to be done. Documentation of test procedures can range from nonexistent (informal, unwritten, traditional ways of doing things) to the elaborate (carefully formated, reviewed, approved, published volumes). In the case of a routine test being done on a typical detector by an experienced operator using proven equipment, the procedure to be followed may warrant no special attention, and the operator may appropriately follow an existing procedure.

If the test is in any way out of the ordinary, new procedures must be developed. The operator, the supervisor, and the customer are all responsible for reaching an agreement on what that procedure will be. If it is decided to leave the procedure to the supervisor's discretion, the customer should concur; if left to the operator's discretion, the supervisor and customer should consciously so agree. The test procedure must include any pretest inspections. It should include any required notification of engineers, customers, or inspectors to witness or participate in one or more of the operations. It should include how data are to be acquired, recorded, reduced, and reported. It should include handling of the unit after the test.

5.1.4. Pretest Operations

Receipt of the Unit. Visually check the unit and any accompanying data: Is it what you expected? Is it in good condition? Is the expected information provided? It is a frustration and a waste of time to accept a unit on Monday, only to discover on Wednesday when your schedule allows you to start the pumpdown that the window is cracked or that there is no detector in the dewar.

Room-Temperature Checkout. It is useful to develop and use a quick room-temperature check of the detector and electronics before committing the dewar to a pump or to the cooldown process. The purpose is to discover opens, shorts, or inoperative in-the-dewar electronics. In some cases a simple ohmmeter resistance measurement may be adequate. Before doing such a test, confirm that it will not damage the detector, the leads, or the electronics: the current developed by some ohmmeters can burn out the fine wires used for thermal isolation of the detector. Discovering such a problem before the dewar is pumped saves the pumping time. Discovering it before the dewar is cooled saves both the cooldown and warm-up time since the dewar would have to be warm for further diagnostics and repair. In general, the longer the pumping and cooling time, the more effort and time can be allocated to room-temperature functional checks.

Pumping. Vacuum practices and the problem of defining what is an adequate vacuum for detector testing are many and varied; some of them are discussed in Chapter 8. It is especially important to have a clear agreement on the pumping procedure.

Cooling. Many cooling techniques are possible; see Chapter 7. Have a procedure and follow it.

5.2. *V–I* CURVES

A knowledge of the current–voltage relationship for a given detector is essential for the circuit designer, and it is also a good diagnostic tool: it is a measure of the "health" of the device. This current–voltage relationship is routinely measured or plotted for photovoltaic detectors.

For both PV and PC detectors the *V–I* relationship depends on temperature and background. Data taken at an arbitrary operating condition (even room temperature) may give you some information about the health of the detector, but the risks in extrapolating from one set of test conditions to another should be considered carefully before agreeing to the apparent simplification of testing at other than the end-use conditions.

5.2.1. Equipment Used for *V–I* Curves

All that is needed to trace out *V–I* curves is a source of current or voltage that can be varied smoothly, and a way of measuring and displaying current and voltage. Commercial curve tracers or special bias boxes with XY recorders (Figure 5.1) can be used.

A finished detector can be mounted and cooled in a conventional dewar for the curve trace. Wafers of many in-process detectors are sometimes mounted on a block of copper surrounded by liquid nitrogen. This allows the detectors to be probed directly, without the time and expense of mounting them in a dewar, and then pumping and cooling the dewar.

5.2.2. Analysis of *V–I* Curves for Photovoltaic Detectors

A *V–I* curve drawn under known conditions (background and temperature) can tell you much about the detector. Items to look for are illustrated in Figure 5.2.

1. *Short-Circuit Current.* If the background is known accurately, the short-circuit current will allow calculation of the detector quantum efficiency, since $I = \eta Q A e$. Even if the background is known only roughly, it is a good idea to compare the observed and expected short-circuit currents just to make sure that everything is working properly. Indicate I_0 on the curve; calculate and record η.

2. *Open-Circuit Voltage.* The open-circuit voltage is quite independent of background, but does depend on temperature. By checking the open-

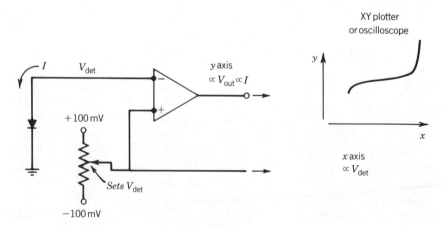

Figure 5.1. *V–I* curve tracer.

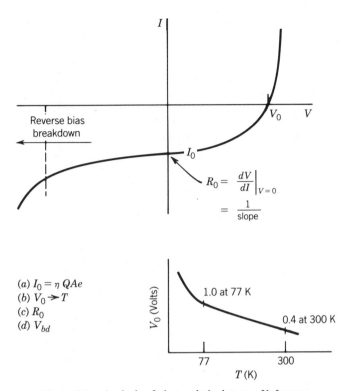

Figure 5.2. Analysis of photovoltaic detector *V–I* curve.

circuit voltage on each curve drawn, obvious problems or blunders can be identified quickly. Indicate V_0 on the curve and compare with the expected value.

3. *Resistance at V = 0.* This is a measure of the "health" of the diode: Diodes with low resistance are "leaky" and are apt to have high noise, or breakdown with time. The product of resistance at $V = 0$ and the area is referred to often as the "R-naught-A Product"—R_0A. Typical values are shown in Figures 2.23 (InSb) and 2.25 (HgCdTe). Calculate R_0 by drawing a line tangent to the curve at $V = 0$ and determining the ratio $\Delta V/\Delta I$. Calculate R_0A. Report both R_0 and R_0A.

4. *Reverse Bias Resistance.* For many applications the detector will be operated slightly reverse biased, primarily to avoid the possibility of forward bias if component values are not quite the nominal ones. In this case the reverse bias resistance is of more interest than is R_0. The slope in the reverse bias range is generally constant (until breakdown begins). To determine R_r, draw a line tangent to the straight portion of the reverse

bias $V–I$ curve and calculate $R_r = \Delta V/\Delta I$ from that tangent line. Report R_r and R_rA.

5. *Reverse Bias Breakdown Voltage.* The maximum voltage a diode will take in the reverse direction is another measure of the "health" of the diode and should be noted. Although it is clear that breakdown is occurring, the definition of V_b is not universal. You may define it as the voltage at which the slope is twice that in the reverse straight-line segment, or where the slope or current reaches some predetermined threshold. Indicate V_b and record it on your report.

5.2.3. *V–I* Data for Photoconductors

Although $V–I$ curves can be drawn for photoconductors, it is not common do so. Instead, one measures or is given one or two voltage, current pairs, or may only have the resistance values: $R = V/I$. These measurements are often made at the blackbody test set while measuring signal and noise; the primary purpose is to determine and document the bias conditions that maximize S/N.

Although not a normal part of testing, it is instructive to acquire and plot $V–I$ data for a photoconductor over a wide range of currents and voltages. The $V–I$ curve is much more nearly linear than for the PV detectors, but deviations from linearity do occur and their presence can be a reassuring indicator that the device is behaving as expected (see Figure 2.11). These variations come about because the mobility and lifetime of semiconductors are somewhat dependent on the electric field. (Ohm's law is, properly speaking, not truly a law. It is a definition of resistance and a statement that for many cases, resistance is independent of electric field.)

5.3. NOISE

Noise measurements are essential ingredients of detector characterization. Noise is defined as the rms deviation of the detector output from its average value. It can be determined from a number of digital samples of the detector output, from an analog meter connected to the output, or even estimated from a visual examination of an oscilloscope trace.

5.3.1. Digital Calculation

Noise can be calculated from digitized samples of the detector output: It is the standard deviation of the population (σ) of the detector output voltage (or current). Several algorithms for computing the standard deviation are available. These are usually implemented in software or firmware.

Direct Method. One way to determine noise is to apply the definition directly. First, calculate the sum of all the values, and the mean. Then go back through the list of values, calculating the deviation from the mean, the squares of the deviations, their sums, and the square root of the sum divided by $J - 1$:

STANDARD DEVIATION OF THE POPULATION

$$\sigma = \sqrt{\frac{\sum_{}^{J} (x_j - \bar{x})^2}{J - 1}} \tag{5.1}$$

where

$$\bar{x} = \frac{\sum_{}^{J} x_j}{J} \tag{5.2}$$

This process could be called a brute-force, or two-pass method. It is straightforward but requires that we save or store all the values, then go through the list of values twice: once for the mean value, once for the deviations. In some situations this can be undesirable; it may take too much time or too much data storage space. Once you have computed the noise, the decision to include a few more values requires you to start over and repeat the entire process with your new (larger) data set. Methods are available that avoid this problem.

Sum and Sum of Squares. The standard deviation definition can be manipulated to provide an algebraic expression that requires only the sum and sum of the squares.

STANDARD DEVIATION FROM "SUM AND SUM OF SQUARES"

$$\sigma^2 = \frac{\sum_{}^{J} x_j^2 - (\sum x_j)^2/J}{J - 1} \tag{5.3}$$

This means that we do not have to store all the individual values—just keep a running total of x and x^2 (and the number of samples summed). Data acquisition systems can do this easily. When enough samples have been summed, calculate the noise directly. (The required number of samples is discussed in Section 5.3.2.)

The advantages of the sum and sum of squares method are that little storage space is required and the calculation is fast. Additional samples can be included without starting over. The only limitation is that the sum of the squares can sometimes become a vary large number, overflowing the available data storage capacity.

Sum and Sum of Squares, with an Offset. If we can estimate ahead of time what the average value will be, we can offset all values by that estimate before calculating the sum and sum of squares (Equation 5.4). This allows us to work with smaller numbers, perhaps avoiding the overflow problem. The offset can be added back to the average value, and the noise needs no correction since the deviation is unaffected by a constant offset.

STANDARD DEVIATION WITH AN OFFSET

$$\sigma(x) = \sigma(x') \qquad (5.4a)$$

$$\bar{x}_j = \bar{x}_j' + C \qquad (5.4b)$$

where

$$x_j' = x_j - C \qquad (5.4c)$$

$$C = \text{offset used} \qquad (5.4d)$$

This is a useful technique, even when calculating means and sigmas by hand. One limitation is that it requires some estimate of the final mean, and that inevitably involves some loss in generality in the process.

5.3.2.   Error in Digital Noise Measurements

When digital techniques are used, the analog detector voltage or current is sampled, then converted to a digital value which is then processed. In addition to any systematic (calibration) error in the analog circuit, error is introduced by the digitization process (quantization error), and in the computer itself (round-off error). In any noise determination (digital or analog) there is error due to the fact that a finite data set provides only an *estimate* of the noise associated with the entire population of output values. In the following sections we discuss these three sources of error.

Round-off Error. Because computers represent all numbers in binary, there is some rounding error in all input, intermediate, and reported values. If many repeated calculations are performed, the rounding errors can accumulate. This can be a limiting contribution to noise accuracy, and in fact, one can reach the point where acquisition and processing of more and more samples, in an attempt to improve accuracy, can actually *reduce* the accuracy. Analysis of these computer related errors and algorithms to minimize them are found in articles by Hanson (1975), Chang and Lewis (1979), and West (1979) in *Communications of the Association of Computing Machinery* (ACM).

Quantization Error. When digital information is required, an analog-to-digital (A–D) converter is used. There is some error introduced in the process—the output of the A–D is a discrete value, not exactly equal to the analog (input) value. A measure of this potential error is the resolution of the digitizer, normally called the *least significant bit* (LSB). If the digitizer covers a range R (either zero to R, or from $-R/2$ to $+R/2$) with K bits, the LSB is given by

$$\text{LSB} = \frac{R}{2^K} \tag{5.5}$$

For example, for 12 bits, 2^K is 4096, so a 12-bit system can represent a 10-V range with a resolution of 2.44 mV:

$$\text{LSB} = \frac{10\ \text{V}}{4096} \simeq 2.44\ \text{mV}$$

Digitization introduces a signal error whose absolute value is between zero and LSB/2.

If the LSB is less than about two times the detector noise being measured, the effect on noise is to add an effective noise of magnitude $\text{LSB}/\sqrt{12}$ (about 0.3 LSB) to the actual noise. This can be derived by calculating the rms deviation in the signal due to the digitization process. If LSB is small compared to the noise, the probablility of any signal between $+\text{LSB}/2$ and $-\text{LSB}/2$ is roughly constant, and the rms error (or added effective noise) is

$$(\text{rms error})^2 \simeq \frac{\displaystyle\int_{-\text{LSB}/2}^{+\text{LSB}/2} \epsilon^2\, d\epsilon}{\displaystyle\int_{-\text{LSB}/2}^{+\text{LSB}/2} d\epsilon} = \frac{\text{LSB}^2}{12}$$

For LSB values that exceed about two times the detector noise, quantization can distort the data in a way that is difficult to predict. The resulting noise can be too small—as small as zero—or too large—as large as LSB/2: If the mean value is near one of the allowed digital values, the observed noise is too small (Figure 5.3a), but if the mean value is near the boundary between two digital ranges, the reported noise will be too large (see Figure 5.3b).

EFFECTIVE NOISE DUE TO QUANTIZATION ERROR

$$\text{Quantization noise} \simeq \frac{\text{LSB}}{\sqrt{12}} \quad \text{if LSB} < 2 \times \text{anticipated basic noise}$$

$$(5.6)$$

Quantization noise becomes more unpredictable as LSB exceeds the basic noise; it can be as large as LSB/2, or can be zero.

Effect of Sample Size on Error in Noise Measurements. If we calculate the noise from J independent (as defined below) signal values, using the rms value of the deviations from the average, as described earlier, then the uncertainty (1 σ, or 1 standard deviation) for that noise value is given by Mandel (1964, p. 236), as seen in Eqs. 5.7 and 5.8.

Table 5.2   Uncertainty in Noise Measurements for Different Numbers of Independent Samples

Number of Samples	Probability (%) That Noise Is within Specified Percentage of "Correct" Value		
	68%	95%	99.7%
10	22	45	67
30	13	26	39
100	7	14	21
300	4	8	12
1000	2	5	7
3000	1	3	4

For 100 samples there is a 68% probability that the result is within 7% of the "correct" values. It requires 3000 samples for 95% confidence that the value is within 3% of the "correct" value.

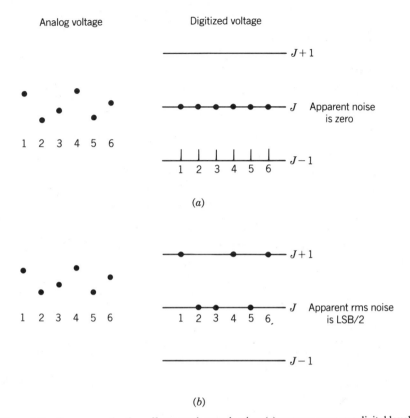

Figure 5.3. Large quantization effects mask actual noise; (a) overage near one digital level; (b) average between digital levels.

UNCERTAINTY IN NOISE DUE TO FINITE NUMBER OF SAMPLES

$$\sigma_{noise} = \frac{noise}{\sqrt{2J}} \tag{5.7}$$

UNCERTAINTY EXPRESSED AS A RELATIVE NOISE ERROR

$$\frac{\sigma_{noise}}{noise} = \sqrt{\frac{1}{2J}} \tag{5.8}$$

The 1, 2, and 3σ values are given in Table 5.2 and Figure 5.4. The table and figure are based on the formula above for the standard deviation of noise values and the normal curve of error (discussed in Chapter 6).

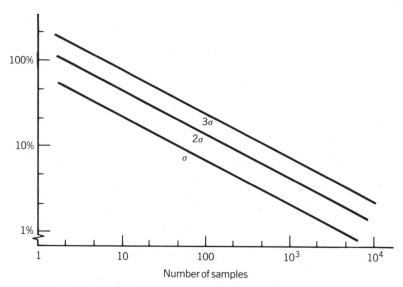

Figure 5.4. Relative uncertainty in noise versus sample size.

Example 1: If noise is determined from the rms deviation of 100 samples, there is a 68% probability (sometimes referred to as 1σ probability) that the result is within 7% (Table 5.2) of the "true" value, and a 95% probability that the result is within 14% of the "true" value (Figure 5.4). ■

Example 2: To be 95% certain (i.e., 2σ) that the noise is within 1% of "correct" requires about 20,000 samples (Figure 5.4). ■

Mandel (1964) comments that "it is a deplorable misconception, based on ignorance or disregard of elementary statistical fact, that small samples are sufficient to obtain satisfactory estimates of variability."

The noise uncertainty expressions given above assume that normal or gaussian statistics apply, and that the samples are independent. White noise obeys gaussian statistics, but other types may not. $1/f$ noise is one that does not.

Samples are independent as long as they are taken at a rate less than $2\Delta f$, where Δf is the noise equivalent bandpass of the system. If the signal is measured by an analog device that integrates for time t, the number of independent samples is $t \times 2 \times \Delta f$, and the integration time required to include n independent samples is $t = n/(2\ \Delta f)$.

Example: With a 20-Hz equivalent noise bandwidth we can take independent data no faster than 40 samples per second. To acquire 20,000 samples would take 500 s, more than 8 min.   ■

5.3.3. Analog Measurements of Noise: Meters and Their Errors

Analog meters with which one could attempt to measure noise fall into three general types:

Meter 1 simply averages—it is called an *average responding meter*. It is of no use to us as a noise meter; it always reads the average value—either just the signal, or zero if there is no signal.

Meter 2 actually averages the square of the wave, then takes the square root—it is a *true rms meter*. It is ideal, and we should always use such a true rms meter if we can get it.

Meter 3 averages a rectified wave. It can be used, but will read gaussian noise low by a factor of $2/\sqrt{\pi}$ (about 1.128). This occurs because a meter like this is generally calibrated so that it displays directly the true rms value of *sinusoids*. To do this, the scale is "fudged" to display a value $\pi \sqrt{2}/4$ (about 1.11) times the average of the full wave rectified deviation. The rms of noise with a gaussian distribution is $\sqrt{2\pi}/2$ (about 1.25) times greater than the average of the rectified deviation. Thus readings gaussian noise made with such a meter must be multiplied by an additional factor of $2/\sqrt{\pi}$ (about 1.128).*

5.3.4. Bandwidth for Noise Calculations

The bandwidth of electrical circuits affects the noise that passes through the circuit. For a perfect square filter the bandwidth is clear, but for a filter that "turns on" and "off" more gradually, the equivalent noise bandwidth must be calculated from the circuit gain as a function of frequency. This equation is given by Hudson (1969) in terms of the power gain $G(f)$, and by Jones et al. (1960) in terms of the voltage gain $g(f)$. It appears here as Eq. 5.9. One way to do the indicated calculation is to plot (on linear paper) the square of the voltage gain versus frequency, determine the area under the curve, and construct a rectangle of height

---

* Jones et al. (1960, Sec. 3.7) comments on this source of error, but states that the meter will read low by 0.9003. This is apparently an error; the 1.128 value is well documented elsewhere (Hudson, 1969, pp. 317, 318; General Radio Company, 1963, p. 4; Marconi Instruments, 1965, pp. 18–22).

g_0 and width such that it includes the same area as the curve. The width of the rectangle is the equivalent noise bandwidth (see Figure 5.5).

EQUIVALENT NOISE BANDWIDTH FROM VOLTAGE GAIN

$$\Delta f = \int_0^\infty \left[\frac{g(f)}{g_0} \right]^2 \bigg/ df \qquad (5.9)$$

where $g(f)$ is the voltage gain at frequency f and g_0 is the maximum voltage gain.

For a circuit that rejects high frequencies at 3 dB per octave (a single-pole filter so that the response to a voltage pulse is an exponential decay with time constant τ) with a corner frequency f_0, the equivalent noise bandwidth is expressed in Equation 5.10. A circuit that integrates for time T averages out the high-frequency effects (both signal and noise). The effective noise bandwidth is given by Lange (1967) and Boyd (1983) and is shown in Equation 5.11.

EQUIVALENT NOISE BANDWIDTH: SINGLE-POLE FILTER

$$\Delta f = \frac{\pi}{2} f_0 = \frac{1}{4\tau} \qquad (5.10)$$

where $f_0 = 1/2\pi\tau$.

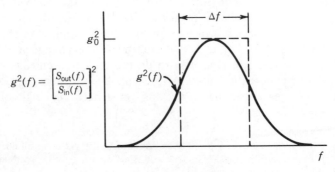

Figure 5.5. Equivalent noise bandwidth.

EQUIVALENT NOISE BANDWIDTH: INTEGRATOR

$$\Delta f = \frac{1}{2T} \qquad\qquad (5.11)$$

5.3.5. Characterization of Low-Frequency Noise

Most semiconductors exhibit increased noise at low frequencies, as shown in Figure 5.6. For some devices this excess noise spectral density fits the following equations:

$$\text{power} \; \propto \frac{1}{f} \qquad\qquad (5.12a)$$

$$\text{voltage} \propto \sqrt{\frac{1}{f}} \qquad\qquad (5.12b)$$

Noise that fits those equations is $1/f$ *noise*. (Note that $1/f$ noise **voltage** varies as one over the **square root** of frequency.) Very often we refer to the increased noise at low frequencies as $1/f$ noise—whether or not it obeys that equation.

The acceptable amount of $1/f$ noise is sometimes specified, or must be measured for other reasons. This can be done using a spectrum analyzer, which generates a plot of spectral density versus frequency. The amount of $1/f$ noise can also be conveyed by a single number—the frequency that divides the excess noise region from the white noise region. The higher this frequency, the higher the low-frequency noise will be (see Figure 5.6).

The critical frequency can be defined and determined in several ways. Since the results will not be the same (unless the noise obeys the assumed $1/f$ relationship exactly) it is important that everyone involved reach agreement on which method will be used.

The real need is *not* for $1/f$ corner frequency. Remember that the customer's ultimate concern is probably not really the $1/f$ corner frequency, but rather some system-level effect (such as flicker, or uniformity drift) which we know is related to low-frequency noise, but whose dependence on low-frequency noise is complex and generally only poorly understood. The corner frequency is just an attempt to get a handle on a more complicated problem. It probably does not make sense to spend a great deal of time refining the $1/f$ corner frequency definition until the real system need is equally well identified and characterized.

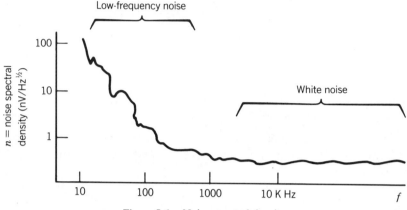

Figure 5.6.   Noise spectral density.

This is particularly true since $1/f$ noise is very difficult to characterize reliably since it is by nature erratic and variable. If noise readings are required between 50 and 300 Hz the presence of 60-Hz line noise will further complicate detector $1/f$ noise characterization.

Noise Spectral Density Plots.   The most complete and accurate way to characterize low-frequency noise is with a plot of noise spectral density (volts/square root hertz) versus frequency (Figure 5.6). This method suffers from two problems: noise spectrals are slow to acquire, and the data are not compact (only one or two curves fit on a page).

3-dB Frequency.   We are sometimes requested to condense the low-frequency noise information into just one number. One way to do that is to provide the *3-dB frequency* f_{3dB}: the frequency below which the excess noise is 3 dB* above the white noise (see Figure 5.7). This definition works even if the noise does not obey the $1/f$ equation.

Determination of f_{3dB} as defined here requires narrowband noise measurements at several frequencies near the expected f_{3dB}, as well as a good fix on n_{white}. First fit a horizontal line through the high-frequency data, and note the corresponding noise spectral density. Draw a smooth curve through the low-frequency data. The frequency at which the noise is 1.413 higher than the horizontal line is the 3-dB frequency.

---

* "3 dB" and "a square root of 2 in voltage ratio" are often used interchangeably. Strictly speaking, 3 dB is a voltage ratio of 10 to the 3/20 power (1.4125 . . .); this is within about 0.1% of the square root of 2. For most purposes this is entirely adequate.

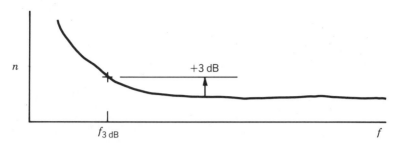

Figure 5.7. 3-dB frequency can be defined even if $n(f)$ does not obey simple equation.

Corner Frequency. Another way to characterize low-frequency noise with one number is to provide the *corner frequency* f_c. To visualize this definition (and to determine f_c graphically), plot the noise spectral density on log paper and draw two straight-line approximations: one for low frequencies and a horizontal one for the high frequency (white noise), as in Figure 5.8. The corner frequency is the frequency at which the straight-line approximations meet. This "works" if the noise can be approximated by straight lines on log paper, which is equivalent to saying that the noise obeys a power law: The square of the noise voltage varies as $1/f$ to some exponent a. (a is 1 for true $1/f$ noise.) If this is not true (see Figure 5.7, for example), the corner frequency is ambiguous and not a very useful figure of merit.

If the corner frequency can be defined, and if the noise varies smoothly, the corner frequency and the 3-dB frequency are equal: At the corner frequency the $1/f$ and white noise are equal, so the total noise spectral

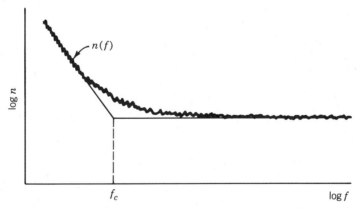

Figure 5.8. Corner frequency can be defined only if log n' versus log f is linear at high and low frequencies.

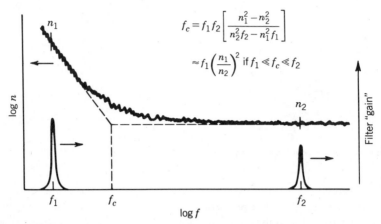

Figure 5.9. Noise measured in two narrow bandpasses allows determination of corner frequency.

density equals the square root of two times the white noise spectral density; this is 3 dB (within acceptable error).

If the foregoing conditions are met, the total noise spectral density is given by

$$n^2 = n_{\text{white}}^2 \left[\left(\frac{f_c}{f} \right)^a + 1 \right] \tag{5.13}$$

where a is the exponent mentioned earlier.

Corner Frequency from Two Data Points. If the noise obeys a known power law (including, but not limited to $1/f$) the corner frequency can be determined without measuring the complete noise spectral density. This saves test time. We need only measure the noise spectral density at two frequencies, or the integrated noise in two narrow bandwidths (see Figure 5.9).

The equations are most easily derived if we know the noise spectral densities at two discrete frequencies (see Equation 5.14a and 5.14b).

$1/f$ CORNER FREQUENCY FROM TWO NOISE SPECTRAL DENSITIES

$$f_c = f_1 f_2 \frac{n_1^2 - n_2^2}{n_2^2 f_2 - n_1^2 f_1} \tag{5.14a}$$

$$f_c \simeq f_1 \left(\frac{n_1}{n_2}\right)^2 \qquad \text{if } f_1 \ll f_c \text{ and } f_2 \gg f_c \qquad (5.14b)$$

where f_1 and f_2 are the center frequencies of the narrowbands, and n_1 and n_2 are the corresponding noise spectral densities.

It is not essential to measure the noise spectral density using narrow-bandpass filters. The bandpasses of Figure 5.10 could be used, for example. In that case the formulas are more complex and depend on the shape of the electrical filter characteristics. Once again, all these methods will yield different values unless the actual noise obeys our "ideal" formula.

Considerations in Selecting the Bandpasses for 1/f Determination

1. Integration times for any desired accuracy are less if wide bandwidths are used.
2. Equations are simpler, derivations are easier, and accuracy is better if:
   a. One measurement is limited to frequencies much less than f_c, and one much greater than f_c
   b. A narrow bandpass is used for the low-frequency filter.
3. If a narrow frequency bandpass is used, noise at discrete frequencies (60 Hz, for example) can contaminate the results easily. The

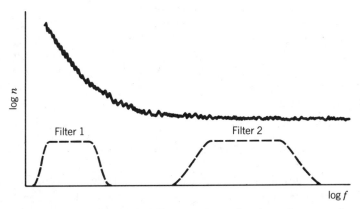

Figure 5.10. Wide bandpass electrical filters can be used to determine corner frequency, but the analysis is more complicated.

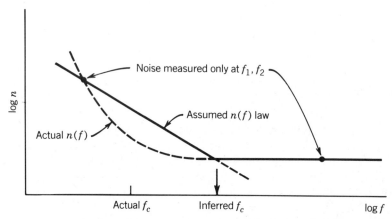

Figure 5.11. Error in determining corner frequency results if the $n(f)$ dependence is not as assumed.

wider the bandpass, the less effect a discrete frequency problem will have.

Figure 5.10 shows a filter combination that works well. The low-frequency filter has a bandpass that is about 25% of the expected corner frequency and is centered well below the expected corner frequency. The high-frequency filter has a broad bandpass but is well above the expected $1/f$ corner frequency and well below the the system cutoff frequency.

The closer the spectral density is to the assumed equation, the better different methods will agree. Generally, detectors of the same type will have similar low-frequency characteristics, and any one method will allow a useful comparison of the goodness of different detectors.

In general, however, we will agree to determine an "effective" corner frequency by applying one of these algorithms to our data, *without questioning the actual shape of the noise spectral density*. The resulting ambiguity or loss of accuracy is the price we pay for attempting to describe a spectral phenomena with just one number. See Figure 5.11 for an example of such an error.

5.4. BLACKBODY SIGNAL AND RESPONSIVITY

The most common measurement of detector performance is the blackbody signal and responsivity. The detector is placed so that it can "see" a blackbody, and the resulting signal is observed. The signal can be either the dc voltage (or current) from the bias circuit, or the ac signal that results when the blackbody source is modulated, generally with a chopper. The ac system is generally chosen because ac signals are more stable and less sensitive to thermal drifts than are dc voltages.

5.4.1. Blackbody Test Set

A typical blackbody test set includes:

Source assembly
    The blackbody itself
    Blackbody temperature controller
    Modulator and controller
    Optical filters
Dewar mounting surface
Electronics
    Bias circuit, controls, and readouts
    Preamplifier
    Electrical filter
    Postamplifier
    Readout and recorders

The design of a blackbody test set and potential problems with such a station are described in Chapter 4.

5.4.2. Blackbody Responsivity

Given a test setup with a known blackbody temperature and aperture diameter, known distance from the blackbody aperture to detector, and characterized intervening optics, the blackbody incidance at the detector plane can be calculated. The necessary calculations are described in Chapter 3.

For convenience in testing, we define and calculate a "blackbody responsivity"—just the signal divided by the radiometric input to the detector. Because the spectral content of the incidance depends on the blackbody temperature, *the blackbody responsivity will depend on the blackbody temperature*. A more fundamental figure of merit—one that does not depend on blackbody temperature—is the responsivity at a specific wavelength, but to obtain this value we need to know the spectral response of the detector. The necessary calculations are described in Section 5.9.3.

Before a great deal of time is spent collecting data the operator should be aware of the potential problems with blackbody setups; these were discussed in Chapter 4. Methods of building one's confidence in the data are described in Section 5.5.3; these also should be considered. Now consider the problem of determining the best bias voltage for a photo-conductive detector.

5.4.3. Optimization of Bias Voltage

The optimum bias is determined by measuring signal and noise at a number of bias values, tabulating or plotting the signal-to-noise ratio as a function of bias, and selecting the bias or range of biases that yields the best performance for a given application. Typical blackbody signal and noise data are shown in Table 5.3. The detector is a photoconductor, and the signal increases more or less linearly with bias voltage until either saturation or "breakdown" occurs.

The detector noise is also quite linear with bias voltage, until at some point it increases dramatically due to breakdown of some sort. At very

Table 5.3   Sample Blackbody Responsivity and D^* Data

V bias (V)	I (μA)	V det (V)	Noise (mV)	Signal (mV)	S/N	Resp	D^*	
0	0.0		0.05	0	0			
10	2.5		0.18	280	1556			
20	4.6		0.27	520	1926			
30	7.6		0.35	840	2400			
40	9.5	39.5	0.41	1100	2683		9.4E+10	
50	11.5		0.52	1350	2596			
60	14.0		0.95	1650	1737			
70	16.5		2.20	1950	886			
80	19.0		3.00	2200	733			
90	21.0		4.50	2450	544			

Remarks: Top contact biased negative (noisier when positive)
Optimum field is about 200 V/cm (39+ V, 0.2 cm)

Material: GeHg   Blackbody Temp: 500 K   Date: 1/28/1970
size: 2 mm x 2 mm   BB Incidance :   0.655 μW/cm^2 FOV: 11 deg
ingot: 94-T   Window: KRS-5
operating temperature: 6 K   Spectral Filter: none
C(BB-Pk):   2.5

load resistor: 100 K   chopping frequency: 1800 Hz
amplifier gain: 600   bandpass: 5.6 Hz

low bias voltages the total noise is limited by Johnson noise or test set noise. Where both signal and total noise are proportional to bias voltage, their ratio is independent of bias. At very low biases, and at very high biases the signal-to-noise ratio will be worse.

5.5. ACCURACY OF BLACKBODY RESPONSIVITY DATA

To assist in identifying the potential errors in your responsivity measurements, try to list potential blunders, systematic, and random errors associated with the three components of the responsivity calculation: signal, incidance, and detector area. Such a list will be unique to your situation, but a typical set of variables follows.

Signal
    Blunders: meter reading
    Systematic errors:
        Gain calibration
        Meter calibration
    Random: effect of sample size
Incidance
    Blunders: calculations
    Systematic errors:
        Reflections
        Vignetting
        Aperture diameter
        Blackbody–detector distance
        Blackbody temperatures
        Emissivity
        Window transmitance
        Modulation factor
    Random: (none)
Detector area
    Blunders: dimension measurement
    Systematic:
        Calibration of toolmaker's microscope
        Photoetching undercut
        Nonuniform response
    Random: variation from element to element

In this section we discuss the effect of sample size on signal accuracy, and ways to increase confidence that the systematic errors have been minimized and the blunders have been identified and eliminated.

5.5.1.   Effect of Sample Size on Signal Uncertainty

For a given S/N ratio, what is the probable relative error in signal as determined from the average of J samples? Stated another way, how many samples must we average to obtain a given probable relative error in signal?

The signal is the mean of a set of samples, so the one sigma uncertainty δS in the signal is the standard deviation of the mean:

$$\delta S = \sigma_m = \frac{\sigma}{\sqrt{J}} \tag{5.15}$$

Here σ is the standard deviation of the population, which we normally refer to as the noise. Thus the uncertainty in the signal is the noise divided by the square root of the number of samples taken.

The *relative* uncertainty in the signal is δS divided by the signal itself, as in Equation 5.16.

RELATIVE UNCERTAINTY IN SIGNAL DUE TO SAMPLE SIZE

$$\frac{\delta S}{S} = \frac{N}{S\sqrt{J}} \tag{5.16}$$

where S = signal
          N = noise
          J = number of independent samples

This is illustrated in Figure 5.12.

Example 1: For a signal-to-noise ratio of 10, with 100 samples, the relative uncertainty (1σ) of the signal is 1%.   ∎

Example 2: For a signal-to-noise ratio of 5, with four samples, the relative uncertainty of the signal is 10%. To reduce that to 1% would require 400 samples.   ∎

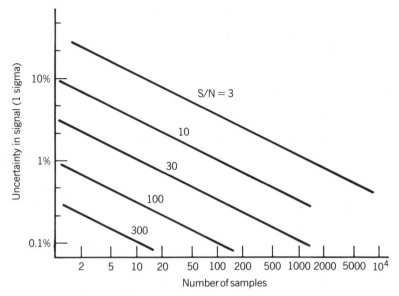

Figure 5.12. Relative uncertainty in signal due to limited sample size.

All of the above is based on the assumptions that we have normal or gaussian statistics, and that the samples are independent, as discussed earlier (see "Effect of Sample Size on Error in Noise Measurements" in Section 5.3.2).

5.5.2. Effective Area

The calculation of quantum efficiency is complicated if the response of the detector varies significantly across its area. In that case we can compute the quantum efficiency at a selected point on the detector, or compute some sort of average. In either case the responsivity versus position must be measured (using a spot scanner) or assumed. The equation for the *effective area* A_{eff} (referred to the point x_0, y_0) is given by Jones et al. (1960) and Eisenmann et al. (1975):

$$A_{\text{eff}} = \int_0^\infty \int_0^\infty \frac{\mathcal{R}(x, y)}{\mathcal{R}(x_0, y_0)} \, dx \, dy \qquad (5.17)$$

Use of the effective area in the normal formulas will result in the quantum efficiency *at the point* x_0, y_0. Error in the effective area is undoubtedly the largest single source of error in determining absolute responsivity and quantum efficiency.

5.5.3. Confidence-Building Techniques

There are many ways in which errors can creep into blackbody data. Checklist 5.2 provides a repertoire of ways to check for such errors. These are techniques used primarily while acquiring and reducing data; Checklist 4.2 in Chapter 4 is a similar list for evaluation and troubleshooting of test sets. An attempt was made to provide two independent lists, both complete for the intended purpose; however, the nature of the problems overlap so much that one should really review both lists if a problem is not quickly resolved.

The extent to which one invokes those techniques (i.e., the number of checks made and the care with which they are made) must be a judgment based on the need for accuracy, the confidence one already has in the data, and the time and personnel available. These can be included in written procedures, but they should also be taught and used automatically, even when doing tests without written procedures. Many of the suggested checks require only visual observation and assessment; with experience these can be accomplished in just a few seconds.

Checklist 5.2: Confidence-Building Techniques

Data acquisition
Check signal variation with:
Position on test set (should be as shown in Figure 5.13a)
Frequency (corner frequency can be predicted)
Aperture diameter (signal should vary roughly as d^2)
Feedback resistance (PV signal should vary as R_f)
Bias voltage (PC signal should be proportional to V_b unless signal saturates)
Is the signal repeatable as these variables are reset?
Double check the amplifier gain settings and meter scales.
Blackbody setup (see also Checklist 4.2)
Visually check the apertures and optical path:
Is the correct aperture in place and centered?
Is there dust in apertures?
Are the folding mirrors dirty?
Is there an obstruction in the optical path?

Are there supports or baffles that might reflect?

Is the dewar window clean and smooth?

Is the aperture or chopper warmer than you assumed?

Verify the distance from the detector to the blackbody aperture.

Verify the blackbody temperature.

Ask some questions:

Has the test set been modified or disassembled recently?

Are other users having any problems?

Have other users verified that the test set is working well?

Assumptions about the dewar

Window and spectral filters:

Check the assumed transmittance versus wavelength, including out-of-band leakage.

Could scratches or uncoated areas allow leakage?

Was transmittance measured (or correctly extrapolated to) the operating temperature and incidence angle?

Are the surfaces around the detector and optical path blackened or baffled to prevent unwanted reflections?

Is the detector at the assumed temperature?

Could icing, condensation, or other contamination on the windows, filters, or detector be reducing the apparent responsivity?

Calculations

For quantum efficiency calculations: How reliable is the assumed effective area? (The definition and accurate determination of quantum efficiency requires an effective area determination unless the responsive area is very uniform and well defined.)

Do the calculations provide blackbody or peak values? Is the conversion factor being calculated properly? Does the report clearly indicate which is being reported?

5.5.4.   Hints for Data Reduction

Application of a few data reduction methods can improve the accuracy and credibility of your results; these methods should be applied consistently in your work.

Timely Data Reduction. A good guideline for *any* experimental data is to check them for "sanity" (reasonableness) and to perform the complete

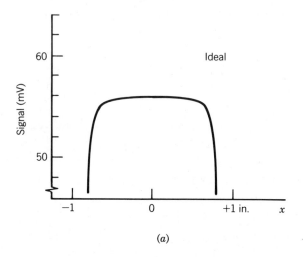

(a)

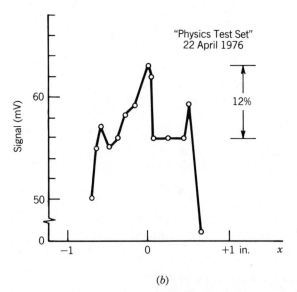

(b)

Figure 5.13. Signal versus position profiles: (a) expected behavior; (b) indicates unwanted reflections.

reduction and analysis as quickly as possible—certainly before the test setup is dissassembled. Postponing data reduction and review has several potential problems:

- We sometimes have deadlines—meetings, presentations, or report due dates—that must be met. If you find an error just before the deadline, it may be difficult to recover and correct errors in time.
- A good deal of time and effort can be wasted taking "garbage" data if problems are not identified quickly.
- It is much easier to find errors or problems while you are still at the test set. Your operating procedure, instruments, and raw data are immediately available. A week later that may not be the case.
- You will be more sensitive to the data required if you reduce them on the spot. The kinds of data, the amount, the accuracy—all these requirements are difficult to quantify, describe, or specify. But if *you* do the data reduction, analysis, and presentation, you will have a pretty good "feel" for the data needed. Reducing the data later means you throw away that advantage and sensitivity until it is too late to be useful.

The fastest method of accomplishing this sanity check is to compute the expected signals and noises *before starting the test*. This can be done days ahead of time, while preparing the test procedure, for example. This not only provides an immediate check for good data at the test set, but it also tells you, in the planning stages, if the test will work.

Look for Trends, Correlations, and Relationships. Any technique that includes and correlates several data points is generally preferable to one that relies on one data point. The most obvious reason for this is the averaging out of random errors. Another is that blunders are more readily identified as such when they are shown next to similar data. Still another reason is the ability to show or test for correct functional relationships. Examples of this will be discussed shortly: $1/r^2$ dependence, and linearity of signal with incidance.

Linearity Data. The apparent responsivity of a detector is the most common parameter used to assess the performance of a blackbody setup. This is often based on the detector signal at one incidance value. One can gain more information, and with better confidence, by plotting signal versus blackbody incidance, as shown in Figure 5.14.

The fit of the points to a smooth curve is a good test of the blackbody

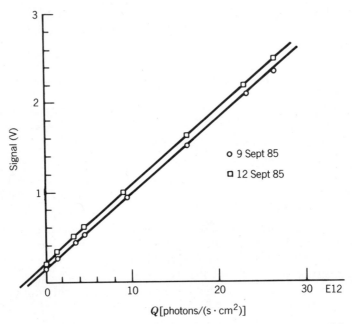

Figure 5.14. Plot of signal versus blackbody incidance is a good check on validity of data and allows an estimate of the background.

apertures used. This does not require that the detector used be linear over the incidance levels used, merely that it be well behaved. Responsivity of the detector can be inferred more accurately from a smooth fit to the data points than from just one or two points, since the errors are averaged out.

If the blackbody incidance is modulated and rms signals are plotted versus rms incidance, a line through the signal versus incidance points should pass through the origin. Failure to do so indicates the presence of an offset of some sort.

If dc signals are plotted against unmodulated blackbody incidance, the intercept on the signal axis is the background signal. The background incidance is the background signal divided by the responsivity. This calculation can be done by direct division, or graphically (by extending the curve down to the incidance axis (see Figure 5.14). Because the Planck's law prediction of the background is based on ideal conditions that are unlikely to occur—perfect shielding and zero out-of-band filter transmittance—an experimental determination of the background as described here should be done whenever possible.

Mapping the Spatial Incidance Profile. If the detector is moved about over the blackbody, by sliding the dewar around on the test set table, for example, the signal will vary. The extent and nature of those variations are a check on the extent to which the setup is "well behaved." Such a check should always be made when working with a new or unfamiliar blackbody or dewar.

The experiment can be done formally by laying out a coordinate system on the table, then recording and plotting signal versus position. Figure 5.13 is representative of actual observed profiles. For a well-behaved system the response should peak smoothly and gradually over a mounting area that is fairly large. If localized and/or asymmetric peaks are observed, reflections are probably present and the Planck's law calculation of incidance will not accurately predict the actual incidance, and responsivities calculated from the Planck's law values are in error.

For a test setup that has been checked out, it is still worthwhile occasionally to do a simple check by moving the dewar around while observing the output signal on a meter or oscilloscope. With a little practice this method will identify major test set problems. The common procedure of adjusting the dewar to maximize the signal is justified only if the incidance profile is well behaved.

$1/r^2$ Dependence. If the detector-to-blackbody distance is much greater than the aperture diameter or detector size, the blackbody incidance should vary as $1/r^2$. By taking data at several distances between the blackbody and detector, one can test to see if the incidance falls off properly with distance. For a linear detector, failure to follow the expected $1/r^2$ law is an indication of the presence of reflections, or of an error in the distance measurements. Whatever the reason for the problem, the incidance calculated from Planck's law cannot be believed if the signal does not follow the $1/r^2$ law.

The data can be presented and compared with theory in several ways:

1. Tabulate distance, signal, and distance squared times signal. This last value should be the same for all the distances.
2. Plot the log of the signal versus log of distance; the resulting points should fall on a straight line with a slope of -2.
3. Plot signal versus distance on Cartesian graph paper, and plot on the same sheet a curve of the form A/r^2, picking A so that the curve goes through one of the observed data points. If the signal varies correctly as $1/r^2$, a curve that goes through one point will go through all the others.

5.6. ANALYSIS OF DETECTOR CURRENT, SIGNAL, AND NOISE DATA

In many situations one will be faced with test data that do not indicate satisfactory performance, or do not agree with expected values. Someone must then try to determine what is wrong. The usual question is: Is the signal too low, or is the noise too high? To complicate matters, the background is generally not known accurately enough to allow the necessary determinations of detector performance. (Calculations of the background using Planck's law and the geometry of the test station neglect internal reflections and are often in error by as much as a factor of 3.)

For PV detectors the answer is fairly easy. The quantum efficiency and responsivity are quite predictable, and the background can be determined from the dc current. Given the quantum efficiency and the background, one can estimate what the noise should be. The signal current can also be predicted with fair accuracy.

A good way to implement those checks of PV data is to plot i_N^2 versus I. For very low currents the noise will probably be dominated by test set noise, and the plot will show that limiting noise level, or *noise floor*. At higher currents the graph should be linear, showing the shot noise limit; the slope should be $2e\Delta f$:

$$i_N^2 = (2e\Delta f)I \qquad\qquad (5.18)$$

This method checks for internal consistancy, agreement with theory, and for excess noise. Considerable confidence can be had in data if the suggested plot behaves in the expected way.

For PC detectors the PC gain complicates matters. Unless you know the material properties very well it is difficult to estimate the PC gain within a factor of 3, and that is not adequate for most assessments of signal and noise. If you have access to reliable data on "good" detectors from the same crystal material and fabrication process, and if you know the background, you can use the equations from Chapter 2 to predict what your detector signal and noise should be.

One method that has proven convenient is indicated in Equation 5.19. We treat the quantum efficiency, the PC gain, and the background as three unknowns and solve the equations for the current, the signal, and the noise simultaneously to determine the three unknowns. Although this model has some shortcomings, we can often derive useful information from it.

BACKGROUND, QUANTUM EFFICIENCY, AND PC GAIN FROM PC BLACKBODY DATA

$$Q_{\text{BK}} = \frac{I}{i_s} Q_s \qquad (5.19a)$$

$$\eta G = \frac{i_s}{Q_s A_d e} \qquad (5.19b)$$

$$G = \frac{i_N^2}{4Ie} \qquad (5.19c)$$

$$\eta = \frac{i_s}{Q_s A_d e} \times \frac{4Ie}{i_N^2} \qquad (5.19d)$$

where

$$Q_s = E_s \frac{\lambda_{\text{PK}}}{hc} = \text{effective signal photon incidance} \qquad (5.20a)$$

$$i_s = \frac{v_s C_{BB-PK}}{R_{\text{parallel}}} = \text{short-circuit signal current} \qquad (5.20b)$$

$$i = \frac{v_N}{R_{\text{parallel}}\sqrt{\Delta f}} = \text{short-circuit noise current} \qquad (5.20c)$$

The analysis above assumes the ideal behavior:

$$I = \eta Q A_d e G \qquad (5.21a)$$

$$i_s = \eta Q_s A_d e G \qquad (5.21b)$$

$$i_N^2 = 4\eta Q A_d e G \, \Delta f \qquad (5.21c)$$

If one or more of the parameters i_N, i_s, or I do not obey the assumed expressions, the quantities Q, G, and η derived from them will be inaccurate or misleading. Noise in particular tends to be higher than that predicted by the assumed expression, so G and η derived this way are suspect. To remind ourselves about this possibility, it would be wise to refer to quantities derived this way as the *apparent* background, or quantum efficiency, or gain.

The resulting values can be compared against some expected or reasonable values:

- The background should be between one and three times the value calculated from Planck's law and the usual perfect shielding, perfect spectral filtering assumptions.
- The quantum efficiency can be estimated to be $1 - R$, where R is the reflectance calculated from the index of refraction (for an uncoated detector), or from reflectance measurements for an anti-reflection coated detector.
- The photoconductive gain can be calculated from the electric field used and the mobility lifetime product. This latter quantity comes best from a similar analysis on other detectors—the more, the better.

If the calculated values are close to the expected values, all is well. If not, the results give a clue to the problem. The three equations assumed ideal BLIP noise. If noise is excessive, the model will yield PC gain values that are too large, and the calculated quantum efficiency will be too low. This does not mean that the PC gain is really large, or that the quantum efficiency is low; it is just a symptom of using our simple model on a noisy detector. In any case, the low apparent quantum efficiency and high apparent PC gain tell us that the noise is too high.

As an example of this method, the 40-V data of Table 5.3 were reduced using the foregoing formulas. The resulting values were:

$$Q = 1.13 \times 10^{16} \text{ photons/(cm}^2\text{·s)}$$

$$\eta G = 0.132$$

$$G = 1.2$$

$$\eta = 0.11 \ (11\%)$$

$$\mu\tau = \frac{Gs}{E} = 1.2 \times 10^{-3} \text{ cm}^2/\text{V}$$

The apparent background is about 20% above the Planck's law prediction for 12-μm cutoff detector viewing a 300-K background through an 11° field of view. The apparent quantum efficiency is lower (by almost a factor of 2) than that predicted by the absorption, length, and reflectance values. Excess noise could well be responsible for some of this difference. The $\mu\tau$ product is consistent with other detectors of this type, although the overall spread from detector to detector is on the order of a factor of 2.

5.7. FREQUENCY RESPONSE

The frequency response of a detector (or detector and electronic circuit) can be determined and reported either in the form of continuous plots of response versus frequency, or condensed to the single frequency at which the response drops off by 3 decibels or some other predetermined factor. The continuous plots provide more information, but data acquisition is slower and reporting takes more space.

5.7.1. Response versus Frequency

To obtain a plot of response versus frequency, we measure the signal from the detector or circuit as the input radiation is modulated at different frequencies. This requires a chopper whose frequency can be varied from the lowest to the highest frequencies of interest, and an amplifier whose gain is independent of frequency (or whose frequency dependence is known with adequate accuracy, so that that final data can be corrected for amplifier roll-off).

The modulation is generally done in one of two ways. A pure sinusoid is generated using a special chopper. (To verify that the sinusoid is pure, observe the resulting signal with a spectrum analyzer.) The output voltage can then be measured using a simple ac meter, without any special electrical filtering. (This assumes that the meter is calibrated over all the frequencies of interest; this is generally a good assumption.) Instead of an rms meter, it is easier to use a spectrum analyzer for all the signal measurements, measuring only the signal associated with the fundamental chopping frequency. This allows the use of a chopper that generates an arbitrary wave shape. That wave shape contains (can be thought of as being composed of) sinusoids at many frequencies. The spectrum analyzer measures only the desired component. Since the fraction of the radiated power at the fundamental frequency and harmonics does not change as the frequency of the chopper is increased, the output of the detector is proportional to the responsivity at each frequency. If the frequencies are too high for available choppers, one can sometimes gain at least a factor of 3 in frequency by measuring and plotting the signal at a harmonic as the chopping frequency is varied.

Another technique for obtaining continuous data is to measure the noise spectral density over the frequency range of interest. At frequencies above the $1/f$ noise region, the noise and the signal often have the same frequency dependence, so the measurement of the noise spectral density provides the desired signal response. This method eliminates the need for a high-speed chopper, but one must still compensate for any amplifier roll-off. In any case, it is usual to normalize this curve by dividing all values by the largest value.

5.7.2. Time Constant and Corner Frequency

For many detectors and circuits, the frequency response obeys an equation that can be written in terms of the time constant τ or the corner frequency f_c:

$$S(f) = \frac{S(f = 0)}{[1 + (2\pi f\tau)^2]^{1/2}} \qquad (5.22a)$$

$$S(f) = \frac{S(f = 0)}{[1 + (f/f_c)^2]^{1/2}} \qquad (5.22b)$$

To characterize the frequency response, plot the signal (or noise) versus the frequency on logarithmic paper, and draw two lines through the data points: a horizontal line through the lowest frequency data, and a line with slope -1 through the highest-frequency roll-off data. The corner frequency is the frequency at which two straight lines intersect (see Figure 5.15). The time constant is given by

$$\tau = \frac{1}{2\pi f_c} \qquad (5.23)$$

The corner frequency can also be determined from two data points, one below and one above the corner frequency. If the two frequencies chosen are f_1 and f_2, and the corresponding signals (or noise spectral densities) are s_1 and s_2, then

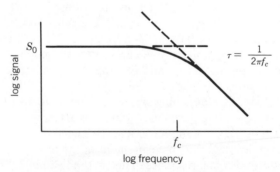

Figure 5.15. Determination of time constant from signal versus frequency data.

SIGNAL ROLLOFF CORNER FREQUENCY

$$f_c = \sqrt{\frac{s_2^2 f_2^2 - s_1^2 f_1^2}{s_1^2 - s_2^2}} \qquad (5.24a)$$

$$f_c \simeq \frac{f_2}{\sqrt{(s_1/s_2)^2 - 1}} \qquad \text{if } f_1 \ll f_c \ll f_2 \qquad (5.24b)$$

The first is an exact equation, but its use may result in unnacceptable errors due to measurement and computational errors if f_1 or f_2 are close to f_c. One can compare the signal at the fundamental frequency and at the harmonics provided that the appropriate modulation factors are included in the data reduction.

Still another method is to apply an incidance step function to the detector, an incidance that increases to a maximum, constant value in a time small compared to the expected response time of the detector. If the detector response can be characterized by the formula above [equation (5.22)], the response to a fast stepped source will be of the form

$$S = S_{\max} \left[1 - \exp\left(\frac{-t}{\tau}\right) \right] \qquad (5.25)$$

and the response time τ can be determined from an oscilloscope trace of signal versus time, as illustrated in Figure 5.16.

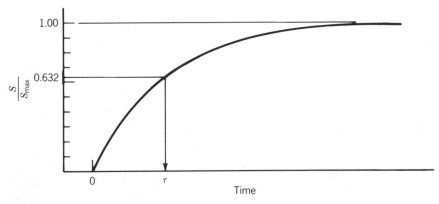

Figure 5.16. Determination of time constant from signal response to step function incidance.

One problem with this method is that there are few fast sources available at the wavelengths of interest. Thus one is forced to make measurements under one set of conditions (wavelength and background), and extrapolate to different conditions. Sometimes one can make a theoretical or experimental argument that the frequency response is independent of the wavelength, but this may not be acceptable to your customer. This skepticism is justified by the number of times that the statement "It doesn't make a difference" has turned out to be wrong.

If the detector frequency response does *not* obey the first equation above, determination of τ or f_c using the methods described will yield inconsistent results, and other definitions are required. (This problem is very much like that of characterizing low-frequency noise.)

5.8. LOW-BACKGROUND TESTING*

As indicated earlier, the performance of most IR detectors is strongly influenced by the background incidance. Space-based detectors require tests at low backgrounds to simulate their end-use environment. Background photon flux levels of 10^8 photons/(cm^2·s) are not uncommon. This is eight orders of magnitude below an unfiltered, unshielded room-temperature background.

The use of extrinsic photoconductors in low-background environments requires the development and application of special low-background test conditions and techniques. Special care is required in preparing dewars and focal planes to avoid shorting these high impedances. Finally, the high impedances impose special electronic requirements.

Background is by definition the incidance level that remains when the source (usually blackbody) incidance is set to zero. In low-background testing, we may use a blackbody to cause an incidance level that simulates the background at which the detector will operate in another application. In this case we would use the apparently contradictory expression "the background due to the blackbody."

The temperature requirements for detectors were discussed in Chapter 2. As the background is reduced, lower operating temperatures are required. Most extrinsic Si and Ge detectors are operated at temperatures that lie between 2 and 50 K.

* Much of the material on low-background testing was taken from a Santa Barbara Research Center document *A Practical Guide to Low Background Testing of Extrinsic Photoconductor Detectors*. That document was written by Bob Kvaas, with input from Richard Nielsen, Jim Fulton, Michael Slonaker, Eric Woodbury, and David Vincent. Used by permission of SBRC.

5.8.1. Background Reduction Methods

Methods used to obtain and calculate low-background incidances are extensions and refinements of the methods discussed in Chapter 3. The angular extent can be limited by cooled apertures, and optical filters can be used to limit the spectral content of the viewed scene. Total attenuations below 10^{-10} are possible if all of these techniques are used together to reduce the background.

For low-background testing one often uses a longer cold shield and smaller cooled aperture than is used for intermediate and high background work. This aperture serves both as a field of view defining and as the blackbody limiting aperture. A 0.010 inch diameter aperture 5 inches from the detector reduces the background by 6 orders of magnitude. Pinholes cannot be used if wide fields of view are required. This might be necessary, for example, if one needed to perform spot scans of arrays or large-area detectors. In such a case one would use spectral filtering and/ or neutral density filters to reduce the background further.

Neutral density (ND) filters are made of a window material coated with an evaporated metal film. Their low transmission reduces the blackbody and background radiation. Germanium is often used as a substrate since it has a relatively constant transmission over a broad spectral range. The germanium window substrate also has the advantage of blocking visible radiation. Nichrome is commonly chosen for the coating metal. Attenuation by a factor of 100 is possible with a ND filter.

If two ND filters are used together, and if interference effects due to reflections can be prevented (by mounting them at a slight angle to each other for example), the transmission of two filters in series is just the product of their individual transmissions. Be cautious about such an approach, however; calibrate the resulting setup whenever possible. Spectral filtering might utilize narrow ("spike") bandpass (BP) filters or short-wavelength filters that transmit only to some wavelength shorter than the usual detector cutoff. These filters can be multilayer coated filters obtained "off the shelf" or built to order. Simple substrates with desirable cutoff properties can also be used.

Corning Blue Glass (Corning 4-76 about 5 mm thick) at cryogenic temperatures has a short wavelength passband which cuts off around 2.7 μm and a transmission of about 10^{-5} beyond 2.7 μm. Thus it passes only short-wavelength infrared. Use of this glass is a common way of obtaining a total background reduction of about 10^{-6}. Sapphire also works as a useful filter.

It is necessary to characterize carefully the out-of-band transmission of the filter. To see how serious such transmission can be, consider a spike filter at 3.5 μm with a bandpass of 0.2 μm and 0.1% out-of-band leakage. The in-band exitance is 2.7×10^{14} photons/(cm²·s). If used with a Ge:Hg detector (10-μm cutoff wavelength), the out-of-band exitance is 4.2×10^{14} photons/(cm²·s). The observed background would be 2.5 times higher than the predicted background, just because of a 0.1% out-of-band leakage!

Since the transmission characteristics of any optical filter can vary strongly with temperature, it is imperative that they be tested and calibrated at their operating temperature. Otherwise, unpredictable changes in the transmission characteristics can go unnoticed. The article "Problems in using cold spectral filters with LWIR detectors" (Stierwalt and Eisenman, 1978) should be reviewed by anyone doing low-background testing.

Use of spectral filters to reduce the background also reduces the optical signal levels. Because the long-wavelength portion of the Planck energy distribution is not a strong function of temperature, raising the temperature of the blackbody does not improve the S/N ratio very much unless only short wavelengths are involved. If, however, a short-wavelength spike filter is incorporated, the background can be reduced substantially without decreasing the signal incidance very much. This increases the optical signal-to-noise ratio. Consequently, testing with short-wavelength narrowband filters provides better signal-to-background ratios than if long wavelengths are used. An associated penalty is the need to extrapolate from the testing wavelengths to those at which the system will be used.

5.8.2. Background Leaks

Because the desired photon flux levels are so low in low-background testing, it is important to maintain rigorous shielding in test dewars to prevent light leaks and unintentional sources of infrared energy from reaching the detectors. Connectors and cables have a high thermal impedance, so they are often relatively warm. Thus they can be an unacceptable photon source unless they are baffled from the detector or carefully heat sunk. Electronic components and any dewar heaters must be mounted where waste heat cannot be "seen" by the detectors.

The discussion of Section 4.5.3 is particularly important to low-background work—reflections inside the dewar can increase the background two to five times the value predicted by line-of-sight Planck's law calculations. Baffles like those in Figure 4.9b and 4.10 must be used.

5.8.3. Low-Signal Problems

To achieve a low-background test setup, we must often restrict the spectral bandpass or introduce ND filters. This results in low signal irradiances. Even when the optical signal is adequate to provide a good optical signal-to-background ratio, it is not uncommon to have both of them so low that electronic noise makes detection impossible. For small detectors operating at low-background conditions, it is common to have a signal-to-noise ratio of only 10:1.

The most accurate and convenient way of measuring the low signal levels is to use a spectrum analyzer. These instruments typically can reduce the bandwidth to 1 Hz, and have a sensitivity of 1 μV/Hz.

Spurious signals like those due to 60-Hz pickup in amplifiers, heterodyne effects in the electronics, or microphonics coming from vibrations of wiring or structural members in the dewar can often be detected and identified by their frequency spectrum. Once identified, they can be eliminated.

Radiometeric Troubleshooting. Measured signals that are lower than theoretical predictions can result from foreign matter in pinholes or on detector array elements, shorted detector or feedback resistors, or feedback resistors whose value changes unexpectedly due to cryogenic cooling. Higher-than-expected optical backgrounds can occur due to leaks in spectral filters, due to either spectral leakage or scratches. Leakage may be present at room temperature or may appear only when the filter is cold.

It is also possible for filters to leak around their mountings if they are not supported properly. In fact, all the shielding must be checked for tiny holes or gaps that can leak light and raise backgrounds. This is a particular problem for vent holes in the shielding caps of the dewar, which are designed to allow the air to be removed during the evacuation of the dewar. They must allow air to be pumped out without allowing IR to leak in.

To determine the background experimentally, divide the "shutter-closed dc current" (zero blackbody incidance) by the responsivity. A graphical equivalent is to plot dc signal versus blackbody incidance, as shown in Figure 5.14. The slope is the responsivity, the signal intercept is the background caused signal, and the intercept on the incidance axis is the negative of the background. These methods rely on the correctness of the blackbody incidance, so errors in blackbody temperature, filter transmission, bandwidths, and so on, will cause an error in the inferred background. On the other hand, if the blackbody problems have all been resolved, this is a reliable way to determine the background.

5.8.4. High-Impedance Electronics

The impedance of a low background detector might be as high as 10^{12} Ω. Similar resistance values are used in the "front-end" electronics that convert the change in detector conductivity into an electrical signal. These high impedances are used to maximize the output signal and to minimize Johnson noise. Such extreme impedance values require extremely clean circuit components. Solder flux residues, oils, and fingerprints act like short circuits. In such a high-impedance area of the electronics, even 10^9 Ω is effectively a short circuit. Extra care is required to avoid such sources of contamination.

The feedback resistors are cooled to low temperatures to minimize Johnson noise. A source-follower FET is often incorporated into the cryogenic part of the circuit to lower the signal impedance as soon as possible so that the high-resistance precautions can be largely confined to the detector, feedback resistor, and source-follower leads. (This is discussed in more detail in Chapter 10.) Once the signal impedance has been reduced with the source follower, it can be handled in a more conventional manner by feeding it into a transimpedance amplifier (TIA) or capacitive transimpedance amplifier (CTIA) for further amplification and signal processing.

New components for a cooled circuit should be tested at the anticipated operating temperature. Although many electronics components designed for room temperature can operate satisfactorily at cryogenic temperatures, some of them display anomalies in their characteristics. Feedback resistors are often very temperature-sensitive at low temperatures. Changes in resistance as a function of current are also common.

Electronic Troubleshooting. FETs can be checked at room temperature before the dewar is cooled by applying bias to the FET when the gate is grounded. A source voltage of 2 to 3 V should be obtained. The addition of a bias to the gate through the low-resistance, room-temperature detector (without detector bias applied) should give ≈80% gain at the source of the FET.

5.8.5. Frequency Response

The frequency response of the signal depends on the type of readout circuit employed. The extremely high impedance values of the detector and feedback resistor tend to generate large RC time constants and poor high-frequency response. Figure 5.17 shows a schematic of the detector, feedback resistor, source follower FET, and TIA. The components within the dashed box are near the focal plane, so that they are at cryogenic temperatures. When a TIA is used (as shown) the frequency response is determined by the feedback resistor (R_{FB}) and the capacitance across the feedback resistor (C_s). In the TIA mode, the upper corner frequency is

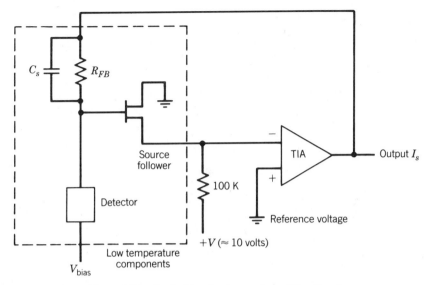

Figure 5.17. Typical low-background amplifier circuit.

$f = 1/(2\pi R_{\mathrm{FB}} C_s)$. If a TIA is not used, the frequency response is determined by the feedback resistor (R_{FB}) and the total capacitance at the input of the source follower, including any stray capacitance in front of the source follower. One can use the small gate capacitance to provide the feedback impedance, and no feedback resistor is used. Dielectric relaxation time effects (see Equation 2.51) further complicate prediction of the frequency response.

5.9. SPECTRAL MEASUREMENTS

5.9.1. Relative Spectral Response

The spectral response of a detector is most often measured and reported as a *relative response*. By this we mean that the data are normalized—all adjusted so that the maximum value shown is unity (1). Such a curve does not tell you the absolute responsivity at any given wavelength; it shows only how the responsivity at one wavelength compares to the responsivity at other wavelengths.

This is done for two reasons: the absolute responsivity may vary from one detector to another, but the relative spectral responsivity is generally the same for all detectors of a given type, or at least for a given processing

Table 5.4   Sample Spectral Response Data

Spectral Response Data

Material: Si:Ga   Date: 6/28/1978
Size: 2 mm × 2 mm
Ingot: 99-Z   Window: KRS-5
                                                            Spectral filter: None
Operating temperature: 5 K   C(BB-Pk):

Load resistance: 100 K
Bias voltage: 40 V   (16.14 V across load;   Noise = 0.11 mV
                        current = 161 μA)

λ	S_{det} (mV)	S'_{det} (mV)	S_{TC} (mV)	Ratio	Factor	Corrected	Normalized
3.5	2.6	2.60	86.00	3.02		3.02	0.09
4.0	2.90	2.90	69.00	4.20		4.20	0.12
4.0	8.95	8.95	245.00	3.65	0.87	4.20	0.12
4.5	10.02	10.02	220.00	4.55		5.24	0.15
5.0	10.00	10.00	180.00	5.56		6.39	0.18
5.5	8.40	8.40	135.00	6.22		7.15	0.20
6.0	5.30	5.30	81.00	6.54		7.52	0.21
6.0	4.10	4.10	61.00	6.72	0.89	7.52	0.21
6.5	2.80	2.80	49.00	5.71		6.39	0.18
7.0	4.00	4.00	40.00	10.00		11.19	0.32
7.5	3.40	3.40	21.80	15.59		17.45	0.49
8.0	2.75	2.75	22.00	12.49		13.98	0.39
8.5	1.90	1.90	15.50	12.24		13.70	0.39
9.0	1.42	1.42	7.30	19.39		21.71	0.61
9.5	1.32	1.32	7.60	17.31		19.38	0.55
10.0	1.28	1.28	6.90	18.48		20.69	0.58
10.0	8.60	8.60	54.00	15.92	0.77	20.69	0.58
10.5	8.35	8.35	44.00	18.98		24.65	0.69
11.0	8.40	8.40	43.00	19.53		25.38	0.71
11.5	7.90	7.90	38.00	20.79		27.01	0.76
12.0	6.30	6.30	27.00	23.33		30.31	0.85
12.5	4.80	4.80	20.00	23.99		31.17	0.88
13.0	4.40	4.40	18.00	24.44		31.75	0.89
13.5	4.40	4.40	18.00	24.44		31.75	0.89
14.0	4.10	4.10	15.00	27.32		35.50	1.00
14.0	4.40	4.40	17.00	25.87	0.73	35.50	1.00
14.5	4.20	4.20	16.00	26.24		30.17	0.85
15.0	3.50	3.50	12.80	27.33		31.42	0.89
15.5	3.00	3.00	11.80	25.41		29.21	0.82
16.0	2.15	2.15	10.00	21.47		24.69	0.70

Table 5.4 (*Continued*)

λ	S_{det} (mV)	S'_{det} (mV)	S_{TC} (mV)	Ratio	Factor	Corrected	Normalized
16.5	1.85	1.85	8.20	22.52		25.89	0.73
17.0	1.70	1.70	7.30	23.24		26.72	0.75
17.5	1.25	1.25	5.80	21.47		24.68	0.70
18.0	0.72	0.71	4.90	14.52		16.70	0.47
18.5	0.20	0.17	4.20	3.98		4.57	0.13
19.0	0.12	0.05	3.50	1.37		1.58	0.04

lot. Thus one relative spectral response curve can describe the whole group.

Another reason is that most spectrometers allow convenient determination of relative response curves, but they do not allow convenient measurement of absolute spectral response values. To determine absolute spectral response data requires that the relative spectral data be combined with blackbody responsivity data. That process and calculation are described in sections 3.2.4 and 5.9.3.

The measurement is done using a spectrometer, a device that passes radiation of a known, narrowband of wavelengths onto the detector. The energy in a narrow wavelength interval is quite small, so care must be taken to focus that energy onto the detector. Part of the setup involves getting the detector centered at the focal point.

To determine how much energy is available at each wavelength, a mirror or beam splitter is provided so that the spot can be shifted over to another (standard) detector whose relative spectral response is known. The relative spectral response of the detector under test is determined by comparing the signal from the two detectors, and including the known spectral response of the standard detector, then dividing all the resulting response values by the largest one. Usually, the standard detector is a thermal detector whose response per watt is independent of wavelength. In that case the spectral response of the unit under test (UUT) is given by

$$\mathcal{R}'_{UUT}(\lambda) = \frac{S_{UUT}(\lambda)}{S_{standard}(\lambda)} \tag{5.26}$$

As noted earlier, spectral response values can be defined in terms of power or photon flux. Make sure that you know which is desired, or meant. Table 5.4 is a sample data sheet showing data recorded and reduced manually, including corrections for noise and discontinuities.

Accurate spectral measurements are not easy to make; there are many potential sources of error, and no convenient way to identify them. Thus special efforts must be made and maintained to obtain an acceptable level of accuracy in spectral measurements. Frequent comparison of spectral data acquired on different instruments and by different laboratories is one way to identify problems and to quantify potential errors. Such a comparison will probably show differences that exceed the accuracy that either laboratory would believe.

There is another complication: Even though accuracy and reproducibility of spectral measurements are real problems, one must not assume that lack of reproducibility is an instrumental problem; observed variations may be due to subtle detector phenomena. Sayre et al. (1977) report that the spectral response of certain photoconductors varies as the radiation is focused close to or away from the contacts. Such variations, or variations of spectral response with bias, background, signal level, or frequency could all suggest to the experimenter that the equipment is not reproducible!

5.9.2. Absolute Spectral Response Values

To determine the *absolute* spectral response values, we multiply the *relative* values (described in Section 5.9.1) times the responsivity *at the wavelength* of *peak response*. We obtain this peak responsivity by combining the blackbody responsivity and the relative spectral data. This is described in the next section.

5.9.3. Effective Incidance and Blackbody-to-Peak Conversion Factor

The responsivity is generally measured using a source that has its energy spread over the infrared spectrum. Not all of the source energy is at the wavelength at which the detector is most sensitive. In fact, the detector is almost certainly "blind" to some of the wavelengths from the source. If all the blackbody energy could be concentrated at the wavelength where the detector is most sensitive, how much greater would the responsivity be? That factor is the *blackbody-to-peak conversion factor*, sometimes referred to as the *cell response factor*. Hudson (1969, p. 344) uses the reciprocal, calling it the effectiveness factor.

The formula for the blackbody-to-peak conversion factor is derived in Section 3.2.4. The result is

$$C = \frac{\displaystyle\int_0^\infty M(\lambda)\, d\lambda}{\displaystyle\int_0^\infty \mathscr{R}'(\lambda)M(\lambda)\, d\lambda} = \frac{M_{BB}}{M_{\text{effective}}} \qquad (5.27)$$

The quantity "M effective" is the *effective* exitance; it is less than M_{BB}, the exitance from the blackbody. This smaller amount of energy—if concentrated at the wavelength of peak response—would generate the same signal as did the blackbody.

Once calculated, the blackbody-to-peak conversion factor can be used to determine all the "peak" parameters (D^*, NEP, responsivity, etc., at the wavelength of peak response):

$$\text{peak responsivity} = \text{BB responsivity} \times C \qquad (5.28a)$$

$$\text{peak } D^* \quad = \text{BB } D^* \times C \qquad (5.28b)$$

$$\text{peak NEP} \quad = \frac{\text{BB NEP}}{C} \qquad (5.28c)$$

The conversion factor can be thought of as the ratio of the blackbody incidance to the useful incidance. Responsivities or detectivities calculated using the total blackbody incidance can be converted to peak values by multiplying by the conversion factor [as indicated in equations (5.28a)–(5.28c)], or the effective incidance can be calculated first from the blackbody value:

$$E_{\text{effective}} = \frac{E_{BB}}{C} \qquad (5.29)$$

Different conversion factors must be calculated for use with incidance in radiant (watts) and photon—flux units. The "type" (radiant or photon) of the incidance and conversion factor must agree.

For an ideal photon detector with an ideal filter, we obtain:

BLACKBODY-TO-PEAK CONVERSION FACTOR: IDEAL PHOTON DETECTOR

$$C_q = \frac{M_q(0 \text{ to } \infty)}{M_q(0 \text{ to } \lambda_{\text{co}})} \qquad (5.30a)$$

$$C_e = \frac{M_e(0 \text{ to } \infty)}{M_q(0 \text{ to } \lambda_{\text{co}})} \frac{\lambda_{\text{co}}}{hc} \qquad (5.30b)$$

5.9.4. Reflectance and Transmittance Measurements

One way to characterize the effectiveness of AR coatings is to measure the fraction of the incident power that is reflected from the surface of the detector. For small detectors this requires that the incidance be focused so that the spot does not extend beyond the surface of interest. Similarly, to determine how much of the energy is not absorbed in the detector, the power transmitted through a detector could be measured.

To eliminate the tight focusing and alignment required if the detectors are small, the measurements can be made on samples that are processed in the same way as the detectors, but have a larger surface area. Use of these *witness samples* is common in coating applications. The potential danger is that the witness samples will in some way not be representative of the actual detectors they are meant to simulate. Their larger size may cause some physical differences, or may cause them to be handled differently.

5.10. SPATIAL MEASUREMENTS

Up to this point we have implied that a detector has a well-defined sensitive area, that is, is uniformly sensitive over that area and does not detect radiation falling outside that area. As usual, these are only approximately true in real life, and the characterization of the spatial sensitivity of the detector is sometimes required.

Spatial measurements are done with focused spot or slit. Optical crosstalk measurements are a special case of spatial measurements: a spot is focued on one element; any electrical signal from other elements is due to crosstalk. (Since this method uses both an optical input and electrical output, it does not discriminate between optical and electrical crosstalk.)

5.10.1. Spot Scan

To characterize the spatial response of a detector, a small spot is focued on the detector and moved over its surface while the resulting signals are recorded. By plotting signal versus spot position, the spatial sensitivity of the detector can be visualized. The resulting plot is called the *point spread function*. A variant is to scan the detector with a thin line or slit of radiation; the resulting plot is the *line spread function* (LSF).

If the spot diameter or slit width is much smaller than the detector size, the resulting plots characterize the spatial sensitivity of the detector. If the spot is *not* much smaller than the detector, the resulting plot is a combination—referred to as the *convolution**—of the actual detector re-

---

* Strictly speaking, the LSF is the mathematical *correlation* of the detector response profile and the slit profile. If either the detector response or the slit profile is symmetric, the mathematical correlation and convolution are identical. This is generally the case. Convolution is more easily done than correlation, and the combination of the two effects is normally treated and referred to as a *convolution* problem.

sponse profile and the incidence in the focused beam or slit. To obtain the actual detector response profile, we must remove the effect of the finite slit width from the LSF; this process is referred to as *deconvolution*. These operations require enough computation effort that they are usually done by computer; they are done using Fourier transforms and modulation transfer functions (MTF).

Before discussing deconvolution and MTF, consider the practical problem of positioning an invisible spot on a very small detector element, then focusing the spot. The process can take hours unless a systematic procedure is followed. One procedure is to begin with the detector intentionally too far from the optics—but only slightly too far. This yields an oversized spot, which facilitates "finding" the spot with the detector and leaves no doubt as to which way the focus should be shifted to improve the focus once the spot is found. If detectors of several sizes are available, begin with the largest one. Move the detector around in a systematic pattern until an appreciable signal is observed. Then maximize the signal by alternately moving the detector horizontally and toward the optics. When the signal cannot be increased further, more a smaller detector over the spot, and repeat the process, or move the array so that the spot is between two closely spaced detectors, and move the detectors and focus to *minimize* the signal from the detectors.

5.10.2. Modulation Transfer Function

The modulation transfer function (MTF) is the spatial equivalent of the frequency response of a system—it tells how the modulated signal decreases as the modulation spatial frequency increases. It is sometimes called the spatial frequency response. MTF and LSFs are equivalent representations of the sa:.. phenomena.

The MTF of the detector/test set combination is calculated from the LSF by a process called the Fourier transform, as described in Equation 5.31.

The MTF is used to describe the spatial frequency response instead of simple LSFs because the effect on MTF of two phenomena (detector and optical blur, for example) can easily be predicted if the individual MTFs are know, but the combination of the two LSFs is much more difficult. To determine the MTF of a detector, we will divide the "raw" (as measured) MTF by the MTF of the optics in the test set. This requires that one calculate or estimate with satisfactory accuracy the MTF of the test set optics. That calculation is *not* described here; instead the reader is referred to books on intermediate and advanced optics.

MTF (AT SPATIAL FREQUENCY k) FROM LSF

$$\text{MTF}(k) = \frac{C(k)}{D} \qquad (5.31a)$$

where

$$C(k) = \sqrt{A^2(k) + B^2(k)} \qquad (5.31b)$$

$$A(k) = \int_{-\infty}^{\infty} \text{LSF}(x) \cos(2\pi kx)\, dx \qquad (5.31c)$$

$$B(k) = \int_{-\infty}^{\infty} \text{LSF}(x) \sin(2\pi kx)\, dx \qquad (5.31d)$$

$$D = \int_{-\infty}^{\infty} \text{LSF}(x)\, dx \qquad (5.31e)$$

If by choice of the x origin, the LSF can be made even [$\text{LSF}(x) = \text{LSF}(-x)$], B will be zero, $C = A$, and $\text{MTF}(k) = A(k)/D$. Because of noise in the LSF data or nonuniformities in the detector and optics, this will never be accomplished perfectly, but it is still worthwhile to select the origin in that way because it facilitates sanity checks.

5.10.3. Crosstalk

Electrical crosstalk can be measured by injecting an electrical signal on one channel of an array and measuring the resulting signal on other channels. Because crosstalk may be affected by subtle factors in the setup, it is advisable to do the crosstalk measurements in the system configuration—or as close to the system configuration as possible. Crosstalk due to inductive, resistive, and capacitive coupling will all have different frequency response, so the frequency used to make the crosstalk measurements should match that to be used in the system, or data should be taken at enough frequencies to allow extrapolation to system conditions.

Since signal readouts are done electrically, optical crosstalk is difficult to isolate from electrical crosstalk. Instead, one measures the total crosstalk (optical plus electrical) by focusing IR radiation on one channel and measuring the resulting electrical signal on other channels.

Because crosstalk values are often very small, noise may affect the readings. It is common to correct for noise by subtracting it (in an rss way).

CROSSTALK FROM ELEMENT j TO ELEMENT k

$$\text{crosstalk} = \frac{\sqrt{S_k^2 - N_k^2}}{\sqrt{S_j^2 - N_j^2}} \tag{5.32}$$

where only element j is intentionally illuminated; S_j and N_j are the signal and noise from element j, and S_k and N_k are the signal and noise from element k.

Example

	Signals (mV)	
	Element 50	Element 55
Spot on element 50	54.3	4.8
Shutter closed	3.7	3.8

The corrected signals are   54.2   2.93

The crosstalk from element 50 to element 55 is 2.93/54.2 = 5.4%.   ■

5.11. TEST REPORTS

Careful documentation of test results is desirable in order to effectively communicate the information gained to everyone interested. Preparation of a good test report also preserves the information for later use—the information may be valuable months or even years later, for reasons that may be unforseen when the original work was done.

A good report is accurate, unambiguous, and complete, without being unnecessarily long or hard to absorb. To design such a report requires care and creativity. If the report describes a specialized or unique test, the report is dominated by description, with a small amount of data. For more routine characterization work, the bulk of the report is tabular and graphical data, with the description covered in separate documents—the test plan or the test procedure. A set of "Test Report Explanatory Notes" may be written to allow abbreviation in the report itself without ambiguity. To avoid ambiguity, use standard notation and nomenclature whenever possible, and define any new or nonstandard terms that must be used.

```
Test Date:    1 Jan 89 12:32                    PROGRAM XYZ                         Array No:  234-56A
Test Seq. No:          1756                 PERFORMANCE SUMMARY
Report Date:  6 Jan 89 04:27                   sheet 1 - MAPS

Responsivity

        1        11        21        31        41        51        61        71        81        91       101       111       121       128
        v         v         v         v         v         v         v         v         v         v         v         v         v         v
A . . . . . ++ . ++ . + . . . . . . . . . . . . . . . . . . . . . . . . . . . . . . . . . . . . . . . . . . . . . . . . . . . . . . .
B . . . : + ++.+++ . . . . . . . . . . . . . . . . . . . . . . . . . . . . . . . . . . . . . - . . . . . . . . . . . . . . . . . . .
C . . . ++ - -++ . ++ . . . . . . . . . . . . . . . . . . . . . . . . . . . . . . . . . . . . . . . . . . . . . . . . . . . . . . . .
D . . . - - : +- . . . . . . . . . . . . . . . . . . . . . . . . . . . . . . . . . . . . . - . . . . . . . . . . . . . . . . . . . .

D*

        1        11        21        31        41        51        61        71        81        91       101       111       121       128
        v         v         v         v         v         v         v         v         v         v         v         v         v         v
A . . . . : - . = . : : = . . . . . . . . . . . . . . . . . . . . . . . . . . . . . . . . . . . . . . . . . . . . . . . . . . . . .
B . . . : : = . : : == . . . . . . . . . . . . - . . . . . . . . . . . . . . . . . . . - . . . . . . . . . . . . . . . . . . . . .
C . . . : : : === . . . . . . . . . . . . . . . . . . . . . . . . . . . . . . . . . . . . . . . . . . . . . . . . . . . . . . . . .
D . . . : : - : -== . . . . . . . . . . . . - . . . . . . . . . . . . - . . . . . . . . . . . . . . . . . . . . . . . . . . . . .

Resistance

        1        11        21        31        41        51        61        71        81        91       101       111       121       128
        v         v         v         v         v         v         v         v         v         v         v         v         v         v
A 1226899537787532221234423302524421102333434442344423423305243124350443135024224134502442220442211142244
B 58658*8658795869721123442114335842114354254423225244211233433253252044243322443322243423533045304335
C 66585879497688*3429*46538411435844221144325442332123443325236210233433433333523433523345122543104344
D 78768834558565859687687687659463442221123442334423025024411230343332215042332250243335233450256462553

legend for responsivity and D*                               legend for resistance
   *  exceeds spec by more than 10 %                           0    R < 41.00 ohms
   +  exceeds spec by       10 % or less                       1    R = 41.xx ohms
   .  in spec                                                  2    R = 42.xx ohms
   -  below spec by  10 % or less                             ...
   =  below spec by more than 10 %                             *    R > 49.99 ohms
```

Figure 5.18. Performance "map" used to summarize array data.

266

To make the report easy to read and absorb, group similar quantities together. Use graphs and histograms in addition to the usual tabulated values. When large amounts of data are involved it is important to provide a brief executive summary. This can be done with tabulated summaries and data "maps" laid out in the array configuration, with a one-character symbol showing the performance for each element in the array. Figure 5.18 is such a map for a hypothetical array of 512 elements arranged in 4 rows of 128 elements each.

The report should include the serial number of the unit tested, the purpose of test, the test date, report date and operator's name. It should define the test setup, either by reference to a controlled drawing set or configuration log, or by direct description. By the "test setup" we mean the chamber or dewar, the electronics, and any software used for acquisition, reduction, and reporting. The report should include test conditions (bias voltages, clock rates and phasing, chopper speeds, blackbody temperatures, etc.). An often-cited reference for reporting is "Standard Procedures for Testing Infrared Detectors and Describing Their Performance." The latest version of this document (Eisenman et al., 1975) has not received as wide a distribution as the earlier version (Jones et al., 1960) and government contracts often cite the earlier version.

REFERENCES

Boyd, Robert W. (1983). *Radiometry and the Detection of Optical Radiation,* Wiley, New York.

Chang, T. F., and J. G. Lewis (1979). "Computing Standard Deviations: Accuracy," *Commun. ACM* 22, 526.

Eisenman, W. L., E. H. Putley, and J. L. Lachambre (1975). *Standard Procedures for Testing Infrared Detectors and Describing Their Performance—Revised 1975,* The Technical Cooperation Program, Subcommittee on Non-atomic Military Research and Development, Washington, D.C.

General Radio Company (1963). *Engineering Department Instrument Notes IN-103,* General Radio Company, West Concord, Massachusetts. p. 4.

Hanson, R. J. (1975). "Stably Updating Mean and Standard Deviation of Data," *Commun. ACM* 18, 57.

Hudson, Richard D., Jr. (1969). *Infrared System Engineering,* Wiley, New York.

Jones, R. C., D. Goodwin, and G. Pullan (1960). *Standard Procedure for Testing Infrared Detectors and Describing their Performance,* Office of Director of Defense Research and Engineering, Washington, D.C.

Lange, F. H. (1967). *Correlation Techniques,* Van Nostrand Reinhold, New York.

Mandel, John (1964). *The Statistical Analysis of Experimental Data*, Dover, New York.

Marconi Instruments Limited (1965). *"Marconi Instrumentation,"* 10(2), 18–22 (Marconi Instruments Limited, St. Albans England).

Sayre, C., D. Arrington, W. Eisenman, and J. Merriam (1977). "Characteristics of Detectors Having Partially Illuminated Sensitive Areas," *Opt. Eng.* 16, 44.

Stierwalt, D. L., and W. L. Eisenman (1978). "Problems in Using Cold Spectral Filters with LWIR Detectors," *Proc. SPIE* 132, 134.

West, D. H. D. (1979). "Updating Mean and Variance Estimates: An Improved Method," *Commun. ACM,* 22, 532.

SUGGESTED READINGS

Dereniak, Eustace L., and Devon G. Crowe (1984). *Optical Radiation Detectors,* Wiley, New York.   Excellent and up-to-date reference; covers radiometry, detectors, and charge transfer devices. Recommended.

Eisenman, W. L., E. H. Putley, and J. L. Lachambre (1975). *Standard Procedures for Testing Infrared Detectors and Describing Their Performance—Revised 1975,* The Technical Cooperation Program, Subcomitee on Non-atomic Military Research and Development, Washington, D.C.

Holter, Marvin R., et al. (1962). *Fundamentals of Infrared Technology,* Macmillan, New York.

Hudson, Richard D., Jr. (1969). *Infrared System Engineering,* Wiley, New York.

Jameson, John, Raymond McFee, Gilbert Plass, Robert Grube, and Robert Richards (1963). *Infrared Physics and Engineering,* Inter-University Electronics Series, McGraw-Hill, New York.

Jones, R. C., D. Goodwin, and G. Pullan (1960). *Standard Procedure for Testing Infrared Detectors and Describing Their Performance,* Office of Director of Defense Research and Engineering, Washington, D.C.

Kruse, P. W., L. D. McGlaughlin, and R. B. McQuistan (1962). *Elements of Infrared Technology,* Wiley, New York.

Smith, R. A., F. E. Jones, and R. P. Chasmer (1957). *The Detection and Measurement of Infrared Radiation,* Oxford University Press, Oxford.

Wolfe, William L., and George J. Zissis, eds. (1978). *The Infrared Handbook,* Environmental Research Institute of Michigan, Ann Arbor, Michigan.

PROBLEMS

1. Given the input and output data of Table 5.5, calculate the noise equivalent electrical bandpass of the filter.

Table 5.5 Electrical Filter Data for Problem 1[a]

Frequency (Hz)	Output Voltage (mV)
60	70
100	93
150	146
200	200
250	249
300	355
350	505
400	683
450	790
500	805
600	795
700	604
800	560
900	460
1000	308
1300	195
2000	120
3000	82
5000	75
7000	70

[a] Input voltage was 800 mV at all frequencies. Noise was about 70 mV referred to the output.

2. A 2-mm-diameter InSb detector with a 60° field of view yielded the $V-I$ curve shown in Figure 5.19. What can you infer about the detector?

3. Radiometric data for the InSb detector of Problem 2 is given in Table 5.6. Use it to find:
   (a) The "best" feedback resistance
   (b) The short-circuit current responsivity
   (c) The detectivity
   (d) The NEP

4. How do the noise and responsivity of Problem 3 compare with the values expected from the curve of Problem 2?

5. Use the Ge:Hg data from Table 5.3 to find:
   (a) The optimum bias voltage

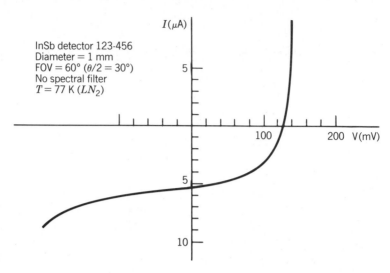

InSb detector 123-456
Diameter = 1 mm
FOV = 60° ($\theta/2 = 30°$)
No spectral filter
$T = 77$ K (LN_2)

$I(\mu A)$

5

100   200   V(mV)

5

10

Figure 5.19. *V–I* curve for Problem 2.

Table 5.6 Blackbody Data for Problem 3

Blackbody Test Data

Material: InSb (PV)   Blackbody   Date 12/25/1986
Size: 1 mm diameter   temperature: 500 K   FOV: 60°
Ingot: 1234–456   Bb incidance: Window: sapphire
                         12.64 μW/cm   Spectral filter: none
                                              C(BB-Pk): 5.5

Operating temperature: 6 K

Op-amp + voltage mode postamp
Feedback resistance: varies—see below
Amplifier gain: varies—see below
Chopping frequency: 780 Hz
Bandpass: 50 Hz

Gain	Feedback (Ω)	V_{det} (V)	Noise (μV)	Signal (mV)	S/N	Resp.	D^*
1000	10 k		140.00	630	4500		
1000	20 k		245.00	1260	5143		
1000	50 k		540.00	3150	5833		
100	50 k		54.00	320	5926		
100	100 k		110.00	640	5818		
100	200 k		205.00	1250	6098		
10	500 k		54.00	101	1870		
10	1 M		100.00	30	300		

Table 5.7 Spectral Response Data for Problem 7

Spectral Response Data

Material: Si:As

Size: 2 mm × 2 mm

λ	S_{det}	STC (mV)
4.00	0.91	245.00
4.50	1.07	220.00
5.00	1.04	180.00
5.50	0.85	135.00
6.00	0.51	81.00
6.50	0.24	49.00
7.00	0.41	40.00
7.50	0.38	21.80
8.00	0.35	22.00
8.50	0.22	15.50
9.00	0.17	7.30
9.50	0.17	7.60
10.00	0.16	6.90
10.50	1.01	44.00
11.00	0.94	43.00
11.50	0.89	38.00
12.00	0.77	27.00
12.50	0.59	20.00
13.00	0.54	18.00
13.50	0.56	18.00
14.00	0.55	15.00
14.50	0.55	16.00
15.00	0.46	12.80
15.50	0.45	11.80
16.00	0.38	10.00
16.50	0.36	8.20
17.00	0.36	7.30
17.50	0.28	5.80
18.00	0.24	4.90
18.50	0.22	4.20
19.00	0.17	3.50
19.50	0.16	3.20
20.00	0.12	2.48
20.50	0.08	1.98
21.00	0.03	1.55

Table 5.8   Frequency Response Data for Problem 8

Frequency (Hz)	Signal (mV)
100	30
200	29
500	26
1 k	20
2 k	12
5 k	4.7
10 k	1.7
20 k	0.78

(b) The short-circuit current responsivity

(c) The short-circuit noise spectral density

(d) The detectivity

(e) The NEP

6. Use the results of Problem 5 to determine the background, quantum efficiency, and photoconductive gain of the detector.

7. Reduce the raw spectrometer data of Table 5.7 to find the relative spectral response of the detector. Plot your results. Indicate the peak wavelength and the cutoff wavelength.

8. Plot the frequency response data in Table 5.8. Indicate the "corner frequency." Find the time constant from the corner frequency and from the 20-Hz and 10-kHz data points.

Unit 2
Related Skills

6

Science and Measurements

In this chapter we attempt to convey the essence of the measurement process. It cannot be stressed strongly enough that this material is essential to good experimental and analytical work—in short, to "doing science." We focus on the "ingredients" of measurements and a few other guidelines for doing science. Not only is this material basic to good science, it is also rich in opportunities for creative, interesting work. Given the same instrument, one person may turn out data with poor credibility, another may generate competent data, while a third person may be able to report values with higher accuracy, and discover trends, effects, or phenomena that escaped the first two. Not only will the last person do better work, but he or she will be happier, and probably better paid!

The ability to improve the quality of data extracted in a particular situation depends strongly on an understanding of the concepts of precision and uncertainty. These are truly fundamental to the measurement sciences, yet they are—in my opinion—rarely taught. There is always a way to improve the data in a given setup, but it will happen only if the observer has a good understanding about precision and the conventions for reporting it. The objective of this chapter is to present those concepts and to promote their use.

6.1. MEASUREMENTS*

Every time we report a measurement, there are three components either stated or implied: the number itself, the units, and the uncertainty associated with the number. Of these, the uncertainty is probably the least familiar concept, so we will devote more time to it than to the units or

---

* Much of the material on measurement, uncertainty and errors in this chapter is taken from a laboratory manual prepared by Dr. George L. Appleton (1981) for his physics classes at California State University–Long Beach. I am indebted to Dr. Appleton for introducing me to these concepts many years ago and for his permission to reproduce some of his ideas here.

best estimate. We will talk about uncertainty in measurements, how to estimate it, how to work with it, and how to express it. First, let's get the units out of the way.

6.1.1. Units

The units are important, and confusion can occur if we are careless. It is quite common for someone to say "I've set the blackbody to 800 degrees, and it seems to be within about 3 degrees of the set point." That's great, unless the person means 800 degrees Celsius, and you think he means 800 kelvin. This may sound too simple or obvious to happen, but it has happened. Be specific and clear about the units you use! Now, on to the number itself, and the uncertainty associated with it.

6.1.2. Best Estimate

The values we report are actually estimates of the values of a particular parameter. The methods of arriving at these estimates are generally well known and are mentioned here primarily for completeness. The real work comes under the heading of uncertainty. Reporting the uncertainty will increase your awareness of the source of the uncertainty, and suggest to you how to reduce the uncertainties. Better estimates will follow almost automatically.

Common methods of arriving at the best estimate are the mean, the weighted mean, the median, and the least-squares fit, as described below.

1. *The Mean.* Make many measurements, and report the average (mean) value. Let x_j (where j is an integer from 1 to J) represent the J individual values. Then the mean value $\langle x \rangle$ is given by Equation 6.1. The mean value is the best estimate one can make if the data are all acquired under similar conditions and have an equal probability of being accurate.

SIMPLE MEAN

$$\langle x \rangle = \frac{\sum\limits_{J} x_j}{J} \tag{6.1}$$

---

2. *Weighted Mean.* If you have reason to believe that some of the data are more accurate than others, they can be weighted before averaging,

as described in Equation 6.2. Determining the weighting factors is a judgment on the part of the investigator, and can be easy to criticize and difficult to defend.

WEIGHTED MEAN

$$\langle x \rangle = \frac{\sum\limits^{J} w_j x_j}{\sum\limits^{J} w_j} \tag{6.2}$$

where w_j are the weighting factors.

---

3. *Median.* The median is that value which exceeds half of the measurements and is smaller than the other half. Although this is not as good an estimate as the mean, it is not contaminated seriously by a few bad data points.

Often one relies on computer reduction of data for which "bad data points" may be common, but rejection criteria have not been implemented. This can yield embarrassing results. Imagine that you are automatically measuring resistances that should be about 25 Ω, and occasionally have an "open" or near open. Inclusion of a 1-MΩ resistance value along with many 20- to 30-Ω values will yield a reported average that is clearly higher than the expected 25 Ω. You will be asked repeatedly for the "real" average, you may spend hours with a hand-held calculator providing a more useful average, and your software will be criticized.

Technically, the mean was computed correctly, and you will not be able to get your customers to agree on acceptable criteria for rejecting the near opens. The ideal solution is to obtain agreement on this subject before the software is written. In practice, this agreement seldom happens because many people regard it as a trivial problem—"Just give us the 'real' average." A more practical solution is to report the median along with the mean.

4. *Least-Squares Fit to a Straight Line.* Often data can be arranged in such a way that a relationship of the form $y = a + bx$ is expected, and either the intercept a or the slope b is a parameter of interest. In this case plotting several sets of (x, y) data on linear graph paper will suggest a straight line. One can fit a straight line through the points either "by eye" (simply drawing a line that looks like it fits as well as possible) or

determine the intercept a and slope b numerically (Beers, 1957):

LEAST-SQUARES STRAIGHT-LINE FIT: $y = a + bx$

$$a = \frac{\sum x_j^2 \sum y_j - \sum x_j \sum (x_j y_j)}{J \sum x_j^2 - (\sum x_j)^2} \qquad (6.3a)$$

$$b = \frac{J \sum (x_j y_j) - \sum x_j \sum y_j}{J \sum x_j^2 - (\sum x_j)^2} \qquad (6.3b)$$

where x_j and y_j ($j = 1$ through J) represent the J data pairs

---

This method is useful in IR detector work to determine the responsivity and background from signal versus blackbody incidance values using the relation $S = (Q_{BK}\mathcal{R}) + \mathcal{R}Q_{BB}$. Even if the formulas are used to determine the slope and intercept, it is worthwhile to plot the data since the plot allows the data to be interpreted quickly and makes problems stand out.

6.1.3. Rejecting Data

Rejecting some data points is the equivalent of assigning them a weight of zero. Data should be rejected if you observed an anomaly or error in the data acquisition process. This is true even if the resulting value "looks okay."

Rejection of data simply because it docs not agree with other data must be handled carefully. Some people will choose to reject one or two data points that appear statistically to be erroneous; others feel that they should be included, but enough additional data taken so that the "wild" values do not contaminate the average significantly. It is agreed that you reject data only once—do not reject some data, then look at the new statistics and reject more data, and so on.

Acknowledge in your report that you have excluded some data; tell how much was excluded, and why. Exclusion of data because they do not support one's point of view is a serious breach of the ethics of science; consistently disclosing any exclusions is one way to remind us to be scrupulously correct in this regard.

6.1.4. Uncertainty

Except for the counting of a set of discrete objects, no physical quantity can be known exactly. This concept is surprising and even disturbing to many people: They expect science to be *exact*, and any amount of uncertainty is somehow "wrong." If pressed to explain, you might hear:

"Science is the business of measuring and predicting and knowing things exactly." Try to divorce yourself from that attitude. Precision and uncertainty are not two mutually exclusive states, but are related words for one mathematical idea: They represent a progression in a never-ending continuum. Our job is to provide the best possible estimate of a particular parameter, to minimize our uncertainties, and to characterize them. That is all we can do. It is unrealistic to expect to eliminate uncertainties, and people who do not understand that are unlikely to characterize the precision of their measurements. This and other principles of uncertainty in measurement are summarized in Table 6.1. The source of uncertainty is discussed in the following sections.

6.2. EXPERIMENTAL ERROR

In the context of science and engineering, *error* generally means an experimental uncertainty. The word uncertainty conveys our meaning more

Table 6.1 Principles of Uncertainty in Measurement

1. No physical quantity can be known exactly (except for the counting of a set of discrete objects).
2. Precision and uncertainty are:
   - not two mutually exclusive states, but rather
   - related words for one mathematical idea
   - a progression in a continuum:

$$\text{less precise} \longleftrightarrow \text{more precise}$$

3. Our job is to
   a. Determine the best estimate of a given parameter
   b. Minimize the uncertainties in that estimate
   c. Report the uncertainty

Experimental "error": In the context of science and engineering

$$\text{error} = \text{experimental uncertainty}$$

Sources of uncertainty
1. Systematic errors: Potentially identifiable, correctable
2. Random errors
   Averaging many measurements minimizes the effect
   No correction possible
3. Illegitimate errors
   Blunders, computational errors, chaotic conditions
   Should be eliminated from all competent work
4. Inadequate model

accurately, but "error" has been around for a long time, and "experimental error" is used much more often than "uncertainty." It is common to separate "errors" into three types: systematic errors, random errors, and what Beers (1957) calls illegitimate errors. In addition to these, lack of validity of the model may be a source of error and must be considered.

6.2.1. Systematic Errors

Systematic errors are repeatable, consistent errors. Examples include a plastic centimeter ruler that has been distorted by sitting in the sun, and the fabled butcher with his thumb on the scale. Systematic errors are always present to some extent in our measuring instruments. They may be due to consistently faulty technique: for example, the parallax effect in reading a scale. The word "accurate" in its technical sense refers to measurements with small systematic errors (Beers, 1957; Leaver and Thomas, 1974).

6.2.2. Random Errors

Random errors are not repeatable, nor are they predictable except in a general, statistical way. No corrections are possible, but they can be reduced by using instruments with better resolution and by making many measurements. A measurement with small random errors is said to be *precise*, or of high precision. That alone is not enough to minimize its uncertainty. To do that we must also minimize the systematic and illegitimate errors.

6.2.3. Illegitimate Errors

Beers (1957) includes in *illegitimate errors* those errors that are avoidable and should be eliminated from competent work before it is reported. These include blunders, computational errors, and chaotic errors.

1. *Blunders or Mistakes.* These are "human errors." They include reading the wrong scale, transposing of digits, and computational blunders. They are unavoidable in the sense that humans make mistakes. They are avoidable in the sense that they can be discovered and corrected. One way to state the problem is: "How often do they occur, and when are they discovered and corrected?" Part of good experimental technique is to include checks and double checks so that any blunders are discovered and eliminated very early in the data-taking process.

2. *Computational Errors.* These errors result when the computational method is less accurate than the data justify. An example is the reduction of five-significant-figure data with a slide rule or with an algorithm that is accurate to only three significant figures.

3. *Chaotic Errors.* These are the last illegitimate errors mentioned by Beers; they are due to disturbances that cause unreasonably large variations in the data. Examples include the effect of temperature variations in an air-conditioned room because many people are entering and leaving, or the electrical or accoustical noise associated with the presence of heavy equipment. In these cases the experiment should be discontinued until the disturbance is removed.

6.2.4. Validity of the Model

This source of error is distinct from the other three. It is an error in the way we visualize and analyze the physical phenomena we are measuring.

Virtually every reduction and report is based on a simplified model. A *model* is a mental image that simulates reality but is simpler and easier to grasp. Before calculating the area of a table top, one probably makes the tacit agreement that the tabletop will be treated as a perfect rectangle (or perhaps a perfect circle).

We may not even mentally acknowledge that we are using a model, but one is always present. The resulting value is meaningful only to the extent that the model is valid.

In the case of our tabletop area calculation, we could discuss the precision of the length measurements, the calibration of the ruler used, and the number of significant digits carried in the multiplication process. That is all necessary and good. At some point, however, we need to acknowledge that we are treating the table as a rectangle, and since it is not perfectly rectangular, we are up against a fundamental problem that may or may not limit the accuracy of our results. We need to decide how valid our model is and whether we want to devise a better one.

Calculations of quantum efficiency can be misleading if we fail to remember that our models often assume that the responsivity of the detector is constant within a defined area, and zero outside that area. If (as is usually the case) the responsivity varies from point to point on the surface and is not zero outside the nominal detector area, we will arrive at results whose error exceeds that predicted from our analysis of meter repeatability and blackbody temperature errors.

Once we acknowledge that we have used a model, we can consider the consequences of using that particular model. We may determine that our model is adequate for our purposes. If not, we can consider using different, more faithful models.

6.3.   ESTIMATING UNCERTAINTIES

The estimating and reporting of uncertainties in quantities you measure is often very important. Decisions to commit money, talent, and time to a particular plan may be based on your numbers. Your reported uncertainties give others a sense of the confidence they can place in the conclusions they draw from your results, and improves the odds of making good decisions. When you record data, you rarely know what decisions will be made on the basis of those data, so it is important that you report the uncertainty of the data for all data you record.

Now—how can we estimate the uncertainties in our data? The methods used are different for each type of errors. These will be described in a moment, but one subjective method can be used to estimate the overall accuracy of the data.

6.3.1.   Subjective or Intuitive Estimate of Uncertainty

Experimenters can often estimate the uncertainty in their data in either a conscious recollection of spreads they have seen, or in a subconscious, almost intuitive way. You will probably encounter experimenters who are reluctant or unable to estimate the uncertainty associated with their values or results. This may be because they are unfamiliar with the ideas of uncertainty, or perhaps because they believe that admitting *any* "error" is a reflection on their competence.

Extracting this information from them can be very frustrating. One way to force such information is to ask the question: Would you be surprised if your results were in error by 0.03 cm? If the answer is "no," proceed with "...0.1 cm?" "...0.3 cm?" and so on. When the answers become hesitant or uncertain you have an estimate of the uncertainty. Despite the subjective nature of this information, it can be a good estimate. It is certainly better than no estimate at all.

6.3.2.   The Probability of Blunders

Uncertainties due to potential blunders are themselves not quantified— a blunder can lead to any "size" error, and the real trick is to eliminate the blunder. Instead of asking how large the error is, we ask: "What is the probability that we have made a mistake?" The key questions here are "How familiar are you with the test and equipment?" and "How much double checking have you done?" The probability of blunders is greatest when the experiment is first done because the operator is unfamiliar with the equipment. By repeating a test several times, and comparing the results from different methods, the probability of an error of this type can be reduced to an acceptable level. I find that I need to "do" any new experiment at least three times before most of the "bugs" are worked out. The third trial is much more reliable than the first, or even the second.

If I had to select a moralistic message for a brass plaque over the entry to a test facility, my choice would be: "What I tell you three times is true" (from *Hunting of the Snark*, by Lewis Carroll). We believe data only to the extent that successive trials agree. If at all possible, two people should perform the tests independently and compare results. In addition, it is worthwhile to perform the test using a completely different method. If you have not had the opportunity to do this double checking, you should say so in your report.

6.3.3. Estimating the Magnitude of Systematic Errors

Errors due to systematic errors in equipment are estimated by the process of calibrating the equipment. This process compares a given piece of equipment with another (the calibration standard) that has itself been calibrated. The possible residual systematic error is the accuracy of the calibration standard plus the uncertainties in making the comparison. This method works only if a standard is available, and if the accuracy of the available standard is adequate for your work. If normal calibration is inadequate or not possible, make your measurements several ways with different sets of equipment and compare the results from those methods. This also allows you to estimate the effects of possible systematic errors due to your technique.

6.3.4. Estimating the Magnitude of Random Errors

The best way to estimate the uncertainty due to random errors is to perform the measurement several times. The uncertainty can be sensed subjectively by the experimenter, estimated from the spread in the data, or determined statistically from appropriate standard deviations. The *spread* is simply the difference between the largest and smallest reported value. It provides a quick sense of the magnitude of the random errors, but is not a rigorous or precise measure of uncertainty.

The *standard deviation of the sample* is the root mean square (rms) of the individual deviations from the mean. (Formulas are provided in Section 6.4.2.) In the next few pages we discuss the interpretation and use of the standard deviation.

6.4. PROBABILITY

A thorough discussion of error analysis requires a rigorous discussion of statistics and probability that is beyond the scope of this book. Fortunately, to provide an accurate response to questions about the uncertainty of your measurements, you need not be an expert on statistics and probability. You need only provide the standard deviation associated with the measurement, and a clear statement of how you define and calculate the standard deviation. This information will allow the reader to interpet the probabilities. We first provide a very simplified introduction to probability, and then describe the standard deviation formulas that you will need.

6.4.1. An Introduction to Probability: Normal Curve of Error

A natural question to ask is: "What is the probability that our result is *right*?" The answer is: "There is zero probability that we are right—there will always be error." This is a frustrating, unsatisfactory answer, and it may seem that we are being difficult and splitting hairs, but it is essential to recognize that truth before proceeding: "There is zero probability that we are exactly right."

We can provide an answer that is satisfactory, yet still accurate and not too difficult if we rephrase the question: "What is the probability that we are within ____ % of being right?" This question can be answered easily if you know the appropriate standard deviation [call it σ (sigma) for now] and a standard format for the reply: "There is a 68% probability that we are within (σ) of the 'actual'* value, 95% that we are within (2σ) and 99.7% that we are within (3σ)." For example, if the best estimate of a signal is 200 mV and it has a standard deviation (σ) of 7 mV, there is a 68% probability that the "actual" signal is within 7 mV of the best

---

* The quotes around "actual" are to assist us in remembering that the "correct" or "actual" value for a physical parameter will never be found or known by mortals. It would be better to say "There is a 68% probability associated with a 1σ error, 95% probability associated with a 2σ error, and so on." That seems a little vague, and we can settle for the more specific words above if we will remind ourselves to retain the quotes—at least mentally.

estimate, 95% that it is within 14 mV of the estimate, and 99.7% that it is within 21 mV of the estimate.

These probabilities are inferred from the *normal curve of error* (Figure 6.1). This is a distribution function—a sort of fine-scale histogram—showing how many random readings will occur with given values x. x is normally measured in terms of the appropriate standard deviation above or below the mean. For example, if we are interested in values of x less than the mean by 2.4 cm, and the standard deviation is 0.8 cm, we would say that x is 3σ or more below the mean.

The probability of readings with values between x_1 and x_2 is proportional to the area under the normal curve between x_1 and x_2. For example, 34% of the values of an infinite sample will exceed the mean by between zero and 1σ, and 68% will be between $\pm 1\sigma$ of the mean. A few such areas are shown on Figure 6.1 for illustration, but they can be read directly from the integrated values (Figure 6.2) or from "normal curve of error" tables. The notation used for such tables is not always consistent, so it is suggested that when you first use a new table, you compare a few values with those in Figures 6.1 and 6.2 to ensure that your understanding of the table is correct.

6.4.2. Standard Deviation for Common Best Estimates

We will need the standard deviation for several situations: a single measurement, the average (mean) of J samples, the noise associated with a

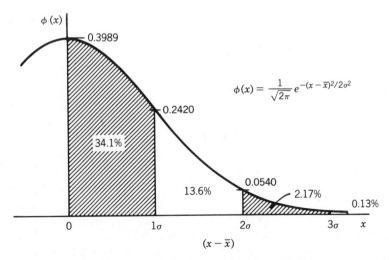

Figure 6.1. Normal distribution of random variables.

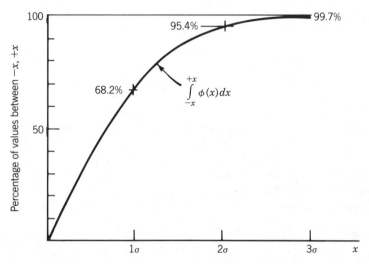

Figure 6.2. Probability that random variables will have a value between $\bar{x} - x$ and $\bar{x} + x$.

set of J samples, and the intercept and slope determined from a least-squares linear fit. Examples of the first three are given in Section 6.4.3. The standard deviations and their applications are as follows.

1. *The Standard Deviation of the Sample—s.* The word *sample* here refers to our finite set of J readings. J may be 20, or a few hundred, or even a few thousand (see Equation 6.4).

STANDARD DEVIATION OF THE SAMPLE

$$s = \sqrt{\frac{\sum (x_j - \bar{x})^2}{J}} \tag{6.4}$$

---

2. *The Standard Deviation of the Population—σ.* The word *population* refers to the infinite possible values of a parameter (length of a table, or output voltage) that might result from a given measurement technique. The standard deviation of the population is slightly greater than the standard deviation of the sample, (see Equation 6.5). The standard deviation of the population is the appropriate standard deviation for use in estimating the uncertainty of any *one* such reading. If the readings are de-

tector output values, the standard deviation of the population is the electrical noise.

STANDARD DEVIATION OF THE POPULATION

$$\sigma = \sqrt{\frac{\sum_{j}^{J} (x_j - \bar{x})^2}{J - 1}} \tag{6.5}$$

In practice we often use s and σ interchangeably. Partly this is due to carelessness, and partly because they differ by such a small amount in practical situations: J will be large because the statistical methods work best for large number of samples. If J is 10 or greater, the difference between the two formulas is less than 5%, and we seldom need better accuracy in estimating the magnitude of errors.

3. *The Standard Deviation of the Mean—σ_m*. The uncertainty associated with the average of a set of J readings is the standard deviation of the mean (Equation 6.6); it is the standard deviation of the population divided by the square root of J. The uncertainty of the mean is less than the uncertainty of any one of the readings (for a given test setup) and it decreases (improves) as we take more samples: the standard deviation of the mean of 20 samples is less than that for five samples.

STANDARD DEVIATION OF THE MEAN

$$\sigma_M = \frac{\sigma}{\sqrt{J}} = \sqrt{\frac{\sum_{j}^{J} (x_j - \bar{x})^2}{J \cdot (J - 1)}} \tag{6.6}$$

---

4. *The Standard Deviation of Noise Values—σ_N*. This is the noise itself divided by the square root of $2J$, where J is the number of independent samples used in determining the noise (Mandel, 1964); see Equation 6.7. (The samples are independent as long as they are taken at a rate $<2\Delta f$, where Δf is the noise equivalent bandpass of the system.)

STANDARD DEVIATION OF THE NOISE

$$\sigma \text{ for noise } = \frac{\text{noise}}{\sqrt{2J}} \tag{6.7a}$$

$$= \sqrt{\frac{\sum (x_j - \bar{x})^2}{2J(J-1)}} \tag{6.7b}$$

5. *Standard Deviations Associated with a Least-Squares Line Fit—* s_a, s_b, s_0. Given the intercept and slope determined by fitting a line through J data points (x_j, y_j), the standard deviations associated with a and b are determined from Equation 6.8–6.11 (Beers, 1957; Leaver and Thomas, 1974).

STANDARD DEVIATION OF LEAST-SQUARES FIT

Let

$$\delta y_j = y_j - (a + bx_j) \tag{6.8a}$$

and

$$s_y = \sqrt{\frac{\sum (\delta y_j)^2}{J - 2}} \tag{6.8b}$$

Then

$$s_a = s_y \sqrt{\frac{\sum x_j^2}{J \sum x_j^2 - (\sum x_j)^2}} \tag{6.9a}$$

and

$$s_b = s_y \sqrt{\frac{J}{J \sum x_j^2 - (\sum x_j)^2}} \tag{6.9b}$$

Further, if we use a and b to determine y associated with some value x_0:

$$y = y_0 = a + bx_0 \tag{6.10}$$

then the standard deviation for y_0 is

$$s_0 = s_y \sqrt{\frac{\sum x_j^2 - 2x_0 \sum x_j + Jx_0^2}{J \sum x_j^2 - (\sum x_j)^2}} \qquad (6.11)$$

Table 6.2 Estimating the Magnitude of Random Error: An Example[a]

	Data Set A: 5 Measurements			Data Set B: 20 Measurements		
	x_j	$x_j - \bar{x}$	$(x_j - \bar{x})^2$	x_j	$x_j - \bar{x}$	$(x_j - \bar{x})^2$
	9.78	-0.02	0.0004	9.78	-0.02	0.0004
	9.80	0.00	0.0000	9.80	0.00	0.0000
	9.79	-0.01	0.0001	9.79	-0.01	0.0001
	9.81	$+0.01$	0.0001	9.81	$+0.01$	0.0001
	9.82	$+0.02$	0.0004	9.82	$+0.02$	0.0004
				9.79	-0.01	0.0001
				9.80	-0.00	0.0000
				9.80	0.00	0.0000
				9.81	$+0.01$	0.0001
				9.78	-0.02	0.0004
				9.83	$+0.03$	0.0009
				9.80	0.00	0.0000
				9.79	-0.01	0.0001
				9.77	-0.03	0.0009
				9.82	$+0.02$	0.0004
				9.80	0.00	0.0000
				9.81	$+0.01$	0.0001
				9.79	-0.01	0.0001
				9.81	$+0.01$	0.0001
				9.80	$+0.00$	0.0000
Total	49.00	0.00	0.0010	19.60	0.00	0.0042
Average	9.80	0.00	0.00020	9.80	0.00	0.00021
s = std dev of sample		=	0.0141			0.0145
σ = std dev of population		=	0.0158			0.0149
σ_M = std dev of mean		=	0.0071			0.0033
σ_N = std dev of noise		=	0.0050			0.0024

[a] x_j, measured or sample values; $\bar{x}$, mean or average value; $x_j - \bar{x}$, deviation from the mean.

S_0 is a minimum if the x_0 value is at the mean of the x_j used to determine the line, and becomes worse as x_0 moves toward the extremes and outside the original data set.

6.4.3. Example of Standard Deviations for Two Data Sets

The standard deviation is widely used but often not properly interpreted. Table 6.2 show two data sets, and the deviations, means, and three types of standard deviations of interest to us. If x_j is the detector electrical output (millivolts) at a particular instant t_j, then $\bar{x}$ is the best estimate of the signal: 9.8 mV and σ is the best estimate of the noise: about 0.015 mV. Note that the mean $\bar{x}$ is about the same in both examples.

The standard deviations *of the sample* and *of the population* (about 0.014 or 0.015 in the examples) are also about the same in both sets. They are a measure of the repeatability *of the individual measurements*; they tell us something about the equipment used, but by themselves do *not* tell us about the precision of the resulting *mean* values. They are about the same since the same equipment was used.

The standard deviation of the mean (σ_m) is the uncertainty in the signal value: 0.0071 mV for set A; 0.0033 mV for set B. The standard deviation associated with the noise (σ_N) is the uncertainty in the noise value: 0.0050 mV for set A; 0.0024 mV for set B.

The standard deviation of the mean and of the noise are better (lower) with 20 samples than with 5 samples. The standard deviation of the mean tells us about the precision of the resulting means. The standard deviations of the mean and of the noise for data set B (based on 20 measurements) are about twice as good as those for data set A (based on 5 measurements).

6.5. PROPAGATION OF ERRORS

If we know the uncertainty or possible error in each of three values used in computing a fourth parameter, can we estimate how accurately we will know that fourth parameter? For example, imagine that we want to calculate a parameter X given by the formula

$$X = A \frac{B - C}{D^2}$$

and that we know that

$$A = 2.0 \pm 0.2$$
$$B = 15.2 \pm 0.1$$
$$C = 4.8 \pm 0.1$$
$$D = 1.2 \pm 0.1$$

The best estimate of X is 14.44, but what is the uncertainty in X? The process that we are discussing is referred to as determining the *propagation of the component errors*. There are two general methods that can be used.

6.5.1. Direct Method

The first method is the direct or brute-force method. It will always work and makes no assumptions or approximations. Simply do the calculation twice, once using the limiting values of the components that will make the result as *large* as possible and once using the values that will make the result as *small* as possible. In the case referred to above, the largest value of X would result if we use the largest possible values of A and B and the smallest values of C and D:

$$A = 2.0 + 0.2 = 2.2$$
$$B = 15.2 + 0.1 = 15.3$$
$$C = 4.8 - 0.1 = 4.7$$
$$D = 1.2 - 0.1 = 1.1$$

The resulting value is

$$X = X_{max} = 19.27$$

For the smallest value of X:

$$A = 2.0 - 0.2 = 1.8$$
$$B = 15.2 - 0.1 = 15.1$$
$$C = 4.8 + 0.1 = 4.9$$
$$D = 1.2 + 0.1 = 1.3$$

Thus

$$X_{min} = 10.86$$

Both these values of X are possible given the uncertainties in A, B, C, and D. X_{max} is 33% above the nominal value and X_{min} is 25% below the nominal value. (For smaller errors, the nominal value will be more nearly centered between the high and low extremes.)

6.5.2. Shortcuts in Estimating Error Propagation

Instead of using the brute-force method to estimate the propagation of errors there are some shortcuts we can take. To use them we will need to state our errors in a relative way—as a fraction or percentage of the best estimate. For example, if x is about 20 cm and the error is 0.2 cm, we could say that the error is 1%. Errors described this way are called *relative errors*. The following methods (Equation 6.12) work well if the relative errors of the components are 10% or less.

Let the absolute errors be ΔA, ΔB, ΔC, and so on, and the relative errors be $\Delta A/A$, $\Delta B/B$, $\Delta C/C$, and so on.

ERROR PROPAGATION: WORST CASE

(a) If $W = A - B$, the worst-case error

$$\Delta W \simeq \Delta A + \Delta B \qquad (6.12a)$$

(b) If $X = CD$, or if $X = C/D$,

$$\frac{\Delta X}{X} \simeq \frac{\Delta C}{C} + \frac{\Delta D}{D} \qquad (6.12b)$$

(c) If $Y = E^n$,

$$\frac{\Delta Y}{Y} \simeq n\frac{\Delta E}{E} \qquad (6.12c)$$

(d) If $Z = \exp(FG)$,

$$\frac{\Delta Z}{Z} \simeq FG\left(\frac{\Delta F}{F} + \frac{\Delta G}{G}\right) \qquad (6.12d)$$

The total exitance M from a blackbody provides an example of case c: Since M varies as the fourth power of temperature, a 2% error in temperature would cause an 8% error in exitance. The spectral exitance $M(\lambda)$ from a blackbody provides an example of case d: For short wavelengths, $M(\lambda)$ varies as e^{-x}, where x is $C_2/\lambda T$. If x is 10, a 2% error in T causes a 20% error in M.

These rules can be combined to handle more complex cases. For the example discussed earlier:

$$X = A\frac{B - C}{D^2}$$

$$\frac{\Delta x}{x} \simeq \frac{\Delta A}{A} + \frac{\Delta B + \Delta C}{B - C} + 2\frac{\Delta D}{D}$$

$$= \frac{0.2}{2.0} + \frac{0.1 + 0.1}{15.2 - 4.8} + 2\left(\frac{0.1}{1.2}\right)$$

$$\cong 10\% + 2\% + 16\%$$

$$= 28\%$$

6.5.3. Probable Total Error

In all of the preceding discussion, we considered how errors could combine in the *worst case*. It is unlikely that the errors will combine in that worst possible way. If the sources of the errors are uncorrelated, the probable resulting error is the rss (root sum of the squares) of the component errors. For example, errors of 10, 2, and 16% yield a total (rss) 19% error:

$$\sqrt{10^2 + 2^2 + 16^2} = \sqrt{360} \simeq 19$$

Standard deviations due to unrelated sources of error are "rss'd" together to determine the composite standard deviation.

6.6. REPORTING UNCERTAINTIES

Uncertainties can be reported explicitly by including either the range of probable values observed, or the standard deviation of the means. Even if neither of these is done, we *imply* the uncertainty of our measurements by the number of significant digits we use. Consider two fractions:

$$\frac{1}{3} = 0.333333... \quad \text{(threes going on forever)}$$

$$\frac{1}{7} = 0.142857142857142857... \quad \text{(repeating forever)}$$

It is not true that one-third *equals* 0.333, nor is it true that one-seventh *equals* 0.1428571. We may *say* that one-third is 0.333, but what we *mean* is: "To the level of precision of three significant figures, one-third is approximated by 0.333."

We will use a *convention* that unstated digits beyond the decimal place are not known. This gives us a shorthand way to report uncertainties: If we say that a distance is 3.56 km, we are implying that it might be 3.564 km, or 3.561 km, or 3.557 km.

This convention is at odds with, and must replace, a convention that we learned in school. There we learned that one-half was *equal* to the decimal 0.5, and that 0.5 was *equal* to 0.50. From now on, let us agree that one-half, 0.5, 0.50, and 0.500 all mean something slightly different: They each convey a different sense of exactness. Similarly, 1.2 MΩ, 1200 kΩ, and 1,200,000 Ω all convey a different sense of exactness.

6.7. THE ETHICS OF SCIENTIFIC WORK

Scientific work is judged by a number of rules of conduct that are not often written down or discussed. The book *False Prophets: Fraud and Error in Science and Medicine* by A. Kohn (1986) is a needed reminder that we can stray from accepted principles and that the community is not tolerant of those that do. In addition to describing many interesting cases of science gone wrong, Kohn discusses the proper conduct of science and provides references to other studies.

The rules have a long-term practical benefit; their need is summarized by Kenneth S. Norris: "Science is a set of rules that keep the scientists from lying to each other" (quoted by Kohn, 1986). A summary of the rules was given by Mohr (1979): "Be honest; never manipulate data; be precise; be fair with regard to priority; be without bias with regard to data and ideas of your rival; do not make compromises in trying to solve a problem."

A little more detail:

Activities and efforts should be directed toward an extension of scientific knowledge, not toward personal or institutional interests.

Judge in terms of intellectual criteria accepted by your branch of science, not in terms of your personal feelings.

Test claims empirically and logically: Don't accept claims on the basis of the authority or stature of the source.

Be both accurate *and complete* in reporting test conditions and results. Failure to report information about the data that might affect their interpretation violates this precept—even if no one asked you for that information. Make sure that the words "those are details that will just confuse the reader" are not an excuse for leaving out unexplained or contradictory data.

Keep a databook with numbered pages, and preserve it for as long as there may be an interest in your work.

If errors are encountered in work already reported, acknowledge them promptly to the same community who received the original report.

Be tolerant of the ideas of others. Accept the fact that you may be wrong.

Particular "sins" that have been noted include:

Recording of data for experiments not done
Manipulating data to make them look better
Reporting only data that fit the researcher's hypothesis
Failure to give proper credit for work done by others
Multiple reporting of the same work
Inflating authorship lists (listing "honorary" authors)

6.7.1. Independent Checks

Check your own work, encourage others to check you, and check the work of others. There is no slur, insult, or discredit intended in this process. It should be recognized by all as an essential part of the scientific process.

Describe your experiments completely enough and accurately enough that they can be repeated by others. The reproducing of experiments provides a cross-check that tends both to identify blunders and to enhance growth and understanding by increasing the combinations of experiments with new viewpoints, environments, and backgrounds. The requirement that experimental conditions be made available also limits the ability of charlatans to continue undiscovered.

6.7.2.   Agressive Learning

Be curious about the data you are taking. Ask questions of those who assign you a task. Make sure that you know why you are doing an experiment—what is expected and how is it different from what has been done before. If the data do not do what you expect, double check them. (But do not just discredit or discard surprising results as "bad data.")

6.7.3.   A Note of Caution

We are not (with very few exceptions) pure scientists. We generally work for a commercial business, with costs, profits, schedules, and other commitments as fundamental constraints. This can run counter to the guidelines of science in two important ways:

1. At least some of the information associated with our work is proprietary—owned by the company that supported the work—so we may be limited in our ability to share test conditions and results.
2. The extent to which we can satisfy our curiosity is limited by the dictates of schedule and budget: We must complete tasks assigned by others. There is, however, always *some* of our time and resources available for discretionary research, and part of our personal growth depends on how we use that discretionary time.

6.8.   OVERALL TEST TASK

Before measurements are made, someone must plan the data acquisition. Afterward comes interpretation of the data. Still later comes presenting the data, either in the form of a written report or in a verbal presentation. All of these are necessary tasks, and they can be very satisfying.

6.8.1.   Designing the Test

The job of planning what data will be taken and how they will be taken is called *test design*. As part of the test design, one should estimate what the general results and uncertainty will be. This helps decide how much data need be taken, at what intervals, and so on. Test design or planning can be a point of contention between managers and the scientists or engineers who will perform the tests. The manager is charged with getting the most work accomplished in a given time and budget, and wants to know that the proposed test will accomplish whatever is expected of it. He knows from experience that this sometimes does not happen, so he is anxious to see that the work is thoughtfully planned.

Engineers may see the plan as a waste of time or as a paperwork task that only slows them down. It provides management with something specific to criticize and to be reworked before engineers can start "the real work." In addition, many diagnostic tests do not seem amenable to specific plans: "If we knew what to do we probably would know what was wrong, and we wouldn't be doing the test." "I need to poke around and see what the situation is before I can decide what to do next."

The possible result of proceeding without a plan (or without a consensus on the plan) is that the test will not yield meaningful results. "In retrospect we should have done it differently." Sometimes this cannot be avoided. If the team can agree ahead of time on how to proceed, everyone will be happier and a better job will probably be done. An obvious rule (but one that is easy to forget) is: "Define your objective carefully before you attempt to design the test." Many conflicts and iterations in test planning would be avoided if agreement were first reached on a clear set of objectives.

6.8.2. Data Review and Interpretation

As soon as the data become available, they should be reviewed. This process has at least two parts:

1. Data should be checked against your expected values or trends. This does not mean that you should discard data that do not agree with your preconceived ideas: It is just to give yourself a chance to be your own critic before others question you. If your results disagree with what you expected to see, you have time—still at the test set—to double check the setup, to verify your readings, and generally to assure yourself that you really believe your results.

2. Data should be checked for internal consistency: Did you get consistent results? Did things vary slowly with time, or were they erratic? If you plotted your data, are the curves smooth or are there unexpected "bumps" or discontinuities? Such behavior may be a symptom of errors in the data, or they may be hints that can lead to useful and exciting new understanding.

In either case it is much more satisfying to confirm or deny your suspicions while you are still in the lab than it is to find out hours or weeks later that you wish you had more data. After you have done your data review, you should strive to interpret it. This can be described as finding a story or message in the data. Other people will not generally have a great interest in all of your data. They want to know the "bottom line": Is there a new effect at work here? Have you confirmed a previous trend? Have you denied a previous hypothesis? Who or what is affected by what you have done? In what way do your data help (or make life more difficult for) your potential audience?

6.8.3. Presentation of Results

If you take data, you will undoubtedly have the opportunity to present them to others. No matter what level of formality is appropriate for your presentation, and whether it is written or verbal, a few guidelines will make it more successful. It is tempting to describe your progress in a chronological or historical way: why you started the project, how it developed, the problems you ran into, how cleverly you solved them, the clues that led up to your latest interpretation, and so on. Unfortunately, no one really cares about all that. What most people want is:

1. The objectives and the bottom line, clearly and succinctly
2. How sure you are about the bottom line
3. What you propose should be done next
4. Your evidence (the level of detail desired here depends on your credibility and the amount of surprise your bottom line produces)

Because of this, good reports (whether written or presented verbally) generally have a rather standard structure:

Introduction. A brief statement of the problem and objectives, and an outline of what will follow

Executive summary. The results, briefly but completely

Discussion. Key detail about the method and results

Appendices. All the gory detail you want to include, or think might be useful some day (a good place to store the data and other information you may need to explain your thinking when the topic is revisited five years later)

In all of this, try to save the audience as much time as you can by making things as clear as possible without insulting their intelligence.

Graphs work better than tables, and tables work better than text. Break your text into digestable sections with an outline and clear headings to get the reader ready for what you are going to do next. Label your figures and graphs clearly. Do not use three different words for one concept.

If you have lots of time, one way to go through the writing process is: think–outline–write–then alternately rethink, get criticism, and rewrite. Finally, get away from it for a few days, do a final editing, and distribute or present your material. If you do not have the time to do all this, don't be hard on yourself if a few errors slip through.

There is often a competing need for rapid distribution of your material. In a fast-paced program or job, people will be waiting anxiously for your results, and information a week old will not be useful, either because your results have already leaked out, or because the need has vanished. You must balance the need to do a good job in presenting the data with the need to do it quickly.

6.8.4. Graphing

Types of Graph Paper. Be familiar with the many kinds of graph paper (and computer-generated graph formats) that are available.

Cartesian (linear, rectangular): good for data that are expected to obey an equation such as $y = a + bx$.

Semilog (logarithmic on the vertical axis, linear on the horizontal axis): good for data with an exponential dependence because $y = a \exp(bx)$ plots as a straight line on this paper.

Log (sometimes called log-log, with logarithmic scales on both axis): good for $y = x$ raised to a power.

Scale. Selection of the scale is not as easy as one might think, since several objectives should be satisfied. This may not be feasible, in which case you will need to make some compromises.

Use conventional divisions: one, two, or five units per small division on the paper. Consider the level of precision you want to maintain and convey with your graph. For calibration or interpolation, you want the graph to convey the same level of precision as your original data, so make the smallest divisions on the paper represent your measurement precision, or even a little less. If the measurement precision was 0.3 mV, one division on my paper should be 0.1 or 0.2 mV. The size of the resulting graph may be inconvenient, but that cannot be helped unless you are willing to give up some precision. If the graph is to show trends or generic features, size and scale are selected to show the phenomena of interest in a convenient format. Graph paper is most common in 8.5 × 11-in. sheets, but the 11-

× 17-in. size is commonly used, and paper is even available in rolls many feet long.

Labeling. Indicate the data points clearly and include a legend to identify any special symbols or notation that you have used. Always include a reference that will help you locate your raw data later. Consistently supporting your results with raw data is good practice: It reminds you to be honest and accurate, and enhances your credibility.

REFERENCES

Appleton, George L. (1981). *Laboratory Manual for Physics 151*, unpublished notes used at California State College at Long Beach.

Beers, Yardley (1957). *Introduction to the Theory of Error*, Addison-Wesley, Reading, Mass.

Kohn, Alexander (1986). *False Prophets: Fraud and Error in Science and Medicine*, Basil Blackwell, New York.

Leaver, R. H., and T. R. Thomas (1974). *Analysis and Presentation of Experimental Results*, Halstead Press, New York.

Mandel, John (1964). *The Statistical Analysis of Experimental Data*, Dover, New York.

Mohr, M. (1979). "The Ethics of Science," *Int. Sci. Rev.* 4, 45.

SUGGESTED READING

Beers, Yardley (1957). *Introduction to the Theory of Error*, Addison-Wesley, Reading, Mass. This is a short (66 pages) intermediate-level text with a useful mix of rigor and intuitive material. I find it an excellent source.

Kohn, Alexander (1986). *False Prophets: Fraud and Error in Science and Medicine*, Basil Blackwell, New York. Reviewed in *Physics Today*, September 1988, page 107; discusses the Piltdown Man, Hershel, Isaac Newton, Gregor Mendel's peas, and Margaret Mead. Good reminder of the ethics of science.

Leaver, R. H., and T. R. Thomas (1974). *Analysis and Presentation of Experimental Results*, Halstead Press, New York. Another short book (174 pages), this one addressing experiment planning and statistics, and reporting of experimental data. Very readable.

Moffat, Robert J. (1988). "Describing the Uncertainties in Experimental Results," *Experimental Thermal and Fluid Science* 1, 3. Provides some perceptive comments on the analysis and reporting of uncertainities that I have not seen elsewhere.

Tufte, Edward R. (1983). *The Visual Display of Quantitative Information*, Graph-

ics Press, Cheshire, Conn. A comprehensive review of ways of displaying
data, including some guidelines, historical examples, and examples of "chart
junk."

Young, Hugh D. (1962). *Statistical Treatment of Experimental Data*, McGraw-
Hill, New York.

PROBLEMS

1. List the three types of "errors," how to reduce them, and how to
    estimate their magnitude.

2. Using digital voltmeter and current meters that were calibrated six
    days ago by the Secondary Standards Lab of your company, workers
    measured the voltage across a resistor and current through it, and
    determined its resistance to be 37.23426 kΩ.

    Data were taken on four days. Two technicians worked together
    on the first and second days, but only one technician took the data
    the last two days. Data for three of the four days were similar, but
    on the first day the spread in the results was much larger than on
    the other days.

    The reported value is the average of 98 "good" values: It excludes
    the "bad" day's data and a few values from each of the other days.
    Representative data for the second day are shown below.

    Describe the confidence you would have in the result reported.
    Comment on systematic errors, random errors, and blunders.

Tuesday, 27 July   **Ed Barnes and A. Gud Mann**

Time	Temperature (°C)	Current (μA)	Voltage (mV)	Resistance (kΩ)
0834	16.6	12.47	464.3	37.2334
		12.58	468.9	37.2735
		23.17	865.1	37.3371
		25.14	905.3	36.0103*
		23.21	862.9	37.1779
0914	17.3	9.122	339.6	37.2287
		9.236	342.9	37.1265

* Excluded from average; moisture had condensed on resistor terminals.

3. You have been assigned to determine the density of a rare pearl.
    Describe your method, the errors (uncertainties) in each component
    of the process, and the resulting uncertainty in the reported density.

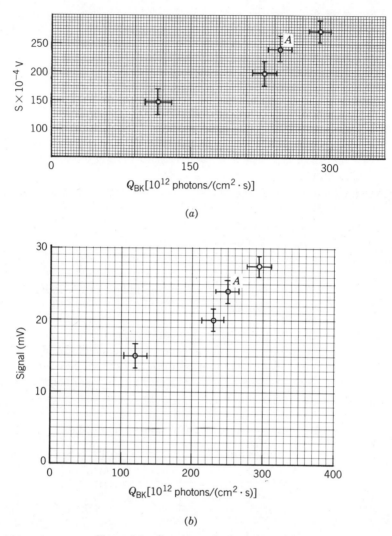

Figure 6.3. Sample graphs for problem 4.

(a) You have 1 minute and no equipment, but you can touch and lift the pearl.

(b) You have 10 minutes, a kitchen scale, and a 6-in. plastic ruler.

(c) You have 1 year, and a budget of $300,000.

4. (a) Critique the two graphs of Figure 6.3. List everything you do not like about them.

    (b) What is the incidance (Q_{bk}) and the signal for data point A?

5. Obtain (or "drylab") some data, select paper, and prepare graphs for the following:

    (a) Weight of an infant measured once a week for the first three months of life.

    (b) Weight of four cable assemblies, each exposed to a different humidity, and very accurately weighed each day for a total of 21 days.

    (c) Price of gold over the last 10 years, *or* the cost of a "typical" three-bedroom house in your area during the last 10 years, *or* a similar problem of your choice.

    (d) Signal from a slightly nonlinear detector over the incidance range of 2×10^{12} to 7.5×10^{15} photons/(cm^2·s) for the purpose of
    (i) Illustrating the general range of signals to a customer
    (ii) A "no-computer needed" method of translating signal to incidance with an accuracy of 1%.

7

Cryogenics

Cryogenics is the branch of science that deals with the production and effects of very low temperatures. In this chapter we discuss the means of cooling detectors to cryogenic temperatures, measuring the temperatures, and some associated safety hazards. For further study we recommend *Experimental Techniques in Low Temperature Physics* by White (1987).

Cryogenic temperatures are almost always measured on the Kelvin scale. Figure 7.1 shows the relationship between the Fahrenheit, Celsius, and Kelvin scales, along with a few temperatures of special interest. The Celsius (sometimes called centigrade) scale is constructed so that the freezing and boiling points of water are at 0 and 100°C; these values are 32 and 212°F, respectively. The Kelvin scale is set up so that the lowest theoretically attainable temperature is at 0 K, and that the difference in temperature between the ice and boiling points of water is 100 K. The result is that 0°C is 273.15 K. The temperature of the sun is about 6000 K, kitchen ovens operate around 477 K, and the comfort level for human beings is about 293 ± 5 K.

7.1. CRYOGENS

The common cryogens for IR detector work are listed in Table 7.1; the most common are liquid nitrogen and liquid helium. Helium is the coldest known cryogen. Under normal atmopsheric pressure it boils at 4.2 K, very close to absolute zero. The boiling point can be reduced by reducing the pressure; this is done by pumping over the helium with large mechanical pumps. Helium has the lowest cooling capacity of the common cryogens; it boils off very quickly, so requires very careful insulation in order to store it for appreciable periods of time.

Nitrogen (77.3 K) is relatively inexpensive and has a relatively high cooling capacity. It is less expensive than Helium and Neon, and is used in large quantities both to thermally insulate the more expensive cryogens,

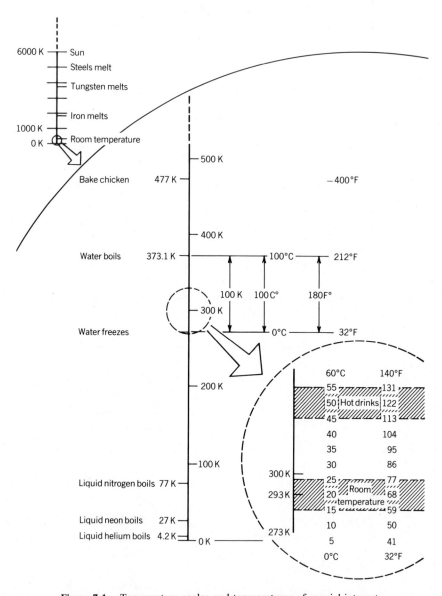

Figure 7.1. Temperature scales and temperatures of special interest.

Table 7.1   Common Cryogens and Their Characteristics[a]

Cryogen	Temp (K)	Cooling Capacity (J/cm³)	Density (g/L)			Cost (dollars/liquid liter)
			Liquid NBP	Gas NBP	Gas STP	
Helium	4.2	2.6	125	16.7	0.1656	7
Hydrogen	20.4	31	70	1.33	0.061	
Neon	27.1	103	1210	9.5	0.810	225
Nitrogen	77.3	161	808	4.41	1.12	0.55
Argon	87.4	229	1400	5.76	1.78	3
Oxygen	90.2	243	1140	4.75	1.29	0.75
Freon-14	145.1	218	1960	7.2		
Freon-13	191.8	220	1505	7.9		
(Water)[b]	373.1	2443	1000			

[a] "Temp." is the boiling temperature at normal atmospheric pressure (760 torr); it is sometimes called the normal boiling point (NBP); cooling capacity is the heat required to vaporize 1 cm³ of the liquid at its NBP; STP is "standard temperature and pressure" (20°C, 760 torr); cost is an approximate value in quantities of 150 L (1988).

[b] Water is not a cryogen, but is included here to make the point that cryogens are liquids; they differ from water only in that their normal boiling point is below room temperature.

and to cool detectors. Neon (27 K) is expensive because of its rarity, but provides a convenient temperature midway between that of helium and nitrogen.

Oxygen (90.2 K) can be purchased or liquefied in the laboratory by condensing atmospheric oxygen onto tubes chilled with liquid nitrogen. It presents a special safety hazard because materials that are normally only mildly flammable will burn very vigorously in the presence of a high oxygen content. Several Freons are available; they provide a range of temperatures between 100 and 300 K.

7.1.1.   Heat of Vaporization

The energy required to vaporize ("boil off") a given quantity of liquid is called the *heat of vaporization h*. For example, the heat of vaporization of liquid helium is 20.9 J/g. The heat of vaporization times the density ρ is the energy required per volume; for helium this is 2.61 J/cm³ of liquid (see Equation 7.1).

HEAT ABSORBED BY GAS DURING VAPORIZATION

$$Q = hM \qquad (7.1a)$$
$$= h\rho V \qquad (7.1b)$$

where M and V are the mass and volume of liquid vaporized by heat Q.

7.1.2. Enthalpy

When helium or any other cryogen is used in a cooling system, it absorbs heat as it vaporizes (the heat of vaporization), then absorbs more heat as it is warmed to room temperature. The total heat absorbed is the heat of vaporization plus the specific heat integrated from the boiling point to the final temperature of the gas (Equation 7.2). This total heat can be calculated from tables of the *enthalpy H* of the cryogen (see Table 7.2); the total heat is the difference in the enthalpy of the material at the starting and ending temperatures. Note that the heat of vaporization is only a small part of the cooling capacity of cryogens. The vapor can provide a great deal of useful cooling if used efficiently; this cooling capability is lost if the vapor is allowed to vent directly to the atmosphere. For helium, 1 cm^3 of liquid absorbs about 2.6 J when it vaporizes, but will absorb another 192 J before it reaches room temperature.

HEAT ABSORBED BY THE GAS (VAPORIZATION + WARMING)

$$Q = [H(T_2) - H(T_1)]\rho_l V_l \qquad (7.2)$$

7.1.3. Storage of Cryogens: Dewars

To keep cryogens liquid for any period of time it is necessary to isolate them from room temperatures. This is normally done by storing them in *dewars** designed for that purpose. A "thermos bottle" is a household version of a storage dewar. *Storage dewars* normally have a large capacity and small-neck diameters. They are often provided by the vendor of the cryogens, and the user may pay a daily demurage charge. Experimental and test dewars differ from storage dewars in that they include windows,

* Named after its inventor, Sir James Dewar (1842–1923), the Scottish chemist and physicist who first liquefied oxygen.

Table 7.2 Enthalpy of Helium and Nitrogen[a]

Condition	He	N$_2$
Normal boiling point (NBP)	4.2 K	77.4 K
Enthalpy of liquid at NBP	0	0
Heat of vaporization	<u>2.6</u>	<u>160.8</u>
Enthalpy of gas at NBP	2.6	160.8
$\int_{NBP}^{77} c\, Dt$	<u>48.1</u>	<u>0</u>
Enthalpy of gas at 77 K	50.7	160.8
$\int_{77}^{300} c\, Dt$	<u>143.9</u>	<u>189.9</u>
Enthalpy of gas at 300 K	194.6	350.7

Source: Data from Scott (1959) p. 269, 310–313.

[a] Heat of vaporization and enthalpy are in J/cm^3 of liquid. All enthalpies are taken to be zero for the liquid at the normal boiling point.

electrical feedthroughs, temperature sensors, and other features needed for the test. Figure 7.2 shows a storage dewar and a conventional test dewar, along with the tube used to transfer helium from the storage dewar to the test dewar.

7.1.4. Transfer of Cryogens

To move helium and neon from the storage dewar to the experimental dewar without exposing them to room temperatures, special vacuum-insulated *transfer lines* are used. The storage dewar has a fitting to accept a hose with pressurizing helium gas. The pressure forces liquid out of the storage dewar, through the transfer line, and into the experimental dewar. Nitrogen and the Freons have high enough heat of vaporization and are cheap enough that the time and trouble of the cryogenic transfer tube is not justified: They are normally just poured from one container into another.

7.1.5. Continual Transfer Systems

To use the test dewars described earlier, one fills them with cryogen, then disconnects the transfer tube and storage dewar and moves them out of the way while the experiment proceeds. If the cryogen in the test dewar is used up before the experiment is complete, another transfer is made.

Continual transfer systems are gradually replacing the original dewar systems. These newer systems use a special transfer tube that deposits the liquid on the surface to be cooled; it has no real reservoir. The transfer

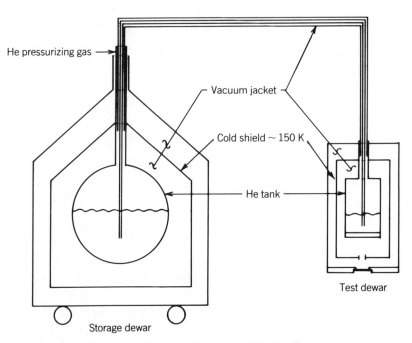

He pressurizing gas

Vacuum jacket

Cold shield ~ 150 K

He tank

Test dewar

Storage dewar

Figure 7.2. Storage and transfer of liquid helium.

tube is left in place and a very slow transfer continues as long as the experiment requires.

Advantages are that the dewar cold end mass is much smaller than that of a "bucket-type" dewar, so that the initial helium losses are much smaller, and cool-down and warm-up are faster. The dewar itself is smaller and thus easier to handle. The dewar can operate in any orientation. One can adjust the transfer rate to help maintain any desired temperature. Cost is a disadvantage—these dewars cost about 50% more than a "bucket"-type dewar.

7.1.6. Liquid Level Sensors

In the early days of cryogenic work, glass dewars were used and a narrow vertical strip was left unsilvered to allow the user to view the liquid level. This was difficult at best: The liquid is nearly invisible because its index of refraction is very close to that of air. Of course, metal dewars prohibit even that method of level detection, so alternative methods are necessary. Electronic liquid-level sensors are available, but for helium and neon another scheme is simpler and more reliable.

A small brass "cup" about 1 in. in diameter and an inch high is soldered to a thin stainless steel tube (a $\frac{1}{8}$-in. outside diameter works well). The top of the cup is closed off with a rubber balloon, a finger cot, or even the palm of the hand, and the tube is inserted slowly into the dewar. As the tube cools, oscillations of the gas (sometimes called *Taconis oscillations**) in the tube begin. These can be felt at the membrane on top of the cup. They change frequency radically when the tip of the tube enters the liquid. Typically, the user inserts the tube slowly into the tank, then marks with a thumb or an alligator clip the point on the tube at which the oscillations change frequency, and compares it to the point at which the tube reaches the bottom of tank. White (1987) discusses this and other types of level indicators.

7.2. THERMAL PROPERTIES OF SOLIDS

Work at low temperatures will require calculations of the time required for components to cool, the ultimate temperature they will reach, and the amount of cryogen used in cooling process. Those calculations depend on the ability of the materials to conduct heat, the amount of heat they store, and the cooling capacity of the cryogens. In this section we discuss the material properties and how the required calculations are done.

7.2.1. Thermal Expansion

As materials change temperature, their linear dimensions change. The change in length (or width, or thickness) is the original dimension times the temperature difference times the *coefficient of expansion* α. The coefficient of expansion has the units of reciprocal degrees. To emphasize the dependence on original length, the units of α are sometimes written as (in./in.)/K, or (cm/cm)/K.

The expansion issue is important because mismatches can stress parts that are bonded together. Great care is required in selecting substrates for detector material, in the mounting of windows that will be cooled, and in solder or glue joints.

The coefficient of expansion α is temperature dependent, going to zero at very low temperatures, as shown in Figure 7.3. Because the coefficient

---

* W. Taconis et al. (1949) attributed the reproducibility of some of their data to effective mixing of their solution by these oscillations. They cite Keesom (1942), who mentioned great difficulties caused by the oscillations. They can be an unexpected source of heat leaks [see White (1987)]. For theoretical explanations see Taconis et al. (1949), Kramers (1949), Wheatley and Cox (1985), and Yazaki et al. (1980).

of expansion is temperature dependent, average or integrated values are required (Equation 7.3) if ΔT is not small. Integrated values are shown in Figure 7.4; more detailed information can be found in the literature: Touloukian (1970), for example. Representative values of the integrated coefficients of expansion are given in Figure 7.4.

THERMAL EXPANSION

$$\Delta x = x_0 \int_{T_1}^{T_2} \alpha(T)\, dT \qquad\qquad (7.3a)$$

$$= x_0 \bar{\alpha} (T_2 - T_1) \qquad\qquad (7.3b)$$

$$\simeq x_0 \alpha(T_1)\, (T_2 - T_1) \qquad \text{if } T_2 - T_1 \text{ is small} \qquad (7.3c)$$

7.2.2. Heat Conduction; Thermal Conductivity

The thermal conductivity of a material is a measure of its ability to transfer heat by conduction. For example, the thermal conductivity of copper and silver are much higher than that of stainless steel or wood. Consider the

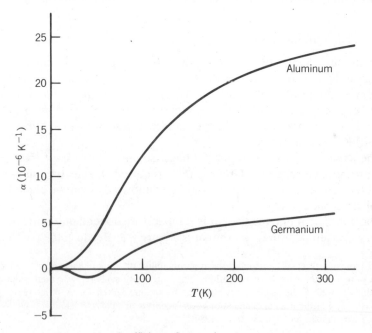

Figure 7.3.   Coefficient of expansion versus temperature.

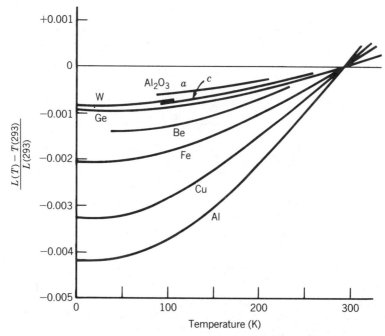

Figure 7.4. Integrated thermal expansion for common cryogenic building materials. [Data from Touloukian and Ho (1970).] Curves labeled Al_2O_3 a and c are for expansion parallel to the a and c crystal axes.

bar of uniform cross-sectional area a and length l shown in Figure 7.5. If the temperature difference ΔT between the ends of the bar is small, the heat transferred by conduction through the bar is given by Equation 7.4:

HEAT CONDUCTION

$$P = k \frac{A}{L} \Delta T \qquad (7.4)$$

where ΔT is the temperature difference across the bar and k is the thermal conductivity. The units of thermal conductivity are W/(cm·K).

The thermal conductivity of most materials varies dramatically with temperature in the cryogenic temperature ranges (see Figure 7.6). Note how much the conductivity varies for a given material from room temperature to its peak value, then how quickly it falls to low values near liquid helium temperature. Note also the wide range of conductivities: The graph covers seven orders of magnitude.

The thermal conductivity at cryogenic temperatures is also a strong

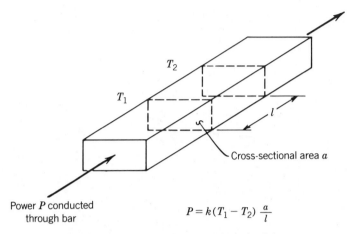

$$P = k(T_1 - T_2)\,\frac{a}{l}$$

Power P conducted
through bar

Figure 7.5. Definition of thermal conductivity.

function of the material purity and heat treating. The curves of Figure 7.7 are all for copper of different varities and condition. Because of the large differences in the conductivity, it is misleading to state a specific value for the conductivity of any material without also providing the temperature at which the value applies, and a description of the purity and treatment of the sample.

Extensive tables of thermal conductivity are available; see, for example, Touloukian (1970). For critical designs use care in selecting and specifying the material to be used and in determining the appropriate conductivity to use in your calculations. Denning (1969) provides the thermal conductivity of some cryogenic adhesives.

If the temperature difference ΔT is small, formula (7.4) can be used directly with conductivity values taken from graphs (such as Figure 7.7) or tables. If the temperature difference is large (300 K to 77 K, for example) we must use an average conductivity in the formulas above, or replace the product of factors k and $(T_2 - T_1)$ by an integrated value:

CONDUCTED HEAT

$$P = \int_{T_1}^{T_2} k(T)\,dT\,\frac{A}{L} \qquad\qquad (7.5a)$$

$$= \bar{k}(T_2 - T_1)\frac{A}{L} \qquad\qquad (7.5b)$$

$$\simeq k(T_1)\,(T_2 - T_1)\frac{A}{L} \qquad \text{if } T_2 - T_1 \text{ is small} \qquad (7.5c)$$

The integrated values between 4.2 K, 77 K, and 300 K for several common cryogenic building materials are given in Table 7.3. Other references provide values for different materials, integrated over different temperature ranges.

The steps in performing thermal conductivity calculations are then: (1)

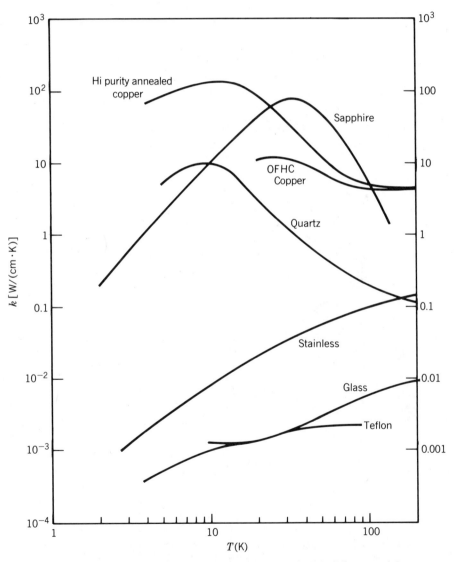

Figure 7.6. Thermal conductivity for common cryogenic building materials.

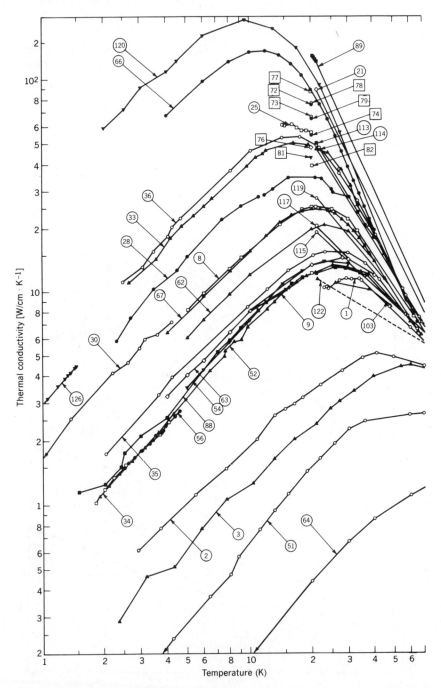

Figure 7.7. Thermal conductivity of copper, showing the wide variations with purity and heat treatment. From Touloukian Y. S., R. W. Powell, C. Y. Ho, and P. G. Klemens, (1970)

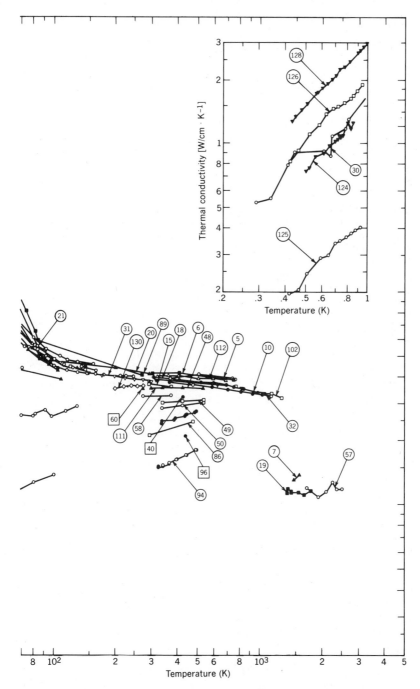

"Thermal Conductivity—Metallic Elements and Alloys," Volume 1 of *Thermophysical Properties of Matter—The TPRC Data Series, IFI*/Plenum Data Co., New York, page 68; used by permission.

Table 7.3 Thermal Conductivity and Integrated Thermal Conductivity for Common Cryogenic Building Materials

	Thermal Conductivity [W/(cm·K)]			Integrated Values (W/cm)		
	4.2 K	77 K	300 K	0–4.2 K	4.2–77 K	77–300 K
Sapphire	2.7	10.	0.46			
Copper						
OFHC	2.5	5.2	4.0	7.5	700	900
ETP				6.4	715	914
Aluminum	0.03	0.57	1.21		223	505
Constantan	0.008	0.18	0.23	0.012	10.2	44.6
Manganin	0.005	0.13	0.22	0.009	6.3	37.7
Stainless	0.0025	0.08	0.15	0.006	3.3	27.4
Pyrex	0.001	0.0045	0.01	0.002	0.18	1.83
Teflon	0.005	0.024	0.027		0.132	0.57
Nylon	0.00013	0.002	0.0003			

Sources: Data from White (1987) p. 133 and Touloukian and Ho (1970).

determine the material and temperature range of interest, and locate a conductivity value or curve appropriate to that material and temperature; and (2) if the temperature range is small enough, or if your accuracy requirements are not demanding, use formula (7.5c) to predict the conducted heat. If more accuracy is desired, obtain the integrated conductivity value or do the integration yourself: The integrated conductivity can be scaled from the area under a graph of thermal conductivity versus temperature.

It is sometimes convenient to combine the conductivity with the area-to-length ratio in a *thermal conductance* (K) or a *thermal impedance* (Z):

$$K = k\frac{A}{l} \tag{7.6a}$$

$$Z = \frac{1}{K} \tag{7.6b}$$

Units of thermal conductance K are W/K. In this form,

$$P = K\,\Delta T = \frac{\Delta T}{Z} \tag{7.7a}$$

$$\Delta T = PZ = \frac{P}{K} \tag{7.7b}$$

Wiedemann–Franz Law. The Wiedemann–Franz law (Kittel, 1966) states that the thermal conductivity of metals is proportional to the electrical conductivity σ times the temperature T:

$$k = L\sigma T \qquad (7.8)$$

The ratio of k to σT is the Lorenz ratio L. For pure metals at not too low temperatures the Lorenz ratio is about 2.5×10^{-8} W·Ω/K^2. Denote this value by L_0 for use in later reference.

If the thermal conductivity cannot be measured or predicted in any other way, this law may be useful to estimate the thermal conductivity. The limitations on its applicability are so stringent that it will not often be accurate, so it should be used only as a last resort, and the assumptions clearly noted. This law holds when the electrons dominate the heat transfer process and when the lifetimes associated with thermal and electronic conduction are the same. The first condition does not hold for insulators and impure metals, and the thermal conductivity of insulators will be higher than that predicted from the Wiedeman–Franz law. The second condition does not hold at low temperatures [see Kittel (1966), p. 180]; at low temperatures the thermal conductivity of good conductors will be less than that predicted by the Wiedeman–Franz law.

Since both the thermal and electrical conductance are given by their conductivities times the same geometrical factor (A/l), the thermal conductance K scales from the electrical resistance R without knowing the geometrical factors:

$$K = L\frac{T}{R} \qquad (7.9)$$

Example: How sensitive would an ohmmeter test be for potential thermal shorts between two pure metal components, one at 175 K and one at 185 K?

Using $R > 1\ \Omega$, $L = 2.5 \times 10^{-8}$ W·Ω/K^2, and $T = 80$ K, formula (7.9) yields $K < 4.5 \times 10^{-6}$ W/K. For $\Delta T = 10$ K, formula (7.7a) yields $Q < 45\ \mu$W. By setting the acceptable minimum resistance at 1 kΩ, one could check for heat leaks of 4.5×10^{-9} W. Remember, however, that this test will not locate thermal shorts due to electrical insulators. ∎

Thermal Conductance through Pressed Contacts. Previously, we discussed the thermal impedance Z associated with a bar of length l and cross-sectional area A. Now consider the joint between two surfaces. Although it has essentially zero thickness, it causes a thermal impedance between the two materials. The heat transfer can be calculated from the conductance formulas in the form $P = K\,\Delta T = \Delta T/Z$. As before the conductance K is proportional to the contact surface area A, but in this case there is no length involved.

The nomenclature is not always consistent—you may find these "contact conductances" stated in units of W/(cm^2·K), in which case the user multiplies the given value by the contact area to arrive at a conductance in the usual units. In other cases the contact conductance will be stated in W/K. Be guided by the units; it is always true that $P = KT$ if K has units of W/K.

The thermal conductance (W/K) between pressed metal contacts is proportional to the total force applied, and the contact conductance [W/(cm^2·K)] is proportional to the pressure applied. The conductance can be enhanced by preparing flat, smooth surfaces. Because most materials exhibit some "creep" under pressure, the thermal conductance may improve with time.

Conductance between dry rough surfaces is much worse in a vacuum than in air. Such a contact can be improved by the use of thermal grease—generally, a grease filled with metal oxide. On the other hand, use of a thermal grease *degrades* the conductance if the surfaces are very flat, smooth, and/or malleable materials (copper to indium, for example).

Values for specific materials and conditions are provided in most recent texts on heat transfer. Older journal references include Berman (1956), Graff (1960), Rogers (1961), Fletcher et al. (1968), and Radebaugh (1977).

7.2.3. Heat Capacity; Specific Heat

The amount of heat required to change the temperature of an object by 1 degree is its *heat capacity*. The usual symbol is C, and the common units are joules per kelvin: J/K. The *specific* heat capacity c (often just called the *specific heat*) is the heat capacity per unit mass, and common units are joules per gram per kelvin: J/(g·K). The amount of heat required to cause an arbitrary temperature change in an object of mass m is described in Equation 7.10. For cryogenic temperatures the heat capacity is a strong function of temperature (see Figure 7.8), so average or integrated values are necessary if large temperature changes are involved.

HEAT INVOLVED IN A TEMPERATURE CHANGE

$$Q = m \int_{T_1}^{T_2} c(T)\, dT \tag{7.10a}$$

$$Q = m\bar{c}(T_2 - T_1) \tag{7.10b}$$

$$Q \approx mc(T_1)(T_2 - T_1) \quad \text{if } T_2 - T_1 \text{ is small} \tag{7.10c}$$

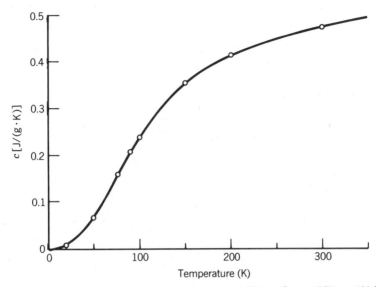

Figure 7.8. Specific heat of type 18-8 stainless steel. [From Scott (1959), p. 130.]

Unlike the thermal conductivity, the specific heat capacity is *not* sensitive to sample purity or preparation technique, so the selection of appropriate values for a particular problem is not difficult. Representative values are given in Table 7.4. If the specific heat capacity for a particular material cannot be located in tables or graphs, it may be possible to obtain the Debye temperature Θ_D, in which case the heat capacity can be estimated from the Debye theory. To predict the specific heat capacity at temperature T from the Debye temperature, first calculate the ratio T/Θ_D,

Table 7.4 Specific Heat Capacity and Integrated Values for Common Cryogenic Building Materials

	Specific Heat [J/(g·K)]			Integrated Values (J/g)		
	4.2 K	77 K	300 K	0–4.2 K	4.2–77 K	77–300 K
Sapphire		0.062	0.80			
Copper	0.0001	0.19	0.39	0.0002	5.4	79.6
Aluminum	0.0003	0.34	0.83	0.0005	8.4	170.4
Stainless		0.16	0.48			
Pyrex	0.0002	0.20	0.76	0.0003	6.2	

Sources: Data from Robert C. Weast, ed. (1986) *Handbook of Chemistry and Physics*, page D-179. CRC Press, Boca Raton, Fl., and Touloukian and Ho (1970).

then refer to tables or graphs of the Debye function—specific heat capacity per mole as a function of T/Θ_D [see Kittel (1968), p. 179; or Kittel and Kroemer (1980), Fig. 4.11; or White (1987), Table D]. Divide that value by the molecular mass M (grams per mole) to obtain the specific heat in J/(g·K):

$$c[\text{J/(g·K)}] = \frac{c[\text{J/(mol·K)}]}{M/(\text{g/mol})} \tag{7.11}$$

Example: The specific heat of copper at 77 K is about 0.17 J/(g·K); for 20 g the heat capacity is 3.4 J/K.

For copper, M is 63.5 g/mol, and the Debye temperature Θ_D is 343 K [Kittel (1968), p. 180, or Kittel and Kroemer (1980), p. 107].

$$\frac{T}{\Theta_D} = \frac{77\text{ K}}{343\text{ K}} = 0.22$$

For that temperature ratio, $c \simeq 11$ J/(mol·K) [from Kittel et al. (1980), p. 108, or White (1987) p. 314].

Finally,

$$c = \frac{11\text{ J/(mol·K)}}{63.5\text{ g/mol}}$$

$$= 0.17\text{ J/(g·K)}$$

$$C = cm = [0.17\text{ J/(g·K)}]\,(20\text{ g}) \simeq 3.4\text{ J/K} \qquad \blacksquare$$

7.2.4. Electrical Analogs

For those with electronic experience, thermal performance and calculations can be visualized in terms of electrical problems: the mathematics are the same if we make the following substitutions:

charge (C)	→ heat energy (J)
current (C/s, A)	→ heat flow (J/s, W)
voltage (V)	→ temperature (K)
impedance (V/A, Ω)	→ thermal impedance (K/W)
capacitance (C/V)	→ heat capacity (J/K)

Once these substitutions are made, the electrical formulas and intuition can be applied to thermal problems. Ohm's law, voltage dividers, parallel and series impedances and conductances, and capacitors charging through a resistor all have thermal analogs.

7.3. DEWAR DESIGN

Conduction from room temperature into the cryogen is minimized by using supports with the smallest possible cross section and as long as possible. Materials with low thermal conductivity are used. Stainless steel is the most common material. Conduction through electrical leads is also important. The thermal conductivity can be reduced by using leads made of manganin or constantan. Unfortunately, materials with low thermal conductivity also have low electrical conductivities, so electrical requirements must be balanced against cryogenic considerations.

Power dissipation in the leads is a source of heat that should be considered; larger-diameter leads reduce the power dissipation but increase the conducted heat load. The optimization problem is discussed by McFee (1959). Walmsley et al. (1972) and subsequent authors in the same journal consider the cooling effect of evaporated gas on the optimization problem.

Convection and *gaseous conduction* can be reduced to negligible values by evacuating the space between the cold chamber and the outside chamber wall. To maintain that vacuum one must use materials that do not outgas and must keep the surfaces clean (even fingerprints can generate enough gas to slow pumpdown or spoil the vacuum shortly after the dewar is removed from the pump). Formulas for thermal conduction through gases are discussed in Chapter 8.

Radiation can be reduced by reducing the emissivity of the cold and warm surfaces that face each other, and by introducing an intermediate temperature surface between the coldest and warmest surfaces. The radiative heat load to a spherical or cylindrical container of area A_1, temperature T_1, and emissivity ϵ_1 from a surrounding container of temperature T_2 and emissivity ϵ_2 is given by Scott (1959):

$$P = \sigma(T_2 - T_1)\epsilon A_1 \qquad (7.12a)$$

where

$$\epsilon = \frac{\epsilon_1 \epsilon_2}{\epsilon_1 + \epsilon_2 - \epsilon_1 \epsilon_2} \qquad (7.12b)$$

The importance of these effects is illustrated by comparing the heat load components for two hypothetical dewar designs (Table 7.5). The first is clearly poorly designed from a thermal viewpoint. The second is perhaps too fragile for practical applications, but it illustrates what can be done to reduce the heat load.

Table 7.5   Heat Load Calculations for Helium Dewar Design

	Dewar A	Heat Load (W)	Dewar B	Heat Load (W)
Conduction				
NECK				
Material	Stainless		Stainless	
$\int k\,dt$ (W/cm)	30.6		30.6	
Diameter (in.)	5/8		1/2	
Wall (in.)	0.060		0.020	
Length (in.)	4.0		6.0	
A/l (cm)	0.075		0.013	
Conduction		2.25W		0.41W
LEADS				
Material	Copper		Manganin	
$\int k\,dt$ (W/cm)	1620		44	
Diameter (in.)	0.010		0.005	
Length (in.)	8.0		12.0	
Number	100		50	
Lead conduction		4.0W		0.009W
Gaseous conduction		Neglect		Neglect
Radiation				
Emissivity	0.20		0.05	
$T_2 - T_1$ (K)	300		200	
Area (cm^2)	400		400	
Radiated power		4.0W		0.4W
Total Heat Load		10.2W		0.8W
Hold time		260 s		3200 s
		4 min		50 min

Additional refinements will improve the hold time and perhaps allow us to add a little strength to the dewar neck: Multilayer insulation (MLI) can be wrapped loosely around the inner wall. It is a "blanket" of alternating layers of aluminized Mylar and a veil-like material to keep the layers from touching. The escaping helium gas can be channeled along the neck walls, cooling them and compensating for some of the conducted heat. We did not consider these improvements in our hypothetical dewars because they complicate the calculations, but they can improve dewar hold time substantially.

The power transferred by radiation if MLI with N independent heat shields float in temperature between the planes at T_1 and T_2 is:

$$P = \frac{\sigma(T_2^4 - T_1^4)}{N + 1} \epsilon A_1 \tag{7.13}$$

7.4. REFRIGERATORS

Cooling to cryogenic temperatures can be done without the direct use of purchased liquid cryogens. Chapter 15 of *The Infrared Handbook* (Donabedian, 1978) provides an overview of the different cooling systems available.

7.4.1. Joule–Thompson Liquefier

This device allows high-pressure gas to expand into an insulated volume. The resulting gas has been cooled by the expansion process and is forced back over the tubing that carried it. This cools the new gas, so when it expands, it is even cooler. Eventually, the exanding gas will emerge as liquid.

These refrigerators are used in miniature glass dewars. A small gas tank can maintain the required operating temperature in a dewar for several hours. As described here, this is an *open-cycle* process; the gas eventually runs out since it leaves the refrigerator. If the gas is collected and compressed, it can be recirculated indefinitely. A device that operates in this way is called a *closed-cycle Joule-Thompson* refrigerator. See Hudson (1969), pages 380 to 384 and 387 to 388, for a more detailed description of these refrigerators.

7.4.2. Thermoelectric Coolers

These coolers use the *Peltier effect*, which can be described as the inverse of the thermocouple effect: When a current flows through a circuit with dissimilar metals, one junction heats and the other cools. By cascading junctions, temperatures of just under 200 K can be achieved with a small heat load.

7.4.3. Radiation Coolers

If a large surface with a high emissivity can be oriented to face away from the sun into cold space, it will emit radiation into space, cooling itself and anything attached to it. This technique requires no cryogens and no moving parts. The only penalty is the size of the required emitter and the complications associated with keeping it facing away from the sun.

7.5. TEMPERATURE MEASUREMENT

For most detector work, temperature measurements with an accuracy of a degree or so are adequate, and thermometry is not too complicated. The common temperature sensors are described here, together with their readouts. The method of attaching leads so that the leads do not significantly alter the temperature of interest or add unwanted power is also discussed. The reader is referred to White (1987) or other specialized sources if accuracies of better than 1 degree are desired. The care required for that kind of work is beyong the scope of this book.

7.5.1. Temperature Sensors

The measurement of temperatures is an important part of cryogenics. The sensors and their use are briefly described here. The characteristics of three common cryogenic temperature sensors are shown in Figures 7.9 and 7.10.

Platinum Resistance Thermometers. The resistance of pure metals decreases as the temperature decreases. If the material is mounted so that it is not stressed or strained during temperature cycling, the resistance is very repeatable and can be used as an accurate temperature indicator. Platinum resistance thermometers (PRTs) are the most common thermometers of this type. Because the resistance is small, it is important to eliminate the lead resistance from the measurements. This is done by using the *four-lead method*: The leads used to measure the voltage are not those used to supply the current (Figure 7.11). In this way the voltage drop measured is only across the PRT; it does not include the IR drop in the leads.

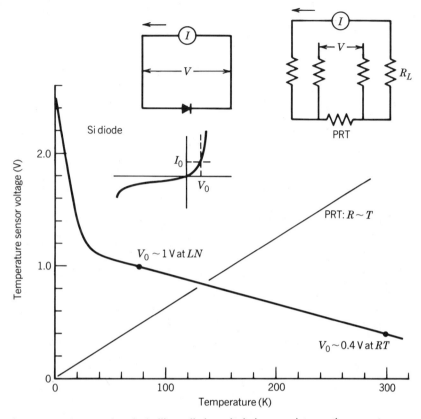

Figure 7.9. Output of typical silicon diode and platinum resistance thermometer versus temperature.

PRTs are generally very accurate but other types of temperature sensors are smaller and have larger relative sensitivities, hence cheaper readouts. The calibration of PRTs is described by White (1987).

Diodes. If a forward current is driven through a diode, the resulting voltage is relatively insensitive to the current but very sensitive to the temperature. A typical voltage–temperature curve is shown in Figure 7.9. These temperature-sensing diodes are available commercially, along with current supplies and readouts that can be adjusted for a given diode so that they read out directly in temperature. The effect on accuracy of voltage drops in the leads should be evaluated; a four-wire technique is often required.

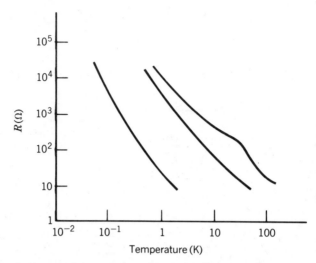

Figure 7.10.   Resistance of three representative germanium resistance thermometers versus temperature.

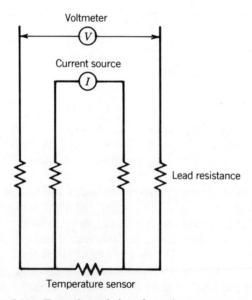

Figure 7.11.   Four-wire technique for resistance measurements.

Carbon and Germanium Resistance Thermometers. The resistance of semiconductor materials increases rapidly as the temperature is decreased. Germanium resistance thermometers with several levels of calibration accuracy are available commercially. Vendors are a good source of information. These thermometers are small and come in several mounting configurations. Since the resistances are large the lead resistance is not a serious problem, and the four-lead technique is not necessary.

Ordinary carbon resistors can be used as temperature sensors. References to carbon resistance thermometry, in chronological order, are: Clement and Quinnell (1952), Corruccini (1963), Borcherds (1969), Kopp and Ashworth (1972), Swartz and Swartz (1974), Kes et al. (1974), and Lawless (1977).

Thermocouples. Whenever two different materials are joined together, a battery is created that generates a *thermal EMF* (electromotive force). The magnitude of the thermal EMF depends on the materials used, but always increases with temperature. If care is taken in the selection and handling of the materials, the resulting EMF can be a reliable and sensitive temperature indicator.

To avoid or minimize unwanted thermal EMFs in other parts of the circuit, it is common to use thermocouple junctions in pairs (Figure 7.12). One of the junctions is maintained at a known temperature—often an ice bath. The temperature of the other junction can then be inferred from the

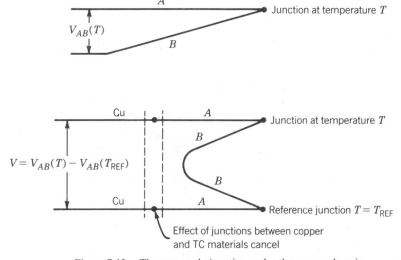

Figure 7.12. Thermocouple junction and a thermocouple pair.

resulting EMF using standard thermocouple tables or special calibration tables prepared for your specific thermocouple.

Another use for thermocouple pairs is to measure temperature *differences* accurately. The EMF from a thermocouple pair is a direct measurement of the temperature difference between the two junctions. Such a determination is more accurate than determining first one temperature, then the other, and finally subtracting the two.

For general instruction on the use of thermocouples, see ASTM *Manual on the Use of Thermocouples in Temperature Measurement* (ASTM, 1981). The National Bureau of Standards publishes a complete set of thermocouple tables (Powell et al., 1974) and White (1987) provides a table of output voltage versus temperature for copper–constantan and two gold–iron thermocouple materials for cryogenic work.

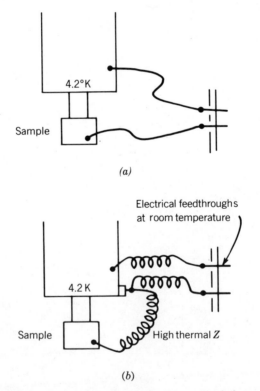

Figure 7.13. Temperature sensor lead attachment: (*a*) poor technique: short leads, poor anchoring; (*b*) good technique: high thermal impedance and good anchoring.

7.5.2. Mounting of Cryogenic Thermometers

Unless care is taken in the mounting of the sensor and the choice and attachment of its leads, the sensor will not be at the temperature of the device we intend to monitor. Good thermal contact must be obtained with the object whose temperature is to be measured. This can be achieved by using threaded metal mounts and contacting across a large surface area.

Great care must be taken to prevent conducting heat to the test object through the temperature sensor leads. The leads should have a high thermal impedance, either by choice of the material (constantan, manganin) or long lengths with a small cross section. Heat leaks to the temperature sensor can be further reduced by thermally connecting the leads to a noncritical point whose temperature is close to that which is to be measured before bringing them to the test object; see Figure 7.13. This technique is referred to as thermally *anchoring*, or *tempering*, the leads.

Whenever two different lead materials are joined, a thermal EMF will be generated. Thus care should be taken to keep those "non-measuring" junctions symmetric and at the same temperature so that the thermal EMFs will cancel out.

7.6. CRYOGENICS AND SAFETY

Several potential safety hazards exist when working with cryogens, but simple precautions are available to protect against all of them.

Skin "Burns". All of the cryogens are cold enough that they can cause skin damage if they come into contact with the skin or eyes for any length of time. Hudson (1969, pp. 376, 377) discusses the physiology of these burns. Nitrogen is more apt to cause burns than is helium because nitrogen boils off less quickly.

Prevent splashing when withdrawing liquid nitrogen from a pressurized source by using a "phase separator" or a good splash shield.

Wear clothing such that cryogens cannot splash against bare skin, or be trapped against your skin. Wear heavy leather gloves, glasses, safety goggles, or a face shield.

Suffocation. If working in a closed area that could be filled with nitrogen, use a buddy system and keep an escape route open.

Fire Hazards of Oxygen. Oxygen itself does not burn, but its presence can promote burning. Materials that are relatively inert may burn fiercely in an oxygen-rich environment. When using oxygen or potentially flammable liquids:

Post signs and instruct others so that those around you know of the danger.

Have plenty of ventilation.

Do not allow any flames in the area.

Dispose of oxygen only in an area free of flames, sparks, and flammable materials where the oxygen can be dispersed rapidly. An empty, open cement parking area is ideal. A fume hood may be used to disperse oxygen slowly if there are no sparks or flammable materials in the hood.

Oxygen can be condensed from the air. The boiling point of nitrogen is 13 K colder than the condensation temperature of oxygen. When liquid nitrogen sits in an open vessel for more than a few minutes, oxygen from the air will collect in the nitrogen. Enough oxygen may collect to require that the nitrogen oxygen mixture be treated in accordance with the guidelines for handling oxygen.

This complication can be largely eliminated if the nitrogen container is closed with a one-way valve, or even if the opening is partially closed down so that escaping nitrogen gas prevents the entry of air to the container.

NOTICE TO THE READER: The following discussion describes potentially hazardous procedures and the reader is cautioned to refer to the Publisher's notice found on the copyright page.

Explosion of Dewars. If the necks of liquid helium or neon dewars become plugged with ice, and if the safety relief valves are plugged, rusted, or otherwise inoperative, the pressure will build up as the cryogen evaporates. This will eventually cause the dewar to explode. *This is a very dangerous situation and must be prevented*:

Keep the neck valve closed when not in use. (This prevents water from condensing in the neck.)

Do not tape or obstruct the relief valves.

If the relief valves are not in good condition, return the dewar to the supplier.

To clear ice plugs if they should occur, a $\frac{1}{4}$-in.-diameter hollow copper tube should be prepared ahead of time and kept handy. The tube should be wrapped near the top end so that the tube cannot drop far enough into the dewar that it could puncture the dewar bottom. The top end should be connected to a valved source of helium gas using a flexible hose long enough to allow the tube to reach the storage dewars.

If an ice plug occurs, clear the area of unneeded personnel. Open the valve to allow a slow flow of room temperature helium gas through the tube. Wearing heavy gloves and a face shield, gently insert the tube into the neck of the dewar, and lower it until the tube reaches the ice plug.

The weight of the tube, heat conducted through the tube, and the warm gas will gradually melt through the ice, allowing the pressure to bleed off. While doing this, keep clear of the dewar. When the plug is melted, a rapid release of very cold gas will result, and the tube may be projected rapidly upward. Stay clear!

Cylinders of High-Pressure Gas. Although not unique to cryogenic situations, high-pressure gas cylinders are often used around cryogenic equipment. If the "neck" of these cylinders ever breaks off, the cylinder becomes a powerful "rocket" that will break through walls and bodies. So:

Strap them in place.

Always replace the protective cover before moving them.

Move tanks only with the appropriate "dollies."

Use only appropriate regulators and connectors.

Whipping of Miniature Gas Lines. If miniature gas lines (used to supply gaseous nitrogen to open cycle refrigerators) should break while under pressure, they can whip around, causing serious cuts or eye damage. To prevent this:

Tape them in place or use sandbags to hold them down.

Do not bend the tubes into tight turns or bend repeatedly.

Implosion of Glass Dewars. Although less common now than in the past, dewars made of glass are sometimes used. If broken, these can shatter, sending glass shards through the air at high speeds. To avoid this, wrap the dewar with tape or constrain it in a wire "cage."

REFERENCES

American Society for Testing and Materials (1981). *Manual on the Use of Thermocouples in Temperature Measurement*, ASTM, Philadelphia.

Berman, R. (1956). "Some Experiments on Thermal Contact at Low Temperatures," *J. Appl. Phys.* 27, 318.

Borcherds, P. H. (1969). "Sensitivity of Carbon Resistance Thermometers," *Cryogenics* 9, 138.

Clement, J. R., and E. H. Quinnell (1952). "The Low Temperature Characteristics of Carbon-Composition Thermometers," *Rev. Sci. Instrum.* 23, 213.

Corruccini, R. J. (1963). "Temperature Measurement in Cryogenic Engineering," *Adv. Cryog. Eng.* 8, 315.

Denning, H. (1969). "Thermal Conductivity of Adhesives at Low Temperatures," *Cryogenics* 9, 282–283.

Donabedian, Martin (1978). "Cooling Systems," Chapter 15 in *The Infrared Handbook*, William L. Wolfe and George J. Zissis, eds., Environmental Research Institute of Michigan, Ann Arbor, Michigan.

Fletcher, Leroy S., Paul A. Smuda, and Donald A. Gyorog (1968). "Thermal Contact Resistance of Selected Low Conductance Intersitial Materials," paper 68-31, *AIAA 6th Aerospace Sciences Meeting*, New York, January 1968, American Institute of Aeronautics and Astronautics, New York.

Graff, W. J. (1960). "Thermal Conductance across Metal Joints," *Mach. Des.* September, 166.

Hudson, Richard D., Jr. (1969). *Infrared System Engineering*, Wiley, New York.

Keesom, W. H. (1942). *Helium*, Elsevier, New York.

Kes, P. H., C. A. M. van der Klein, and D. de Klerk (1974). "A New R-T Relation for Allen-Bradley Carbon Resistor Thermometers," *Cryogenics* 14, 168.

Kittel, Charles (1968). *Introduction to Solid State Physics*, Wiley, New York.

Kittel, Charles, and Herbert Kroemer (1980). *Thermal Physics*, W.H. Freeman, San Francisco.

Kopp, F. J., and T. Ashworth (1972). "Carbon Resistors as Low Temperature Thermometers," *Rev. Sci. Instrum.* 43, 327.

Kramers, H. A. (1949). "Vibration of a Gas Column," *Physica* 15, 971.

Lawless, W. N. (1977). "One Point Calibration of Allen-Bradley Resistor Thermometers, 2–20 K," *Rev. Sci. Instrum.* 48, 361.

McFee, Richard (1959). "Optimum Input Leads for Cryogenic Apparatus," *Rev. Sci. Instrum.* 30, 98.

Powell, Robert L., William J. Hall, Clyde H. Hyink, Jr., Larry L. Sparks, George W. Burns, Margaret G. Scroger, and Harmon H. Plumb (1974). *Thermocouple Reference Tables Based on the IPTS-68,* NBS Monograph 125, National Bureau of Standards, Washington, D.C.

Radebaugh, Ray (1977). "Thermal Conductance of Indium Solder Joints at Low Temperatures," *Rev. Sci. Instrum.* 48, 93.

Rogers, G. F. C. (1961). "Heat Transfer at the Interface of Dissimilar Metals," *Int. J. Heat Mass Transfer* 2, 150.

Scott, R. B. (1959). *Cryogenic Engineering,* D. Van Nostrand, Princeton, N.J.

Swartz, D. L., and J. M. Swartz (1974). "Diode and Resistance Cryogenic Thermometry: A Comparison," *Cryogenics* 14, 67.

Taconis, K. W., J. J. M. Beenakker, A. O. C. Nier, and L. T. Aldrich (1949).

"Measurements Concerning the Vapor–Liquid Equilibrium of Solutions of He$_3$ in He$_4$ below 2.19 K," *Physica* 15, 733.

Touloukian, Y. S., and C. Y. Ho, eds. (1970). *Thermophysical Properties of Matter* (13 volumes), IFI/Plenum, Washington, D.C.

Walmsley, D. G., G. R. Clarke, and H. London (1972). "Enhanced Cooling of Cryogenic Leads," *Cryogenics* 12, 54. See also the follow-up articles: *Cryogenics* 12, 393 (1972) and *Cryogenics* 13, 44 (1973).

Wheatley, John, and Arthur Cox (1985). "Natural Engines," *Phys. Today*, August, 50.

White, Guy Kendall (1987). *Experimental Techniques in Low Temperature Physics*, 3rd ed., Clarendon Press, Oxford.

Yazaki, Y., A. Tominaga, and Y. Narahara (1980). "Experiments on Thermally Driven Acoustic Oscillations of Gaseous Helium," *J. Low Temp. Phys.* 41, 45.

SUGGESTED READING

CRYOGENICS: GENERAL AND ENGINEERING

Advances in Cryogenic Engineering, Plenum Press, New York. This is a series of volumes, each reporting on a Cryogenic Engineering Conference. Volume 8, for example, is the proceedings of the 1962 conference.

Holman, J. P. (1981). *Heat Transfer*, McGraw-Hill, New York.

Kreith, Frank (1962). *Radiation Heat Transfer*, International Textbook, Scranton, Pa.

Reed, Richard P., and Alan F. Clark, eds. (1983). *Materials at Low Temperatures*, American Society for Metals, Metals Park, Ohio. A thorough review of cryogenic material properties and techniques.

Scott, R. B. (1959). *Cryogenic Engineering*, D. Van Nostrand, Princeton, N.J.

White, Guy Kendall (1987). *Experimental Techniques in Low Temperature Physics*, 3rd ed., Clarendon Press, Oxford. This is a classic: comprehensive and well written.

Wolf, Helmut (1983). *Heat Transfer*, Harper & Row, New York.

Journals: In the four scientific journals listed below, most of the articles are very specialized, but one or two in each issue will probably relate in some way to things you are familiar with. It is worth becoming familiar with these if you are involved in cryogenics at all: Cryogenics, Journal of Low Temperature Physics, Journal of Scientific Instruments, and Review of Scientific Instruments.

HANDBOOK AND TABULAR MATERIAL

American Society for Testing and Materials (1981). *Manual on the Use of Thermocouples in Temperature Measurement*, ASTM, Philadelphia.

Corruccini, Robert J., and John J. Gniewek (1961). *Thermal Expansion of Technical Solids at Low Temperatures—A Compilation from the Literature*, NBS Monograph 29, National Bureau of Standards, Washington, D.C.

Powell, Robert L., and William A. Blanpied (1954). *Thermal Conductivity of Metals and Alloys at Low Temperatures—A Review of the Literature*, NBS Circular 556, National Bureau of Standards, Washington, D.C.

Powell, Robert L., William J. Hall, Clyde H. Hyink, Jr., Larry L. Sparks, George W. Burns, Margaret G. Scroger, and Harmon H. Plumb (1974). *Thermocouple Reference Tables Based on the IPTS-68*, NBS Monograph 125, National Bureau of Standards, Washington, D.C.

Touloukian, Y. S., and C. Y. Ho, eds. (1970). *Thermophysical Properties of Matter*, IFI/Plenum, Washington, D. C.   This is an encyclopedic reference to the material properties as a function of temperature—an excellent resource—in 13 large volumes:

1. *Thermal Conductivity—Metallic Elements, Alloys*
2. *Thermal Conductivity—Nonmetallic Solids*
3. *Thermal Conductivity—Nonmetallic Liquids, Gases*
4. *Specific Heat—Metallic Elements and Alloys*
5. *Specific Heat—Nonmetallic Solids*
6. *Specific Heat—Nonmetallic Liquids, Gases*
7. *Thermal Radiative Properties—Metallic Elements, Alloys*
8. *Thermal Radiative Properties—Nonmetallic Solids*
9. *Thermal Radiative Properties—Coatings*
10. *Thermal Diffusivity*
11. *Viscosity*
12. *Thermal Expansion—Metallic Elements and Alloys*
13. *Thermal Expansion—Nonmetallic Solids*

PROBLEMS

1. What is 68°F in Celsius and Kelvin?

2. If I want to use my home oven as a 500-K blackbody, what setting (Fahrenheit) should I use?

3. Mention a few characteristics of each of the following cryogens: (a) Neon; (b) Helium; (c) Oxygen; (d) Nitrogen; (e) Freon.

4. Critique the dewar design shown in Figure 7.14, and suggest improvements.

5. (a) Calculate the heat flow through a stainless steel tube 8 in. long, with a 0.500-in. outside diameter and 0.020-in.-thick walls, if one

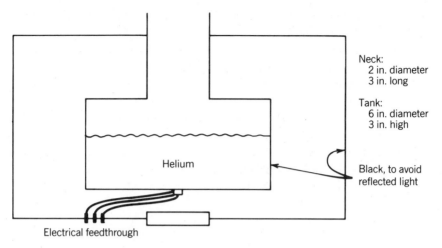

Neck:
2 in. diameter
3 in. long

Tank:
6 in. diameter
3 in. high

Helium

Black, to avoid
reflected light

Electrical feedthrough

Figure 7.14. Hypothetical dewar for Problem 4.

end is immersed in liquid helium and the other is at room temperature.

(b) Same problem as in part (a) except that the tube is made of OFHC copper.

6. How much helium is vaporized to cool a penny from room temperature to liquid helium temperature?

7. What is the heat leak into a dewar that will hold 1 L of helium for 4 h?

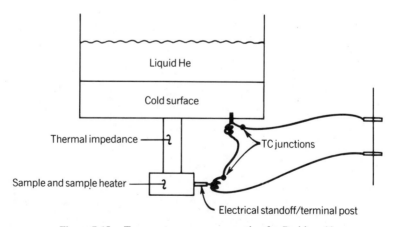

Liquid He

Cold surface

Thermal impedance

TC junctions

Sample and sample heater

Electrical standoff/terminal post

Figure 7.15. Temperature sensor mounting for Problem 11.

8. Imagine that we fill that same dewar with boiling coffee. How long will it take to cool to a drinkable temperature (125°F)?

9. List and describe briefly the three mechanisms for heat transfer.

10. Name some advantages and disadvantages of the following temperature sensors: (a) PRT; (b) Silicon diode; (c) Carbon resistor; (d) Thermocouple junction; (e) Differential thermocouple.

11. Critique the temperature sensor mounting scheme shown in Figure 7.15.

12. List some safety hazards associated with cryogenic work and methods of reducing those hazards.

8

Vacuum Practices

8.1. INTRODUCTION

Vacuum technology is used in the IR detector business for insulation in dewars and to improve adhesion in the deposition of optical coatings. To achieve these vacuums, a nearly leak-free container is required. In addition, the container must be clean: Many materials will vaporize, generating gas that will increase the pressure above usable limits. In this chapter we survey the required pumping, measuring, leak detection, and cleanliness.

Three references are suggested for further study: *Scientific Foundations of Vacuum Technique* by Dushman (1962) is a classic: rigorous, well known, and respected. *Vacuum Technology* by Roth (1982) is comprehensive and a little easier to read. The vacuum chapter in *Building Scientific Apparatus* by Moore et al. (1983) provides a shorter but still comprehensive survey of vacuum techniques.

8.1.1. Units

Early vacuum measurements were made with a manometer—a u-shaped tube filled with mercury. The difference in pressure at the two ends pushed mercury up into the tube. With normal atmospheric pressure at one end, and a vacuum at the other end, a column of mercury about 760 mm high is supported. The lower the pressure, the shorter the column of mercury that the pressure will lift. Pressures can thus be measured in terms of the height of the supported mercury column. A millimeter of mercury—in this context—has been given the name *torr*, (for Evangelista Toricelli, 1608–1647) and that unit of measure has been retained even though mercury manometers have generally been replaced by more sensitive or convenient gauges. Pressures are sometimes stated in "microns" (properly micrometers of mercury): One micron (μm) in this sense is one millitorr. Atmospheric pressure is about 760 torr, test dewars are normally evac-

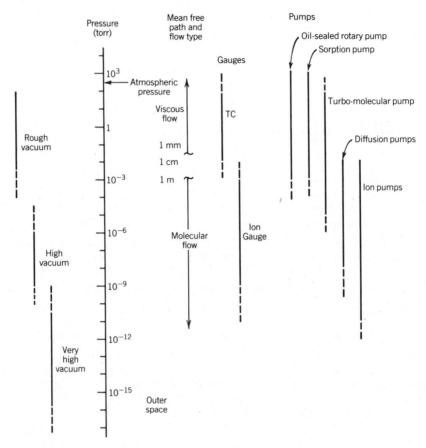

Figure 8.1. Pressure ranges and regimes.

uated to around 1×10^{-6} torr, and special-purpose vacuum pumps can achieve ultrahigh vacuums of 10^{-12} torr (see Figure 8.1).

Occasionally, we need to work in mechanical dimensions—those built on metric units of force and energy. The conversion is done by expressing the weight of a column of mercury in terms of its height h, its density ρ (13.6 g/cm^3), and the acceleration due to gravity g. The result is that

$$1 \text{ torr} = 1.33 \times 10^3 \text{ dyn/cm}^2$$
$$= 133.3 \text{ N/m}^2$$
$$= 133.3 \text{ Pa} \quad \text{(pascal)}$$

Table 2.8 in Roth (1982, p. 42) facilitates conversion between different pressure units.

8.1.2.   Conversion between Molecular Density and Pressure

We will need to determine the effect of a given thickness of absorbed or condensed gas on pressure. The necessary conversions are easily done if the following information is at hand: Avogadro's number (6.023×10^{23} of molecules fills 22.4 L as a gas at room temperature and atmospheric pressure (760 torr) and has a mass equal to the molecular weight of the molecule—18 g for water, as an example. Thus

$$760 \text{ torr} \times 22.4 \text{ L} \rightarrow 6.0 \times 10^{23} \text{ molecules}$$

$$1 \text{ torr-L} \qquad \rightarrow 3.6 \times 10^{19} \text{ molecules}$$

$$18 \text{ g} \qquad \rightarrow 6.0 \times 10^{23} \text{ molecules}$$

$$1 \text{ g} \qquad \rightarrow 33. \times 10^{21} \text{ molecules}$$

Assume that the density of the outgassing material is that of water: 1 g/cm³. A cube 1 cm on a side would have a mass of 1 g, have about 33×10^{21} molecules, 3×10^7 molecules along one edge, and 1×10^{15} molecules on each face. The surface density of molecules is thus about 1×10^{15} molecules per square centimeter. This is shown in Figure 8.2.

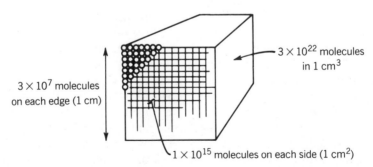

Figure 8.2.   Molecular density in a liquid or solid. 6×10^{23} molecules $H_2O \rightarrow 18$ $g \rightarrow 18$ cm³ liquid; therefore, 1 cm³ liquid contains $\approx 30 \times 10^{21}$ molecules.

8.2. GAUGES

The manometers described above are useful for pressures from atmospheric pressure down to a few millimeters of mercury, but at the pressures required for dewar insulation and coating operations more sensitive gauges are required.

8.2.1. Thermocouple Gauges

Thermocouple (TC) gauges are commonly used; they are rugged and inexpensive. TC gauges are available that work from 10^{-3} to 20 torr, but any one design only covers of range of about 200 (10^{-3} to 0.2 torr, for example). The readout is actually the voltage generated by a thermocouple bead connected to a heated wire. In a good vacuum the wire and thermocouple become very hot. At higher pressures the gas conducts heat away from the bead, cooling it, and changing the output.

8.2.2. Ion Gauges

Ion gauges measure lower pressures than do TC gauges, but they are more expensive and more fragile. The most common ionization gauges work from 10^{-3} to 10^{-8} torr, but specialized designs are available for much lower pressures. They ionize some of the gas present and collect the ions. The resulting current is a measure of the pressure in the tube. The meters generally offer several ranges of readout, and a log scale that allows convenient wide range readout at some loss in accuracy and resolution.

They should be used in conjunction with a TC gauge: the ion gauge filament should not be on until the TC gauge indicates the pressure is below about 10 μm. When not in use, turn off the filament, or at least set the meter to the least sensitive scale. The ionization and collection process actually removes some gas from the vacuum space, and this pumping can improve the vacuum or maintain a vacuum in the presence of a small leak, and much larger versions of this device are made specifically for pumping.

The accumulation of material on the anode and cathode must be removed periodically; this is done with a "degassing current." Usually, this is applied with a momentary switch. As the deposits are driven off, the pressure will rise. When the pressure reaches a steady level, the gauge is clean, and the degassing current should be turned off.

The thermocouple and the ion gauge are the common gauges used.

Other gauges are available; the references cited earlier in this section should be consulted for further information.

8.3. PUMPS

8.3.1. Oil Sealed Rotary Pumps

The most common mechanical pump is the oil-sealed rotary pump. They are commonly used to obtain the rough vacuum (10^{-2} to 10^{-1} torr—10 to 100 μm) required to support other pumps. They are simple, reliable, and common, and can obtain pressures as low as 1×10^{-4} torr in exceptional circumstances.

8.3.2. Turbomolecular Pumps

Turbomolecular pumps rely on rapidly rotating vanes to strike air molecules, literally batting them toward the exit port. Because no oil seal is used, they are very clean. They are also quiet, and are becoming more common.

8.3.3. Diffusion Pumps

Diffusion pumps use high-speed jets of oil or mercury to trap gas molecules and sweep them into a reservoir of fluid. An oil diffusion pump can achieve pressures of about 10^{-6} torr. There are three limitations to the pressures at which they can be used:

1. At the exhaust port, a pressure below about 0.3 torr (300 μm) must be maintained. At higher pressures, hot oil is pushed back through the pump into the high-vacuum system.
2. At the inlet port (the high-vacuum side of the pump) the pressure must be low enough that the mean free path of molecules is less than the distance from the oil jets to the wall. Otherwise, the oil molecules are slowed down by collisions with many gas molecules, and never reach the wall, so they never condense. They simply add to the gas in the volume to be pumped.
3. The ultimate pressure that an untrapped diffusion pump can reach is the room-temperature vapor pressure of the oil being used. Different oils are available; the choice requires a trade-off between expense, resistance to oxidation and cracking, and ultimate pressure.

Mercury can be used as a working fluid, but its toxic nature and ability to combine with brass and other alloys makes it less convenient to use.

8.3.4. Cold Traps

The vapor from a diffusion pump can be condensed by inserting a cold surface between the diffusion pump and the unit to be pumped. These *traps* are generally cooled with liquid nitrogen and reduce the amount of contaminating vapor that could otherwise reach the unit being pumped.

8.3.5. Sorption Pumps

Sorption is a composite of the words *adsorption* (collect on the surface) and *absorbtion* (collect inside). Sorption pumps cool specially prepared materials with very high surface-to-volume ratios to collect gases. When the sorbing capacity of the material is exhausted, it can be rejuvenated: The sorbed material is driven off to the atmosphere by heating.

8.3.6. Ion Pumps

If gas molecules are ionized, they can be drawn to a titanium surface held at a large negative voltage. They strike that cathode with enough energy that they are buried, and also sputter off fresh titanium, which can sorb other gases. Ion pumps are very clean, quiet, and act as their own pressure gauge. They operate from 1×10^{-2} to 1×10^{-11} torr. Primary disadvantages are their cost (particularly in the larger sizes) and the associated stray electric and magnetic fields.

8.3.7. Manifolds

For many applications, provision is made to attach several dewars to one pumping system. The set of valves and the tube that connects them is referred to as the manifold. Care is necessary when connecting or removing a dewar to the system since the pumping of someone else's dewar can be disturbed. If using such a system, be sure you know the agreed-on rules.

8.3.8. Operation of a Typical Multiport Vacuum Station

Figure 8.3 shows a typical pumping station. It includes a mechanical pump, a high-vacuum oil diffusion pump, a manifold, and the associated gauges and valves. Operation of most commercially purchased vacuum

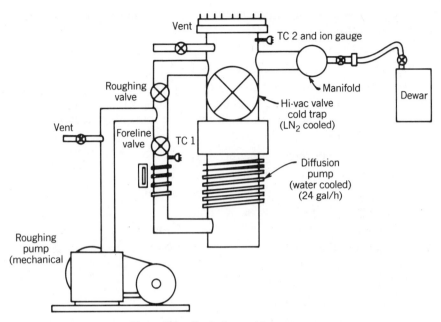

Figure 8.3. Typical pumping system.

systems are now at least partially automated, but older or specialized systems are operated by manual valves. There are a number of steps required to put a dewar on the pump, and if the characteristics of the pumps and gauges are not understood, the units being pumped and the pumping system itself can be contaminated. Recovery from errors may require complete dissasembly, cleaning, replacement of pump oils, reassembly, and a day or more of cleanup pumping.

Because of the potential problems, *it is essential to meet with the pump custodian before using a pump with which you are unfamiliar.* Obtain permission to use the pump, obtain specific directions for its operation, and confirm that you understand what can and cannot be done. All this will be easier to absorb if you understand how such a system works.

Pump Operation: To Pump a Dewar

1. Connect your dewar to the manifold, but do not open any valves.
2. Note the ion gauge pressure.
3. Protect the vacuum in the dewars already on the pump by first closing the valves on the dewars and then on the manifold.

4. Prepare to rough out your dewar by protecting the ion pump and the diffusion pump:
   a. Turn off the ion gauge.
   b. Close the high-vacuum valve.
   c. Close the foreline valve. (At this point the diffusion pump is no longer backed up. It cannot be left that way for long, so move rapidly—but carefully—through the remaining steps.)
5. Rough pump your dewar:
   a. Open the roughing valve.
   b. Open the manifold valve to your dewar.
   c. Pump until TC 2 reaches about 100 μm.
   d. Open the dewar valve slowly.
   e. Pump until TC 2 reaches about 50 μm.
   f. Close the roughing valve. The pressure will begin to rise a little until you get into the high-vacuum mode.
6. Go to the high-vacuum mode:
   a. Open the foreline valve.
   b. Slowly open the high-vacuum valve.
   c. When TC 2 reaches 50 μm, turn on the ion gauge.
7. When you are satisfied that the system is working well and that your dewar does not leak, open the valves to the other dewars:
   a. Open the corresponding manifold valve.
   b. Open the dewar valves. (Do this slowly: if the dewars leak or outgas badly, their pressure could have risen since you closed them off.)

To Remove a Dewar from the Pump

1. Close the valves to your dewar first at the dewar side and then at the manifold side.
2. Disconnect the dewar from the manifold.

8.4. PUMPING DYNAMICS

We will talk later about what an adequate vacuum is. Assuming that we can answer that, we next need to consider how long it takes to achieve that vacuum. To do that we first identify a few parameters used to quantify gas flow.

8.4.1. Vocabulary and Units

Quantity of Gas. One measure of the quantity of a gas is the number of molecules N. The ideal gas law ($PV = NRT$) suggests a more convenient way to express the quantity of gas: At constant temperature, the number of molecules is proportional to the product of pressure and volume. Thus we will express the quantity of gas as pressure times volume, in units of torr·liters, or in "standard cc": that is, 1 atmosphere (760 torr) × 1 cubic centimeter. The two units are not very different:

$$1 \text{ std cm}^3 = (760 \text{ torr}) (1 \times 10^{-3} \text{ L}) = 0.76 \text{ torr·L}$$

Flow Rate. The volume rate of flow of a gas across a plane is the flow rate, measured in L/s.

Throughput. Throughput is the rate at which gas moves across a plane, measured in torr·L/s. If the flow rate is S, the throughput $Q = SP$. The throughput can be the pumping capacity or the outgassing rate of the unit being pumped.

Conductance. A pressure difference across a tube or orifice causes a given throughput of gas. The conductance is the ratio of the throughput (torr·L/s) to the pressure difference (torr). Thus the units of conductance are L/s. The usual symbol for conductance is C:

$$Q = C(P_2 - P_1) \tag{8.1}$$

Pumping Speed. This is the maximum flow rate that a pump or a pumping system can accommodate at a given pressure. If there are no impedances in a system, the speed of the system will be the speed of the pump. The system pumping speed for a pump with conductances in series is

$$S = \left(\frac{1}{S_{\text{pump}}} + \frac{1}{C_1} + \frac{1}{C_2} + \cdots \right)^{-1} \tag{8.2}$$

Pumping Capacity. This is the throughput the pump system can handle. Pumping capacity is a function of pressure: It equals system pumping speed times the pressure.

Mean Free Path. The mean free path is the average distance a molecule moves before colliding with another molecule. This is useful in predicting how a gas will flow. The flow is said to be *viscous* when the mean free path is much less than the dimensions of the pumping system. The gas acts like a crowd leaving a building through a narrow hallway. The flow is said to be *molecular* when the mean free path is much greater than the dimensions of the pumping system. The gas molecules move from one surface to another as individuals, not interacting with their neighbors.

8.4.2. Mean Free Path

To determine whether viscous or molecular equations should be used, we will need to calculate the mean free path and compare it to the dimensions of the system. The mean free path λ is given by Roth (1982, p. 38):

$$\lambda = \frac{kT}{2\pi dP} \tag{8.3}$$

where k = Boltzmann's constant, about 1.38×10^{-23} J/K
        T = temperature (kelvin)
        d = molecular diameter
        P = pressure

To apply this equation, it is necessary to use "mechanical" pressure units—those consistent with the normal force and energy units: 1 torr = $1.33 \times 10^{+3}$ dyne/cm^2. For air, at ambient temperature,

$$\lambda \simeq \frac{5 \times 10^{-3}\ \text{cm·torr}}{P\ (\text{torr})} \tag{8.4}$$

8.4.3. Conductance Formulas

To determine the pumping speed of a system, one must know the speed of the pump itself and the conductance of the connecting system. First determine the mean free path for the pressures of interest and determine whether the flow is viscous (mean free path small compared to the system diameters) or molecular (mean free path large compared to the system diameters).

The appropriate formulas for conductance can then be selected and used. A few of these are provided here (Equations 8.5–8.8); these and others are given in most vacuum texts. These formulas apply for air at room temperature. In these equations the diameters and lengths are given in centimeters, areas in cm^2, and pressures in torr.

CONDUCTANCE OF A LONG CYLINDRICAL TUBE OF DIAMETER D, LENGTH L

$$\text{Viscous: } C = 184 \frac{\text{L/s}}{\text{cm}^3 \cdot \text{torr}} \frac{D^4}{L} \bar{P} \tag{8.5}$$

$$\text{Molecular: } C = 12.1 \frac{\text{L/s}}{\text{cm}^2} \frac{D^3}{L} \tag{8.6}$$

CONDUCTANCE OF AN APERTURE WITH AREA A

$$\text{Viscous: } C = 20 \frac{\text{L/s}}{\text{cm}^2} A \tag{8.7}$$

$$\text{Molecular: } C = 11.6 \frac{\text{L/s}}{\text{cm}^2} A \tag{8.8}$$

Equations 8.5–8.8 are from Roth (1982; formulas 3–54, 3–94, 3–35, and 3–73); they are applications of more general formulas to specialized cases. Roth or Dushman (1962) should be consulted to verify that all the assumptions made in deriving these formulas are applicable to your case before making any detailed analysis.

8.4.4. Time to Pump a Perfectly Clean Volume

It is important to understand—but not obvious to the uninitiated—that the time required to pump a dewar is *not* so much dependent on the volume of the dewar as it is on the time required to remove trapped gases from the surfaces and voids in the dewar. To see this, consider a *perfectly clean* 1-L volume, connected to a pumping system with an overall pumping speed of 1 L/s. Many systems have either larger or smaller volumes and generally higher pumping speeds, but the results here will still be helpful. After this system begins pumping, the pressure will drop according to the following equation:

$$P = P_0 \exp\left(\frac{-t}{\tau}\right) \tag{8.9a}$$

or

$$t = \tau \ln\left(\frac{P}{P_0}\right) \tag{8.9b}$$

$$= 2.3\tau \log\left(\frac{P}{P_0}\right) \tag{8.9c}$$

where τ is the volume/speed and P_0 is the initial pressure. These formulas can be derived from the definitions of pumping speed and gas flow, or inferred by analogy with an electrical circuit, as shown in Figure 8.4.

For our case, $\tau = 1$ s, and $P_0 =$ atmospheric pressure, 760 torr. The equation yields 10^{-3} torr in 15 s, and 10^{-5} torr in 30 s. A perfectly clean dewar could be adequately pumped in 15 to 30 s. Because this is much shorter than the time we are accustomed to pumping*, you may be tempted to suspect an error. The "error" is only in our assumption of a perfectly clean system.

Before we discuss the "clean dewar" assumption, convince yourself that the arithmetic above is reasonable. Consider an empty volume filling with air. The same analysis that yielded the expression above yields

$$P = P_{\text{final}} \left[1 - \exp\left(\frac{-t}{\tau}\right) \right] \qquad (8.10)$$

A volume will be very close to its final pressure in several time constants: say, 2 to 5 s. You can confirm this the next time you let air into an evacuated dewar.

Now return to the problem of the predicted 15- to 30-s pumpdown time versus the hour or longer that we actually pump. The discrepancy between our "theoretical value" and experiment shows that our assumption of a perfectly clean volume is unrealistic. The pumpdown time is *completely dominated* by the outgassing of liquids, solids, and trapped gases in the unit to be pumped.

8.4.5. Outgassing

As indicated above, dewar pumpdown time is dominated by the presence of materials on and trapped in the container. This is true even if one is very careful in handling the container. Because of outgassing we cannot use observed pressures as simple measures of vacuum goodness. This is true in two ways:

1. The pump or manifold pressure is only a poor indicator of the pressure in the dewar itself.
2. Even the *dewar* pressure is not a complete predictor of dewar vacuum life. We really want to pump long enough to get the out-

---

* Dewars used for testing IR detectors must often be pumped for an hour or two. The manufacture of permanently evacuated dewars includes days of high-temperature pumping before being welded shut or "tipped off."

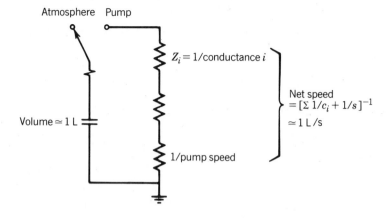

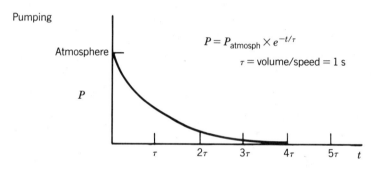

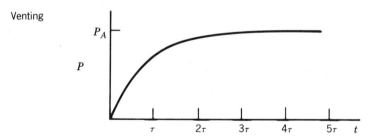

Figure 8.4. Time to pump or vent an ideal clean system.

gassing rate low enough so that—once the dewar is sealed up—
the pressure will not build above some acceptable level until we
are through with our test or other intended use.

To understand how outgassing affects us, consider the effect of one
monolayer of water on the wall of a dewar. There are about 1×10^{15}

molecules/cm² in a surface one molecule deep. If the surface area is 1000 cm², we have about 1×10^{18} molecules. When these evolve they will generate about 3×10^{-2} torr-liters. If these evolve in 1 h, the throughput is about 1×10^{-5} torr·L/s. If the vacuum gauge is in the pump manifold and the dewar valve has a conductance of 1 L/s, the dewar pressure will be 1×10^{-5} torr higher than the manifold pressure: The dewar pressure will be at least 10^{-5} torr, no matter what the pump gauges read. The electrical analog of this situation and the resulting pumpdown curve are shown in Figure 8.5.

If outgassing continues at a rate of 1×10^{-5} torr·L/s after we valve off the dewar, the pressure will build up to 36 μm in 1 h. This is enough to conduct heat appreciably and degrade the hold time of the dewar. The resulting gases can also condense on the detector, contaminating or obscuring it.

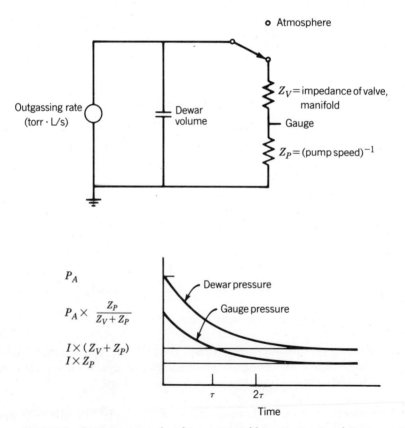

Figure 8.5. Pressure versus time for a system with constant outgassing rate.

Outgassing rates are usually given in torr·L/s. Note, however, that the dimensions are those of power: A pressure of 1 Pa times a volume of 1 m^3 in 1 s is a power of 1 W and 1 torr·L/s (the usual units for outgassing rates) equals 0.133 W. One early paper, for example, gives the outgassing rate of CO from stainless steel as 12×10^{-10} W/m^2; this is equivalent to 1.0×10^{-12} torr·L/(s·cm^2) of surface area.

A thorough analysis of dewar vacuum requires that the leak and outgassing rates be broken down by species (water, nitrogen, methane, etc). Tools for vacuum analysis include:

Helium leak tests
   Standard
   Enhanced sensitivity
Outgassing rate while pumping
   Pressure drop across a known conductance
   Pressure given the pumping speed at the gauge
   Plot of pump pressure versus time
Valved-off pressure versus time ("leak-up" tests)
   Gauge in pump
   Gauge in manifold
   Gauge in dewar
Residual gas analyzer (RGA)

8.4.6. Time to Pump a Real System

To do any quantitative vacuum work, begin by characterizing your pumping system. Obtain from the vendor's literature the speed of the pump itself at different pressures. Calculate the conductance of the connecting tubing and manifold, and then the speed of the system. Plot the pumping capacity (speed times pressure) versus pressure, as shown in Figure 8.6.

For the materials you normally encounter, plot the outgassing rate per unit area versus time. Figure 8.7 is taken from the Consolidated Vacuum Corporation High-Vacuum Calculator. These two plots (pumping capacity versus pressure, and outgassing rate versus time) will allow quick determinations of expected pressure versus time. They help visualize the relationship between outgassing rate, pumping speed, time, and pressure. In the three examples that follow, we assume a pump capacity shown by the bold curve of Figure 8.6.

Example 1: Using the two graphs provided, estimate how long it will take to reach 10^{-6} torr with 1 ft^2 of exposed stainless steel.

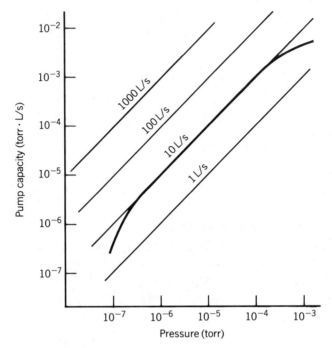

Figure 8.6. Pumping capacity versus pressure.

From Figure 8.6, at 1×10^{-6} torr the pump capacity is about 1×10^{-5} torr·L/s, so the pump can maintain a pressure of 1×10^{-6} torr at that outgassing rate. 1×10^{-5} torr L/s divided by 1 ft² (about 1×10^3 cm²) is 1×10^{-8} torr·L/(s·cm²). Figure 8.7 indicates that it will take about 1 day for the outgassing to drop to that rate.　■

Example 2: What will the pressure be after 3 h if we are pumping on 200 cm of mild steel?

From Figure 8.7 the outgassing rate after 3 h is about 1.5×10^{-7} torr·L/(s·cm²). For an area of 200 cm² this is 1.5×10^{-5} torr·L/s. From Figure 8.6 our pump has that capacity at about 1.5×10^{-6} torr.　■

Example 3: What is the outgassing rate of a processed cable if you can pump it to 10^{-5} torr in 3 h, and 10^{-6} torr in 24 h?

The outgassing rates are just the pumping capacities at the pressures reached: 10^{-4} torr·L/s at 10^{-5} torr (3 h) and 10^{-5} torr·Liter/s at 10^{-6} torr (24 hr).　■

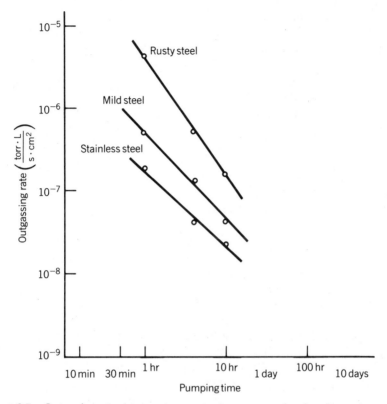

Figure 8.7. Outgassing rates (at room temperature) versus pumping time. Rates shown are from the High Vacuum Calculator, Consolidated Vacuum Corp., Rochester, New York. Roth (1982, p. 190)) gives values for stainless steel that are about 20 times lower than these. Cleaning procedures, including baking at high temperatures, can affect these values drastically.

These examples are not very exciting—we have deliberately not used state-of-the-art pumps or components. However, the method works with your pump and materials—once you have characterized them. The only way to characterize them is to go through this exercise, keeping track of your results. At the same time you will develop a sense for the variability that you will encounter.

8.4.7. Reduction of Outgassing

As we indicated above, the pumpdown time and resulting vacuum life are not limited only by the pumping speed. More often it is the outgassing of molecules that dominates pumpdown time and vacuum integrity. It is thus desirable to minimize the amount of "contaminants" in the dewar.

To minimize water condensation on the dewar walls (and to protect detectors from an unhealthy environment), make sure that the dewar is warm before venting it. Use dry nitrogen gas to fill the dewar when venting it. The nitrogen molecules stick to the dewar surfaces, preventing water molecules from reaching them. The resulting nitrogen–water layer is more quickly removed in the next pump cycle than if the water were directly on the walls. Always wear clean gloves when handling dewar components. Do not use paints, glues, or goos without careful consideration of their effect on vacuum integrity.

8.4.8. Permanent Vacuum

For test applications, it is only necessary to obtain an outgassing rate low enough that the pressure will remain at an acceptable level during the test. Many deliverable packages are permanently sealed, and the vacuum must remain at an acceptable level for periods up to 15 years. The required engineering and processing are fairly complex, and are treated as highly proprietary by most companies.

Consider a small permanently sealed dewar that must have a pressure of 0.05 torr or less for acceptable performance. If the dewar vacuum is to last for 10 years (3.15×10^8 s), and if the dewar contains no getters, it must have a net gas input of less than 0.8×10^{-12} torr·L/s:

$$\frac{(0.05 \text{ torr} \times 5 \text{ cm}^3)/(1000 \text{ cm}^3/\text{L})}{3.15 \times 10^8 \text{ s}} = 0.8 \times 10^{-12} \text{ torr·L/s}$$

This is difficult to do: Even if we can verify by careful leak tests that any leaks are below this level, we cannot reduce the outgassing to this rate.

Long vacuum lives are obtained by the use of getters in addition to careful leak detection, material selection, and vacuum baking before sealing the unit. Materials are selected that do not outgass badly, or can be readily cleaned up. Epoxies and the kapton used in some dewar electrical cables are examples of materials with high outgassing rates. When the use of such materials cannot be avoided, the quantities are carefully limited and controlled. If epoxies are used, good design minimizes the path through which the outgassing occurs and maximizes the surface area to volume ratio so that outgassing will occur early, while units are still on the pump. The outgassing process can be accelerated by heating the unit during the pumpout process. This vacuum baking is done at the highest temperature the unit will tolerate. Strousser (1969) gives an overview of outgassing rates and the effect of vacuum baking on outgassing.

8.5. LEAKS

To prevent leaks, handle parts carefully. Many vacuum assemblies include glass and welded or brazed joints, all of which can be somewhat fragile. If O-ring grooves are designed properly and the rings are in good condition, they require only careful handling to make a good seal. The grooves should be clean, and the O-ring greased very lightly. Be careful when removing, handling, and installing the rings not to pinch or cut them. Once installed the flanges should be tightened uniformly, and need be only finger-tight. Metal gaskets are used without any grease or sealants. They must be scrupulously clean, and the bolts tightened uniformly and to specified torques.

Leaks are a throughput of gas, and have units of torr·L/s or standard cm^3/s. A dewar with a volume of 1 L, that must only keep the pressure below 0.1 torr for 4 h (14.4 × 10^3 s) can tolerate a leak (or outgassing) rate of 7 × 10^{-8} torr·L/s:

$$\frac{0.1 \text{ torr} \times 1 \text{ L}}{14.4 \times 10^3 \text{ s}} = 7 \times 10^{-8} \text{ torr·L/s}$$

8.5.1. Leak Detection

Acetone. Gross leaks can be found by watching the ion gauge on the pump while spraying acetone selectively around the suspected areas: O-rings, welds, and valves. The acetone yields more ions than air does, and the ion gauge will surge when acetone reaches it.

Mass Spectrometers. Leak detectors using mass spectrometers are built to measure the flow rate of helium past the sensor. Helium is relatively rare in the air, so it registers only when there is a leak in the system being tested and when helium is being sprayed near the leak. Using a sensitive system with a very fine and slow helium "probe," leaks as small as 1 × 10^{-10} torr·L/s can be located.

The existence of even smaller leaks can be determined by valving off the unit to be tested, bagging it in helium gas for a few minutes, then opening the unit to the leak detector. If a leak exists, helium will have accumulated in the unit and will register as a burst on the meter. If some care is used, the size of the leak may be estimated, but this method does not lend itself to locating the leak.

Outgassing. If the unit will not pump down but no leak is found, it may be that outgassing is responsible. This is often the case with a new system, which—despite careful cleaning—may contain a large amount of trapped material. Outgassing is a *virtual leak* that may be distinguished from real leaks by monitoring the pressure as a function of time, or as the unit is heated: Outgassing will decrease with time (although it may do so only very slowly if there is a great deal of contaminant material) and will increase with temperature.

8.6. ADEQUATE VACUUM FOR TEST DEWARS

Test personnel are faced daily with a decision about how long (or how well) a test dewar should be pumped. As we discussed earlier, adequate vacuum practice for long-life sealed dewars has been studied thoroughly, and many tools are available to assist in vacuum decisions and predictions: residual gas analysis, outgassing rates, vacuum bake time–temperature schedules, and so on.

Test dewars are different in that we normally face daily (or faster) turnaround times, use fresh (un-outgassed) parts every pumpdown, have frequent dewar modifications, and often do not impose production discipline. A typical scenario is:

Open the dewar, remove the old part.

Leave the dewar standing open for a few hours.

Insert the new unit to be tested.

Pump the dewar for as little time as we think we can get along with (for x hours or until the pump gauge measures y torr or for z hours after the gauge reaches w torr).

Cool the dewar.

Test until the test is done, or until the dewar frosts.

Warm, and vent the dewar.

This scenario is subject to many criticisms: The dewar walls absorb water and other gases while open. The unit being tested may be dirty, and certainly has some water on its surface. We do not know what the pressure was in the dewar when it was removed from the pump. (The gauge on the *pump* does not indicate the true vacuum in the *dewar*.) Minimal checks are made for possible leaks. No consideration is given to outgassing rates. There is no measure of how fast, and where, condensation occurs.

Some improvements are fairly easy or obvious: Increase the conductance of valves on the dewar; add vacuum gauges to the dewars or closer to the manifold on the pump. The main problem remains: We have not provided any practical, quantitative guidelines or rules of thumb that are easily applied to the pumping process.

Answers to the following questions are needed:

What dewar pressure is adequate?

How long must we pump to obtain that pressure?

How long can we maintain that pressure without pumping?

Are proposed improvements necessary, marginal, or overkill?

We cannot develop rules of thumb that will apply to all situations. We will, instead, discuss some of the issues that must be considered if those guidelines are to be developed.

8.6.1. Heat Conduction by Gases

One criterion for adequate vacuum must be the ability of the vacuum to prevent gaseous heat conduction from the warm dewar walls to the cryogen. The material in this section will allow us to predict the gaseous component of the dewar heat load as a function of dewar pressure.

At *high pressures* the mean free path varies as 1 over the pressure, so the thermal conduction is independent of pressure. The heat flow formula is the same used for conduction by solids, with a thermal conductivity K for each gas. Conductivity values are given in Table 8.1 for common gases. Typical heat loads in this regime are not adequate for insulating most dewars.

At *low pressures* the molecular path length is limited by the dewar dimensions, so the thermal conduction is proportional to pressure. Figure 8.8 shows the heat conducted by the gas in the vacuum space of two hypothetical dewars as a function of pressure. Pressures below 10^{-3} or 10^{-4} torr would be needed to make the gaseous conduction negligible compared to other sources of heat loads.

Dushman (1962), Roth (1982), and White (1987) all derive the necessary equations. They are fairly involved but may be simplified for initial estimates, as shown in Equation 8.11.

Table 8.1 Values Needed for Mean Free Path and Thermal Conductivity Calculations[a,b]

Gas	Symbol	Mass	λ_0	k	K_1
Hydrogen	H_2	2	8.4	1.73	60.7
Helium	He	4	13.3	1.43	29.3
Water vapor	H_2O	18	2.90		26.5
Neon	Ne	20	9.44	0.466	13.1
Nitrogen	N_2	28		0.238	16.6
Air	—	—	4.54	0.24	
Carbon monoxide	CO	28			
Oxygen	O_2	32	4.81	0.244	15.6
Argon	Ar	40	4.71	0.160	9.3
Carbon dioxide	CO_2	44	2.95	0.144	17.0
Mercury	Hg	201			4.15
Xenon	Xe	130	2.64		

[a] λ_0, mean free path (10^{-3} cm) for 0°C at 1 torr; k, thermal conductivity (high pressure) 10^{-3} W/(cm·K); K_1, free molecular conductivity, 10^{-3} W/(cm²·K·torr).

[b] Values from Dushman (1962) Tables 1.6, 1.7, and 1.12 (p. 32, 48); see also Roth (1982) Tables 2.9, 2.12, and 2.14 (p. 46, 57, 60). Values taken from most texts must be used with care because similar but inconsistent notation is used for mean free path and for conductivities, and because values are sometimes given in calories per second (1 calorie equals 4.186 joules).

THERMAL CONDUCTANCE THROUGH GAS: HIGH PRESSURE (MEAN FREE PATH < SPACING BETWEEN WALLS)

$$Q = k \frac{A}{d} \Delta T \tag{8.11}$$

where A = area of surfaces facing each other
   d = distance between the hot and cold surfaces
   ΔT = temperature difference between the warm and cold surfaces
   k = thermal conductivity of the gas [values of k are given in Table 8.1; for air, k is about 0.24×10^{-3} W/(cm·K)]

THERMAL CONDUCTANCE THROUGH GAS: LOW PRESSURE (MEAN FREE PATH > SPACING BETWEEN WALLS)

$$Q = K_1 a \sqrt{\frac{273 \text{ K}}{T_i}} PA\Delta T \tag{8.12}$$

where K_1 = *free molecular conductivity*; see Table 8.1 [for air, K_1 is about 16.6 × 10^{-3} W/(cm²·K·torr)]
  a = *accommodation coefficient*
  T_i = temperature of the cold surface.
  P = dewar pressure
  A = area of surfaces facing each other
  ΔT = temperature difference between warm and cold surfaces.

The accommodation coefficient is a measure of how well a molecule comes to thermal equilibrium with a surface before leaving it. For a worst-case calculation take a = 1. Values of 0.8 to 0.5 are common; 0.05 is possible.

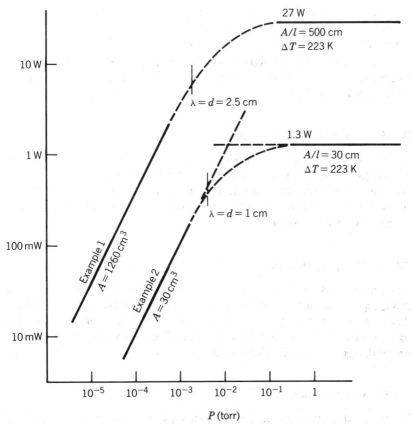

Figure 8.8. Gaseous heat conduction versus pressure for two hypothetical dewars, air or N_2.

8.6.2. Allowable Condensation

Once the dewar is valved off from the pump, the pressure will begin to rise until the cryogen is added. Then cyropumping will begin, and the pressure may drop, but condensation will begin on the cold walls and on the detector. If we are to analyze the vacuum requirements, we must determine in some way how much condensation is allowable. Two effects are of interest: transmission through the condensate, and damage to the detectors.

1. *Absorption.* The transmission depends on the absorption coefficient at the wavelengths of interest on the condensing materials. As an example, consider water—a primary constituent of outgassing. At a wavelength of 4.5 μm, the transmission of 1 μm of ice is about 90% (see Figure 8.9). If we are striving for 10% radiometric accuracy, one limitation on the pressures we can tolerate, then, is that which will allow 1 μm or so of ice to form on our detector in the time we hope to stay cold.

2. *Damage.* The problem of damage is more difficult; it depends on the type of detector and could include short-term (reversible) effects as well as permanent effects.

8.6.3. Leaks and Outgassing

If we know how much pressure or condensation we can tolerate, we need to estimate how fast leaks and outgassing will cause that pressure to build up. This was discussed earlier, and there are many tools available that allow accurate answers to these questions if one is willing to perform the necessary measurements.

8.6.4. Pumping Dynamics

What is the pressure *in the dewar* when you are pumping? Straightforward pumping speed and conductance calculations will allow you to infer the dewar pressure, but the point here is that the pump gauge pressure does not tell you the dewar pressure directly. The necessary calculations and methods are discussed briefly in Section 8.4.

8.6.5. Condensation

When the dewar begins to cool, condensation of gases will begin on the cooled surfaces—including the detector. Assume that the pressure re-

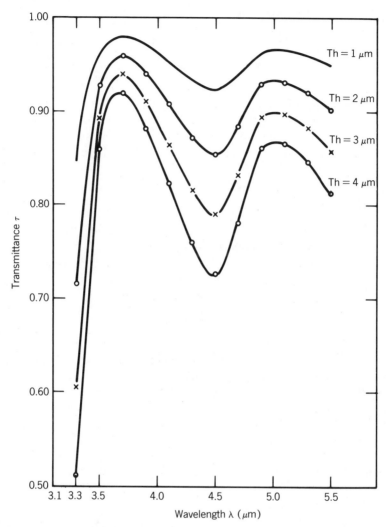

Figure 8.9. Transmittance of ice: at 100 K. $\tau = \exp[-(\alpha)\,(\text{Th})]$. Calculated from absorption coefficients of Bertie et al. (1969). Th is the thickness.

mains uniform within the dewar, and that some fraction η of the gas molecules that collide with the cold detector condense there. The rate at which the condensed solid material builds up on the surface can be predicted from kinetic theory (Equation 8.13).

BUILDUP RATE

$$\dot{t} = \frac{\eta \; P \; (\text{torr}) \times 0.1 \; \text{cm/mm} \times \rho_{Hg} \; g}{\rho_c} \left(\frac{M}{2\pi k T N_0}\right)^{1/2} \qquad (8.13)$$

where $\dot{t}$ = the buildup rate (cm/s)

   η = assumed sticking probability

     = 1.0 for worst-case analysis

   P = pressure (torr) of the gas whose condensation rate is desired.

   ρ_{Hg} = density of mercury (about 13.6 g/cm^3)

   ρ_c = density of the condensed material (about 0.8 gm/cm^3 for ice)

   g = acceleration due to gravity (about 980 cm/s^2)

   M = molecular weight (18 g/mole for water)

   k = Boltzmann's constant (about 1.38 $\times$ 10^{-16} erg/K)

   T = temperature (Kelvin)

   N_0 = Avogadro's number (about 6 $\times$ 10^{23} molecules/mole)

For example, condensation of water onto an 80K surface will be at a rate of 3.5 $\times$ 10^{-2} cm/(s·torr) times the partial pressure of water vapor in torr. At a partial pressure of 10^{-6} torr this yields 3.5 $\times$ 10^{-8} cm/s, or about 1 μm per hour. As shown in Figure 8.9, one micron of ice will attenuate an IR signal by about 10 percent.

One experimental data point: An ion gauge built into a life test dewar shows a pressure of about 5 $\times$ 10^{-8} torr while the dewar is cold, and we observe an apparent water condensation rate of about 1 μm per day (from the responsivity loss with time). This agrees fairly well with the kinetic/ theory calculation above.

8.7.   VACUUM AND SAFETY

There are relatively few hazards associated with vacuum work, and most of those are fairly obvious. They are listed here for completeness.

   Hot Diffusion Pumps. Diffusion pumps are very hot, and anyone working around them should be careful to avoid contact with them and to keep the area around them clear of flammable objects.

Electrical. Any repair or modification to the electronic components associated with gauges or power supplies must be done with the same concern for electrical shock that would be used in any electronic repair. Note that ion gauges operate at a high voltage. The leads are insulated, but it is still unwise to touch them!

Mercury. Mercury is used in some diffusion pumps and in some manometers. Mercury vapor is present whenever mercury is exposed to the air, and prolonged exposure to mercury vapor is harmful. To avoid this exposure, keep unused mercury in tightly sealed containers. Be careful to avoid spills when transferring mercury: Work over a pan so that if mercury is spilled it can be easily collected. Refer disposal of unwanted mercury to hazardous material specialists.

Belts on Mechanical Pumps. Belts and pulleys on mechanical pumps can easily catch fingers, ties, or tools. The results can be broken fingers, abrasions, and worse. To avoid this, keep protective covers over the pulley and belt system, do not wear ties or loose clothing near belts and pulleys, and turn off the pump if you must work near it.

REFERENCES

Bertie, J. E., A. J. Labbe, and E. Whaley (1969). "Absorbtivity of Ice I in the Range 4000–30 cm^{-1}" *J. Chem. Phys.* 50, 4501.

Dushman, Saul (1962). *Scientific Foundations of Vacuum Technique*, 2nd ed., Wiley, New York.

Moore, John H., Christopher C. Davis, and Michael A. Coplan (1983). *Building Scientific Apparatus,* Addison-Wesley, Reading, Mass.

Roth, A. (1982). *Vacuum Technology*, North-Holland, Amsterdam.

Strausser, Yale (1969). "Review of Outgassing Results," *Varian Report VR-51,* Varian Associates, Palo Alto, Calif.

White, Guy Kendall (1987). *Experimental Techniques in Low Temperature Physics*, 3rd ed., Clarendon Press, Oxford.

SUGGESTED READING

Dushman, Saul (1962). *Scientific Foundations of Vacuum Technique,* 2nd ed., Wiley, New York. This is a standard, well-known, respected reference.

Moore, John H., and Michael A. Coplan (1983). *Building Scientific Apparatus* (Advanced Book Program, World Science Division), Addison-Wesley, Reading, Mass. This is an excellent introduction to experimental apparatus. One chapter is devoted to vacuum. Well worth browsing through.

O'Hanlon, John F. (1980). *Users Guide to Vacuum Technology*, Wiley, New York.

Roth, A. (1982). *Vacuum Technology*, 2nd ed., North-Hollland, Amsterdam.

Technical periodicals, available in most scientific libraries: *Vacuum* and *Review of Scientific Instruments*.

PROBLEMS

1. Identify the components of the vacuum system in Figure 8.10.

2. What are some of the advantages and disadvantages of the following vacuum components? (a) Hg manometer; (b) Oil manometer; (c) TC gauge; (d) Ion gauge; (e) Oil-sealed rotary pump; (f) Oil diffusion pump; (g) Mercury diffusion pump; (h) Sorption pump; (i) Ion pump.

3. How long would it take to pump a "perfectly clean" dewar to 1×10^{-5} torr? (Assume a dewar volume, conductance, etc.). (a) Guess the answer; (b) Do the calculations.

4. Consider a pumping system with the following components:

   Pump: 1000 L/s, no lower pressure limit
   Connecting tube: 6 in. diameter, 12 in. long
   Ion gauge in the center of the 6-in. tube
   Valve of 100 L/s conductance
   Connecting tube: 1 in. diameter, 12 in. long
   Dewar valve: 2 L/s conductance
   Dewar volume: 3 L

   (a) What is the conductance of each of the connecting tubes?
   (b) What is the net pump speed?
   (c) If there is no outgassing from the dewar, how long should it take to pump to 1×10^{-2} torr? 1×10^{-5} torr?
   (d) If there is no outgassing from the dewar, find the pressure at $t = 30$ s.
   (e) If the outgassing rate is 1×10^{-4} torr·L/s, find the pressure at the dewar and at the ion gauge after several hours of pumping.

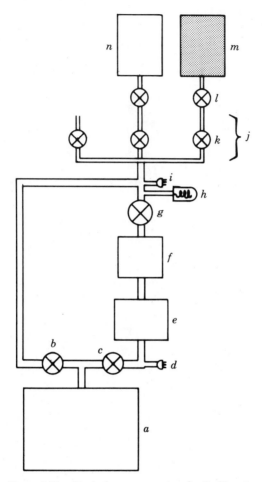

Figure 8.10. Typical vacuum system for Problem 1.

(f) If the outgassing rate is 1×10^{-4} and the dewar is valved off, how long will it take for the pressure to reach 0.03 torr?

5. List the hazards associated with vacuum work and good practices to reduce the dangers.

9

Optics and Optical Materials

9.1. INFRARED OPTICAL MATERIALS

Windows, lenses, and mirrors are used in the IR just as they are in the visible. Only the choice of materials is different: Common glass is opaque beyond about 2 μm, so it is not useful as an IR optical material. Many IR transmitting materials are available for windows and lenses, but we will discuss only window materials here.

9.1.1. Selection of Window Material

Of the many materials available for IR windows, none will be best in all respects. Generally, we must chose the material with the fewest or least objectionable limitations. Factors to consider when selecting a material include:

Desired operating spectral range.

Need for high transmission and low reflection.

Physical strength; thickness required to withstand atmospheric pressure with a vacuum on the other side.

Durability. Some materials are soft and scar easily; others are water soluble.

Desired spectral rejection. The near-zero transmission of Ge and Si in the visible and ultraviolet can be used to advantage to protect InSb and lead salt detectors from undesirable effects of exposure to UV.

Figure 9.1 and Table 9.1 summarize these characteristics for a variety of materials. Table 9.1 is intended to provide a convenient overview, but it is of necessity highly condensed, so refer to original sources for more detailed information before making any final decisions. A discussion of the table data, by columns, follows.

Material, Description. The common name and the chemical symbol are given. Irtran is the name given by the Eastman Kodak company to their

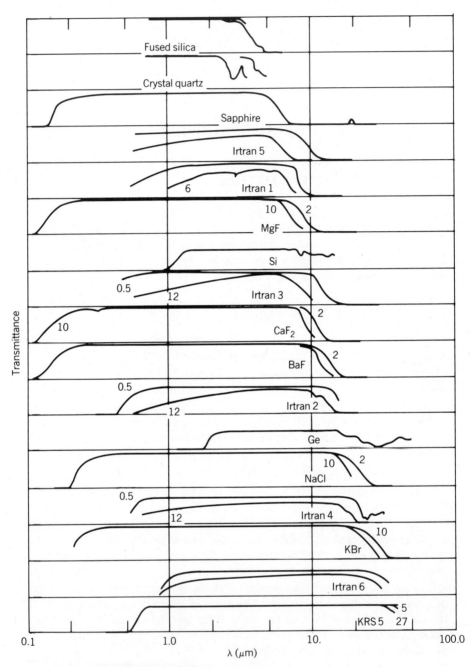

Figure 9.1. Characteristics of window materials.

line of hot-pressed polycrystaline compacts. They heat and compress pure powdered crystaline material to provide relatively inexpensive windows and lenses. At short wavelengths, including the visible, the transmission is generally not as good as for the single-crystal equivalent, but at longer wavelengths there is little difference. (They no longer produce these materials commercially, but many labs have Irtran windows, and similar products are still available.)

n. A nominal value of the index of refraction. Since the index of refraction is wavelength dependent, refer to more detailed sources for the index at your wavelengths.

Transmittance. The transmittance of a window thin enough to be absorption free. Again, consult detailed tables for the transmission at your wavelengths and for a sample of your thickness.

Hardness. These values provide a relative measure of the resistance of the material to scratching. The higher the number, the harder it is.

Solubility. This column gives the solubility of the window in water, at room temperature. The values provided are in grams of window material that will dissolve in 100 g of water. Where that information was not readily available, comments on the relative effect of water are provided. Solubility is of interest since dewars or cryogen transfer lines often condense water from the air, and it may be difficult to avoid getting some water on the windows.

Thickness. Windows are often used on vacuum dewars, with atmospheric pressure forcing the window inward. The thickness here is for a 1-in.-diameter window supported on the edges, and provides a safety factor of 4. Methods of determining the required thickness for other windows is provided in Section 9.1.2.

Modulus of Rupture. Typical values taken from vendors' literature. Window strength is highly variable and is dependent on surface finish.

Cost. Values here are typical for 1-in. samples with nominal tolerances and a "window glass finish," for one window when ordered in lots of five.

Table 9.1 Characteristics of Window Materials

Material[a]		n	Transmittance (%)	Hardness		Solubility	Thickness (in.)	Modulus of Rupture (psi)	Cost	Ref.
				Moh	Knoop					
Fused silica SiO_2	3.3	1.46	93.3		741	Insoluble				
Crystal quartz SiO_2	3.3	1.46	93.3			Insoluble				
Sapphire AlO_2	4.7	1.70	87.4	9.0	1525 to 2200	Near zero	0.019	>65,000		L
Irtran 5 MgO*	6.2	1.65	88.7	6.5		Surface appeared coated	0.029	19,200		K
Irtran 1 MgF*	6.3	1.35	95.7	6.0		No effect	0.028	21,800		K
Magnesium fluoride MgF	6.7	1.38	96.0		415	0.008 g/100 g		7,600	12	H, O
Silicon Si	7.0	3.42	53.9	7.0	1150	Insoluble				
Irtran 3 CaF_2*	8.2	1.40	94.5	4.0		No effect	0.053	5,300		K

Material										
Calcium fluoride CaF_2	8.7	1.40	94.5	4.0	170	0.001 g/100 g	0.053	5,300	4	H, O
Barium fluoride BaF	10.5	1.45	93.5		70	0.02 g/100 g	0.052	3,900	12	H, O
Irtran 2 ZnS*	10.7	2.20	74.9	4.5		No effect	0.034	14,100		K
Germanium Ge	11.0	4.00	47.1			Insoluble				
Sodium chloride NaCl	16.0	1.50	92.3		17	36 g/100 g			1	H, O
Irtran 4 ZnSe*	>15.0	2.41	70.8	3.0		No effect	0.052	6,000		K
Potassium bromide KBr	24.0	1.52	91.8	2.0	6.5	60 g/100 g			2	H, O
Irtran 6 CdTe*	>15.0	2.67	65.7	2.0		No effect	0.060	4,600		K
Thallium bromoiodide (KRS-5)	35.0	2.36	71.8		40	0.05 g/100 g	0.054	3,800	2	H, O

* Asterisk indicates material prepared as a "hot-pressed polycrystalline compact."

Ref. Vendors' handbooks and catalogs used as references. See the Suggested Readings at the end of the chapter for details.

L   Linde
K   Eastman Kodak Irtran Optical Materials, Eastman Kodak Co.
O   Optical Crystal Handbook, Optovac, Inc.
H   Harshaw Optical Crystals, The Harshaw Chemical Co.

9.1.2. Required Thickness for Vacuum Windows

Windows are often used on vacuum dewars, with atmospheric pressure forcing the window inward. Beginning with the formula for the stress on a window supported on the edges, we can derive the thickness necessary to ensure that the stress is less than the modulus of rupture by a safety factor (Equations 9.1*a*, *b*, and *c*).

MINIMUM WINDOW THICKNESS: ROUND WINDOW (DIAMETER *d*)

$$t = dC \tag{9.1a}$$

MINIMUM WINDOW THICKNESS: RECTANGULAR WINDOW (LENGTH *L* × WIDTH *W*)

$$t = 1.5 \left(\frac{L^2 W^2}{L^2 + W^2} \right)^{1/2} C \tag{9.1b}$$

where

$$C = \left(\frac{\Delta P \; \text{SF}}{4M} \right)^{1/2} \tag{9.1c}$$

ΔP = pressure difference across the window

SF = safety factor

M = modulus of rupture

A safety factor of 4 is generally suggested; use of a smaller factor entails real risk, since the modulus of rupture used is an average for many samples and individual samples may be weaker by as much as a factor of 2. Further, any rough spots in the mounting arrangement may increase the stress locally beyond the ideal values that the usual formulas provide.

The stress depends on the method of mounting, so more detailed analysis may be appropriate if unusual mounting methods, higher pressures, or smaller safety factors are used. For the common O-ring mounting often used for dewar windows, with a safety factor of 4, these formulas are adequate.

Values in Table 9.1 are calculated from the equations above for a 1-in.-diameter window, using a 1-atm pressure difference and a safety factor of 4.

9.1.3. Index of Refraction

The wavelength of radiation and its speed both depend on the medium through which the radiation is passing. When we say "the wavelength is 5.5 μm" or "the speed of light c of all wavelengths is about 3×10^{10} cm/s," we mean *in vacuum*. In any other material, including air, electromagnetic radiation travels more slowly and its wavelength is shorter.

The index of refraction of a material is the ratio of c to the speed v of radiation in that material:

$$n = \frac{c}{v} \tag{9.2}$$

An index of refraction near unity means that the radiation moves nearly at the "speed of light." For example, for air, in the visible, n is about 1.0003 (*CRC Handbook*), so the speed of light in air is about 0.03% slower than in a vacuum.

An index of refraction much greater than unity means that the radiation moves much slower than c. For wavelengths of 5 μm (in vacuum), germanium has an index of refraction of about 4: in germanium, those wavelengths move at only one-fourth of their "in-vacuum speed" and have a wavelength of about 1.25 μm.

We are concerned with the index of refraction not because of the speed of the radiation in the material, but because it determines the extent to which the material reflects, transmits, and bends (refracts) incident radiation. If the index of refraction is near unity, the material will not reflect much, nor will it "bend" radiation striking it. If the index is large, the incident radiation will be strongly refracted and reflected and will not be transmitted well.

9.1.4. Dispersion

Dispersion is the rate of change of the index of refraction with wavelength. The term results from the fact that a prism will spread (disperse) a spectrum best if the index changes rapidly with wavelength. It can be shown that near wavelengths which are strongly absorbed, the dispersion is high, and where wavelengths are transmitted well, dispersion is low.

9.1.5. Transmittance of Windows

For consistency and accuracy, the transmittance of window materials should *not* be read from wall charts or other graphs unless absolutely necessary. If possible, a direct measurement is good to have. However, even if that is available, it is a good idea to determine the theoretical or expected transmittance from published data. Tabulated values are available in several handbooks, generally as a function of wavelength. If the transmittance is not listed, it can be calculated from the index of refraction:

TRANSMITTANCE FROM INDEX OF REFRACTION

$$T = \frac{2n}{n^2 + 1} \tag{9.3}$$

(This formula applies only if several assumptions are valid: normal incidence, materials with negligible absorption, spectral content broad enough that interference effects average.)

To assist others in auditing your calculations, include the index of refraction and the source of your information in your calculation notes. Sources of optical data are given in the Suggested Reading section, and data for some common materials are included in Table 9.1.

9.2. GEOMETRICAL OPTICS

Geometrical optics refers to the classical refraction and specular reflection (as opposed to diffraction and interference effects, which are described in *physical optics*). The laws of reflection and refraction are illustrated in Figure 9.2. Consider a ray passing through a material of index n_1, incident on a flat surface of index n_2. The geometry is specified in terms of the angle between the ray and the normal (perpendicular) to the surface (see Figure 9.2).

9.2.1. Angle of Reflection

The reflected beam leaves a smooth surface with an angle equal to that of the incident ray. This is just the way a ball bounces off of a wall.

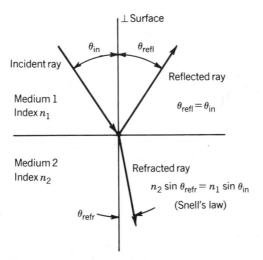

Figure 9.2. Reflection and refraction.

Retroreflectors take advantage of this law, and a little geometry, to bounce incident light directly back at its source. They use three mirrors at right angles to each other (as on three adjoining sides of a cube). The reflection described here is ideal: It assumes smooth surfaces. Such reflection is called *specular reflection*. Specular, or nearly specular, reflection dominates many optical effects of interest. Another important type of reflection is *diffuse reflection*, resulting when light falls on a rough surface. If the surface is sufficiently rough, the resulting reflected rays are uniformly spread over all angles. The surface is said to be *Lambertian*, and the reflectance falls off as the cosine of the angle from the normal.

9.2.2. Refraction and Snell's Law

The path followed by a ray as it passes from one material to another is described by the law of refraction, or *Snell's law*. The angle of the refracted ray depends on the indices of the two materials:

$$n_1 \sin \theta_1 = n_2 \sin \theta_2 \qquad (9.4)$$

9.2.3. Optical Thickness

Imagine a microscope focused on a scratch on the top of a 5-mm-thick glass ($n = 1.67$) window. How far must the microscope be moved down to focus on the bottom of the window? The answer is surprising: it is not

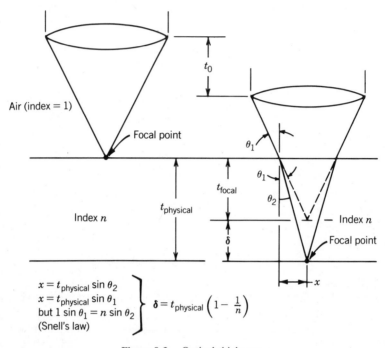

Figure 9.3. Optical thickness.

5 mm, but only about 3 mm. If a traveling microscope is used to determine the distance between the detector and the outside surface of the window, serious errors can result unless the optical thickness of the window is properly accounted for. The explanation is suggested in Figure 9.3. This effect is essential in the design of optical systems but may be ignored for most IR detector work.

9.3. REFLECTANCE AND TRANSMITTANCE CALCULATIONS

9.3.1. Reflectance and Transmittance of Nonabsorbing Materials

The reflectance and transmittance of a particular material at a given wavelength can often be determined from graphs. This is handy for initial selection of materials, but if the transmittance is required for radiometric calculations, more precise values should be obtained by actual measurement or by reference to tabulated values.

If the transmittance cannot be determined directly, it can be calculated from the index of refraction. The formulas for the reflectance and trans-

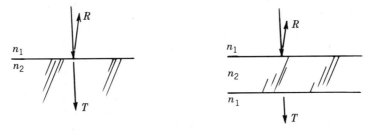

Single surface (semi-infinite slab)   Window (slab with two parallel surfaces)

Figure 9.4. Reflectance and transmittance of nonabsorbing materials.

mittance are illustrated in Figure 9.4 and given in Table 9.2 for two useful geometries:

1. An ideal single surface (a slab of material so thick that no radiation is bounced back from the far side)
2. A window with parallel flat sides

These formulas assume that any absorption in the material is negligible. This is generally true for most window materials. It can be checked by comparing the transmittance curves for different thicknesses of the same material: if the thickness does not change the transmittance, absorption is negligible. (The formulas for absorbing materials are discussed in Section 9.3.3).

These formulas further assume that the radiation strikes the surface at right angles. Small deviations from this will not make a large difference in the reflectance and transmittance, but at larger angles more detailed analysis is appropriate.

9.3.2. Reflection at Other Than Normal Incidence*

Calculation of reflectance at other than normal incidence is fairly involved and requires that we consider the polarization of the waves. An electromagnetic wave contains electric fields that can be broken down into two components. Consider the plane that contains both the incident and reflected rays. The magnitude of the electric fields parallel to the surface is denoted E_p, that for those perpendicular to the surface are denoted E_s (s for the German word *senkrecht*, perpendicular).

Reflection can be thought of as re-radiation by atoms in the reflector

* For a more thorough discussion of reflectance versus incidance angle, see Jenkins and White (1957) pages 509 to 511.

Table 9.2   Reflectance and Transmittance Formulas

Single Surface	Window
General case	
$R = r$ = single surface reflectance	$R = \dfrac{2r}{1 + r}$
$\qquad = \left(\dfrac{n_2 - n_1}{n_2 + n_1}\right)^2$	$\qquad = \dfrac{(n_2 - n_1)^2}{n_2^2 + n_1^2}$
$T = 1 - R = 1 - r$	$T = 1 - R = \dfrac{1 - r}{1 + r}$
$\qquad = \dfrac{4n_1 n_2}{(n_2 + n_1)^2}$	$\qquad = \dfrac{2n_1 n_2}{n_2^2 + n_1^2}$
If $n_1 = 1$ and $n_2 = n$	
$R = \left(\dfrac{n - 1}{n + 1}\right)^2$	$R = \dfrac{(n - 1)^2}{n^2 + 1}$
$T = \dfrac{4n}{(n + 1)^2}$	$T = \dfrac{2n}{n^2 + 1}$
For $n_1 = 1$ and $n_2 = 1.5$	
$R = \ \ 4.0\%$	$R = \ \ 7.7\%$
$T = 96.0\%$	$T = 92.3\%$
For $n_1 = 1$ and $n_2 = 4$	
$R = 36.0\%$	$R = 52.9\%$
$T = 64.0\%$	$T = 47.1\%$

after they are "jostled" by the incident wave. The s polarized wave will always be able to vibrate the atoms in the reflector, but at some incident angles, the p polarized waves are at a geometric disadvantage in vibrating the surface atoms. Thus reflection tends to discriminate against the parallel fields and enhance the proportion of perpendicularly polarized fields. Some sunglass materials use this to advantage to selectively reduce "glare" from the water and roads. Since glare is reflected from water, it is dominated by horizontal polarized light, and the sunglasses screen out these horizontal polarizations.

The reflectance formulas for a single surface are most easily written in terms of r, the angle of refraction* (Equation 9.5).

---

* This is only an intermediate calculation; the reflection is still specular: the reflected angle equals the incident angle.

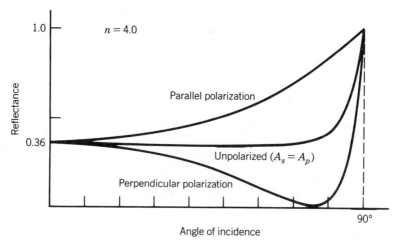

Figure 9.5. Reflectance versus angle of incidance.

REFLECTANCE AS A FUNCTION OF INCIDANCE ANGLE i

$$R_p = \frac{\tan(i - r)}{\tan(i + r)} \qquad (p, \text{ parallel}) \qquad (9.5a)$$

$$R_s = \frac{\sin(i - r)}{\sin(i + r)} \qquad (s, \text{ perpendicular}) \qquad (9.5b)$$

where

$$r = \sin^{-1}\left(\frac{\sin i}{n}\right) \qquad (9.5c)$$

The reflectances from germanium ($n = 4.0$) for pure parallel and pure perpendicular rays are plotted in Figure 9.5. Note the discrimination against reflection of parallel fields mentioned earlier. Figure 25B in Jenkins and White (1957) shows the reflectance for an index of 1.5; Figure 2 in Bennet and Bennet (1978) is a group of curves for various index values and degrees of absorption.

If the incident power contains a fraction A_p of parallel polarized energy and A_s of the perpendicular, the resulting reflectance is

$$R = \frac{A_p R_p + A_s R_s}{A_p + A_s} \qquad (9.6)$$

For unpolarized incident radiation $A_p = A_s = 0.5$, and

$$R = \frac{R_p + R_s}{2} \tag{9.7}$$

At normal incidence ($i = 0$) both R_p and R_s approach the value given earlier*:

$$R = \left(\frac{n - 1}{n + 1}\right)^2 \tag{9.8}$$

At large grazing angles, the reflectance approaches unity for all materials.

9.3.3. Absorption in Optical Materials

Absorption is not often significant in materials used for optical components, but detector operation relies on absorption of the infrared in the detector material. Selection of the detector thickness depends on how well the material absorbs the IR. The absorption in a material is described by the law of exponential absorption:

$$P = P_0 \exp(-\alpha x) \tag{9.9}$$

where α is the absorption coefficient. The total amount absorbed is the initial value less the amount remaining:

$$A = P_0[1 - \exp(-\alpha x)] \tag{9.10}$$

These quantities are plotted in Figure 9.6 for Ge:Hg for both the single surface and the window geometries.

One can never absorb completely all the radiation. In practice, a detector thickness about two absorption lengths is used; this captures about 90% of the energy that enters the material.

9.3.4. Mirrors and Blacks

The discussion of optics to this point has concerned materials with good-to-fair transmittance—window materials and detectors. Reflection and absorption phenomena are also important for materials with high reflec-

---

* This is not clear from the formulas above, since direct subsitution yields 0/0. See Jenkins and White (1957) for the derivation.

Single surface (semi-infinite slab)

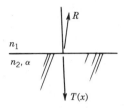

Window (slab with two parallel surfaces)

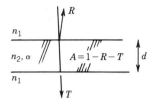

let $r = \left(\dfrac{n_2 - n_1}{n_2 + n_1} \right)^2$

$R = r$

$T(x) = (1 - r)e^{-\alpha x} = \dfrac{4 n_1 n_2}{(n_1 + n_2)^2} e^{-\alpha x}$

$A_{0\ x} = 1 - R - T(x) = (1 - r)(1 - e^{-\alpha x})$

$dA_x = (1 - r)e^{-\alpha x} \alpha\, dx$

let $a = e^{-\alpha d}$

$R = r \cdot \dfrac{1 + a^2 - 2a^2 r}{1 - a^2 r^2}$

$T = a \dfrac{(1 - r)^2}{1 - a^2 r^2}$

$A = \dfrac{(1 - r)(1 - a)}{1 - ar}$

Example: for $n_1 = 1, n_2 = 4, R = 36\%$

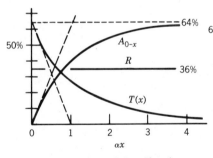

Example: for $n_1 = 1, n_2 = 4$ (Ge, InSb)

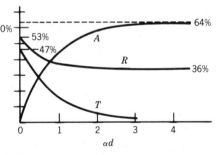

Figure 9.6.   Absorbtance, reflectance, and transmittance.

tivity (mirrors and dewar walls, for example) and low reflectivity (''optical blacks'').

The geometry of reflection we described earlier applies to high-reflec-tivity surfaces as well as to window materials. Good electrical conductivity and a smooth surface are indicators of good reflectance, but specific re-flectance predictions are beyond the scope of this book. Additional in-formation on good reflectors in the IR is provided in Chapter 4.

Low-reflectivity surfaces and materials do *not* reflect in the specular way described for window materials. Instead, radiation is reflected at all angles. This is called diffuse reflection, or scattering (reserving "reflection" for specular reflection). Optical blacks are discussed briefly in Section 4.2.4.

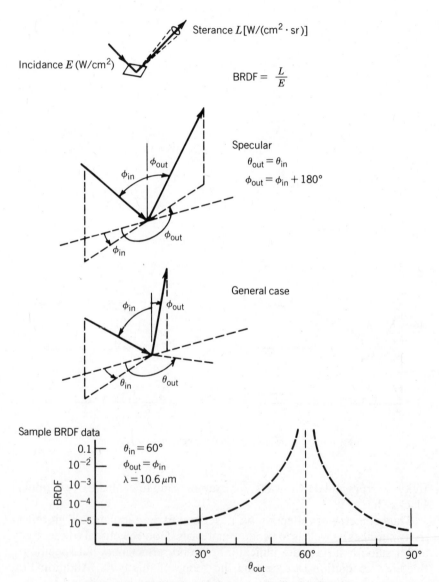

Figure 9.7. Bidirectional reflectance distribution function.

9.3.5. Bidirectional Reflectance Distribution Function

To characterize reflection that is not perfectly specular, we use the *bidirectional reflectance distribution function* (BRDF). This function converts incidence on a surface to the reflected sterance from that surface:

$$L \ [\text{W/(cm}^2\cdot\text{sr)}] = E \ (\text{W/cm}^2) \times \text{BRDF} \qquad (9.11)$$

BRDF has units of $1/\text{sr}$. A perfectly diffuse reflector would have a BRDF of $1/(\pi \ \text{sr})$; about $0.3 \ \text{sr}^{-1}$. A perfectly specular reflector would have zero BRDF everywhere except when exit angle equals the incident angle. At that angle the BRDF would go to infinity (described mathematically by a *delta function*: infinitely high, but zero width). Real surfaces have BRDF somewhere between the two.

BRDF data are a little difficult to present, since BRDF is a function of wavelength and four angles: θ_{in}, θ_{out}, ϕ_{in}, and ϕ_{out} (see Figure 9.7). Typical plots are done:

Versus θ_{out} for a given wavelength, and θ_{in}

With $\phi_{out} = \phi_{in}$ (i.e., the surface normal, the incident and reflected rays in the same plane)

Versus θ_{out} for a given wavelength and θ_{in}

9.4. INTERFERENCE EFFECTS

Optical interference effects are observed when radiation from one source is separated and travels through different paths, then comes together again. Such a situation can be set up with gratings or reflecting layers. Diffraction of light (discussed later) will also provide the necessary situation.

If the peaks of the recombining waves occur at the same place, the resultant electromagnetic wave is large, and we say that the components interfered *constructively*. If the peak of one component and the minimum of the other component come together, the interference is *destructive*, and a small or zero wave results.

Because the phenomenon depends on the wavelengths of the light, it is clearer if the light is monochromatic (one wavelength) or nearly so. Figure 9.8 shows the typical transmittance of a window of thickness d plotted versus the ratio $2d/\lambda$. Table 9.3 includes formulas for the minimum, maximum, and average transmittance.

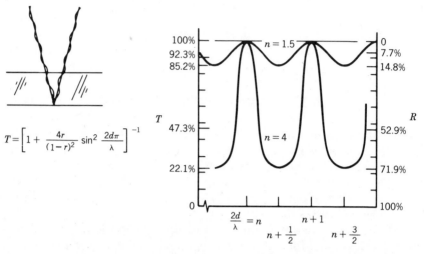

$$T = \left[1 + \frac{4r}{(1-r)^2}\sin^2\frac{2d\pi}{\lambda}\right]^{-1}$$

Figure 9.8. Effect of interferance on reflectance and transmittance. [Data from Jenkins and White (1957, Chaps. 14, 25). *Note*: In the formulas above and in Jenkins and White, Chapter 25, $r = [(n - 1)/(n + 1)]^2$, the reflectance from a single surface. However in Jenkins and White, Chapter 14, $r = [(n - 1)/(n + 1)]$.

9.4.1.   Antireflection Coatings

The reflectance from a surface can be reduced to nearly zero for at least some wavelengths by coating the surface. The simplest coating is a single layer coating. The required index of refraction of the coating and its thickness are determined by the index of refraction of the material to be coated

Table 9.3   Reflection (R) and Transmission (T) Formulas

Minimum reflection, maximum transmission	Maximum reflection, minimum transmission	Average Reflection and Transmission (noncoherent)
$R = 0$	$R_{max} = \dfrac{4r}{(1+r)^2}$	$R_{avg} = \dfrac{2r}{1+r}$
	$= \left(\dfrac{n^2-1}{n^2+1}\right)^2$	$= \dfrac{(n-1)^2}{n^2+1}$
$T_{max} = 1.00$	$T_{min} = \left(\dfrac{1-r}{1+r}\right)^2$	$T_{avg} = \dfrac{1-r}{1+r}$
	$= \left(\dfrac{2n}{n^2+1}\right)^2$	$= \dfrac{2n}{n^2+1}$

and the wavelength at which the minimum reflectance is desired:

$$\text{thickness} = \frac{\lambda}{4} \tag{9.12a}$$

$$n = \sqrt{n_1 n_2} \tag{9.12b}$$

$$\simeq \sqrt{n_2} \quad \text{if } n_i \simeq 1.0 \text{ (air)} \tag{9.12c}$$

The minimum reflection can be obtained for a wider spectral range by the use of multilayer coatings, but for many applications a single-layer coating is adequate.

9.4.2.   Spectral Filters

By using multilayer coatings, filters can be built to very precise spectral specifications. Short wavelength rejection, long wavelength rejection, and narrow and wide bandpass filters are all common.

Such filters often have many layers, and the filter designer must consider the indices of refraction, the ability to deposit materals, and the relative thermal expansion characteristics. It is not surprising then that these filters can be very costly, and their procurement and fabrication can be very time consuming.

Although a filter can be designed to provide desired spectral characteristics in a finite spectral region, one cannot control the transmission at all wavelengths. It may be necessary to use another filter to control the transmission at wavelengths away from the primary transmission region. Such an additional filter is referred to as a blocking filter, and when procuring a bandpass filter, it is necessary to specify how much blocking is needed. The specification for a relatively sophisticated filter is summarized in Figure 9.9.

Effect of Temperature and Angle of Incidence. OCLI puts out a handy booklet on the expected performance of spectral filters as a function of temperature and angle of incidence. The following information is from that booklet, but the original should be consulted if at all possible.

The turn-on and turn-off wavelengths of spectral filters decrease when a filter is cooled. The fractional shift is roughly proportional to the temperature change; typical values range from one to $2\frac{1}{2}\%$ when a filter is cooled from room-temperature to liquid nitrogen temperature. Wavelengths of interest for a spectral filter are shorter by 1 to 2% for a 30° angle of incidence than they are for normal incidence. The shift varies

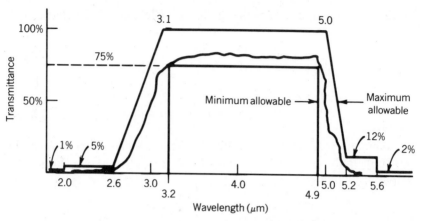

Figure 9.9. Specification of a sophisticated filter.

roughly as the square of the angle of incidence: For 15° angles the shift would be about one-fourth of the values for 30°.

9.4.3. Spectral Transmittance Measurements

Transmittance measurements are made as a function of wavelength using spectrometers. The difficulty in making these measurements depends on the required accuracy (both in transmittance and wavelength) and test conditions (room temperature versus cold, and focused energy versus parallel rays). Small desktop spectrometers can make rapid spectral scans adequate to verify the identity of a filter and determine room-temperature transmittance within 2 to 5% with a wavelengh accuracy of 0.05 μm. For a 1-in. filter, the setup time is less than 5 min and the scan takes about 2 min. Spectrometers with special optical paths that will accommodate cryogenically cooled filters or make very accurate measurements are also available. For these tests the setup may be complex and the total test times may be a day or more.

Wavenumbers. Spectral measurements are often reported in *wavenumbers*:

$$\nu \ (\text{cm}^{-1}) = \frac{10{,}000 \ \mu\text{m}}{\text{cm}} \times \frac{1}{\lambda \ (\mu\text{m})} \tag{9.13}$$

For example, if $\lambda = 5 \ \mu\text{m}$, $\nu = 2000 \ \text{cm}^{-1} = 2000$ wavenumbers. This practice is convenient for many spectroscopic applications since radiation often is associated with energy levels, and

$$E = \frac{hc}{\lambda} \propto \frac{1}{\lambda} \tag{9.14}$$

However, Equation 9.13 cannot be used to directly convert narrow bandwidths and accuracies from wavenumbers to microns: Instead first convert them to *relative* values, then to μm. We take advantage of the relationship

$$\frac{d\nu}{\nu} = -\frac{d\lambda}{\lambda} \tag{9.15}$$

Example: Given λ = 5 μm and an uncertainty of 2 wavenumbers, the uncertainty—in microns—is 0.005 microns.

The wavelength itself (5 μm) corresponds to 2000 wavenumbers, so the uncertainty is 0.1%. Then 0.1% of 5 μm is 0.005 μm. ∎

9.4.4. Neutral Density Filters

Neutral density (ND) filters are generally made by depositing a very thin layer or an only partially covering layer of metal on a suitable substrate. "Neutral" implies that the spectral content of the transmitted radiation is not changed; only the level is changed. ND filters are used to reduce the background in a dewar without restricting the field of view or the spectral bandwidth.

Neutral density filters are never absolutely neutral; they will almost always transmit differently at different wavelengths (see Figure 9.10). The combination of two ND filters in series may yield a transmittance that is greater than just the product of the two component transmittances. Before using two filters in series a check should be made to verify that the expected transmittance is indeed obtained.

ND filters can be made with transmittances from nearly unity to fractions of a percent. Absolute measurement of very low transmittance is very difficult, and measurement of the spectral content is also difficult.

9.4.5. Transmittance of Two Filters in Series

Most neutral density (ND) filters work by reflecting unwanted radiation. This introduces a complication if two such filters are placed in series. The transmittance of two filters in series is commonly assumed to be just the product of their transmittances. This is approximately true if their transmittances are relatively high (i.e., near unity) but is not true in general.

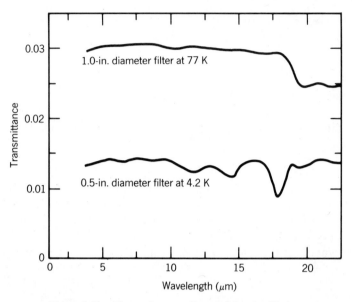

Figure 9.10. Transmittance of neutral density filters.

If the filters are parallel or so nearly parallel that reflections from either filter will reach the detector, the effective transmittance can be greater than $T_1 T_2$.

TRANSMITTANCE OF TWO PARALLEL FILTERS

$$T = \frac{T_1 T_2}{T_1 + T_2 - T_1 T_2} \tag{9.16a}$$

$$\simeq T_1 T_2 \qquad \text{if } T_1, T_2 \simeq 1 \tag{9.16b}$$

$$\simeq \frac{T_1 T_2}{T_1 + T_2} \qquad \text{if } T_1, T_2 \ll 1 \tag{9.16c}$$

Thus two 90% filters yield 81%, but two 2% filters yield 1%. Note that the formulas above are symmetric in T_1 and T_2; it does not matter which filter is in front.

The above formulas assume the two filters are parallel, or nearly so. Any tilt will decrease the net transmittance in a way that is difficult to predict. For that reason the use of ND filters in series is strongly discouraged.

The arguments of the preceding section apply equally well to any two optical components that reflect. If the reflectance is wavelength dependent, the formulas must be applied at each wavelength of interest: We replace T, T_1, and T_2 with $T(\lambda)$, $T_1(\lambda)$, and $T_2(\lambda)$.

9.5. DIFFRACTION

For some applications, radiated electromagnetic energy behaves like waves. As a wave passes a barrier, new waves spread out from that barrier. Thus the energy does not travel in a straight line. The results can be seen when light passes an aperture or knife edge: Instead of a sharp-edged shadow, blurring and more complex patterns are observed if the shadow is observed carefully.

The smaller the aperture used or the longer the wavelength, the more the image is blurred. The image of a point source will appear as a central bright disk (the Airy disk) surrounded by concentric dark and light rings. The diameter of the primary disk can be expressed either in terms of the angle subtended at the lens, or by the linear size on the focal plane (see Figure 9.11 and Equation 9.17.

DIFFRACTION: DIAMETER OF THE AIRY DISK

$$\theta = 2.44 \frac{\lambda}{d} \tag{9.17a}$$

$$x = \theta D = 2.44 \frac{\lambda}{d} D \tag{9.17b}$$

$$= 2.44\lambda \; f/\text{number} \tag{9.17c}$$

where "f/number" is that of the system: D/d.

Note that this blurring occurs even for perfect optics: It is not due to flaws in the optics. The Airy disk contains 84% of the total energy, 7.1% is in the first bright ring, and 2.8, 1.5, and 1.0% in successive rings. Some improvement in the diffraction limit can be made by shading (apodizing) the aperture edges, removing the sharp edges. Generally speaking, however, the diffraction limit as described above is a theoretical limit, difficult to surmount. Analysis of any optical system and prediction of spot size should include a calculation of the diffraction limit. The only practical

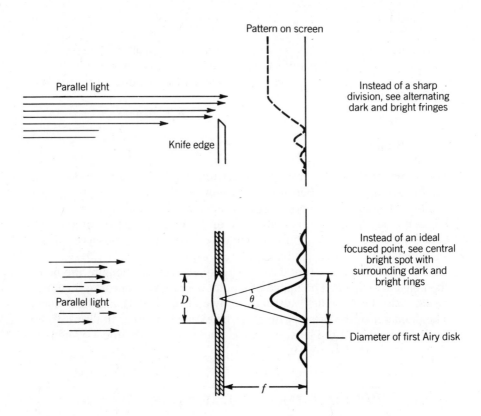

Pattern on screen

Parallel light

Instead of a sharp division, see alternating dark and bright fringes

Knife edge

Parallel light

D θ

Instead of an ideal focused point, see central bright spot with surrounding dark and bright rings

Diameter of first Airy disk

f

Figure 9.11. Diffraction.

$$\theta = 2.44 \frac{\lambda}{D}$$

$$\text{diameter} = 2.44 \frac{\lambda}{D} f = 2.44\lambda \frac{f}{D} = 2.44\lambda \times f/\text{ number}$$

$$= \frac{2.44}{\text{NA}}$$

Example:

$$\lambda = 5\mu\text{m} \quad = 5 \times 10^{-3} \text{ mm}$$

$$D = 2 \text{ cm} \quad = 20 \text{ mm}$$

$$f = 10 \text{ cm} = 100 \text{ mm} \quad \Bigg\} f\, 5, \text{ NA} = 0.2$$

$$\theta = 2.44 \frac{\lambda}{D} = 0.6 \times 10^{-3} \text{ rad}$$

$$D_{\text{Airy}} = \theta \cdot f \quad = 0.06 \text{ mm} = 60 \ \mu$$

$$\simeq 2.4 \times 10^{-3} \text{ in.}$$

This is larger than many small detectors.

way we can reduce the blur diameter is to use larger-diameter apertures or shorter wavelengths. It is for this reason that telescopes use the largest practical diameter lenses and that electron microscopes are used.

9.5.1. Resolution Limits and Criteria

The result of diffraction is that the images of even ideal point sources will have *some* blurring. If two point sources are very close together, the blurring may be enough that you cannot tell whether two images are present, or only one. How close together can the images be and still be able to resolve them? The answer is somewhat subjective, but one criterion is commonly used: *Rayleigh's criterion* states that the two images are resolvable if the minimum from one image falls on the maximum from the other (Figure 9.12). Stated differently, the images are resolvable if they are separated by at least the diffraction blur radius:

$$\theta_{resolvable} > \frac{1.22 \, \lambda}{D} \tag{9.18}$$

An empirical observation about resolution sometimes used in astronomy is *Dawes' limit*: Objects are resolvable (in the visible portion of the spectrum) if they are separated by 4.5 seconds of arc divided by the diameter of the telescope, in inches:

$$\theta_{resolvable} > \frac{4.5 \text{ seconds of arc}}{D \text{ (inches)}} \tag{9.19}$$

Compare this with a prediction made using the Rayleigh criterion: Use a wavelength of 0.55 μm, convert wavelength to inches, and convert angles

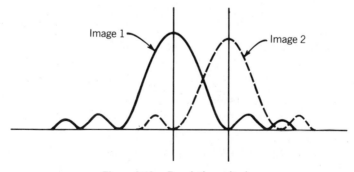

Figure 9.12. Resolution criteria.

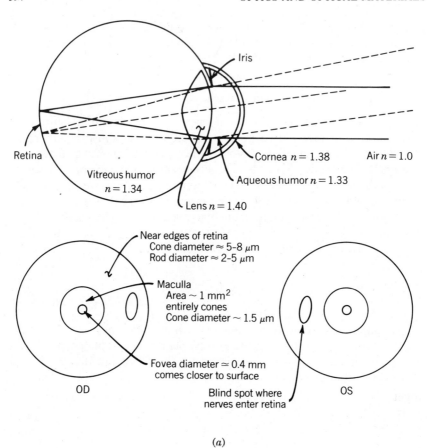

Iris

Retina

Vitreous humor
$n = 1.34$

Lens $n = 1.40$

Cornea $n = 1.38$ Air $n = 1.0$

Aqueous humor $n = 1.33$

Near edges of retina
Cone diameter $\approx$ 5–8 μm
Rod diameter $\approx$ 2–5 μm

Maculla
Area $\sim$ 1 mm^2
entirely cones
Cone diameter $\sim$ 1.5 μm

Fovea diameter $\simeq$ 0.4 mm
comes closer to surface

OD

Blind spot where
nerves enter retina

OS

(a)

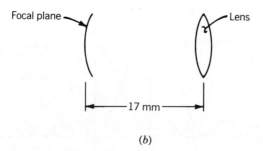

Focal plane

Lens

17 mm

(b)

Figure 9.13.   (a) Structure of the eye; (b) the reduced eye.

to seconds of arc:

$$\theta_{\text{resolvable}} > 1.22 \times \frac{0.55 \times 10^{-4} \text{ cm}}{D} \div 2.54 \text{ cm/in.} \times 57 \text{ deg/rad}$$

$$\times \ 3600 \text{ seconds of arc/degree}$$

$$= \frac{5.4 \text{ seconds of arc·inch}}{D}$$

This agrees quite well with Dawes' limit.

9.5.2. Resolution of the Human Eye

Compared with human-made sensors, even state-of-the-art sensors, the human eye is a marvelous optical system. It has a better dynamic range and better sensitivity in a much smaller package than that of any manufactured system. The eye and its function as a sensor can give us some practice with the optics concepts that we have discussed so far.

The structure of the eye is discussed in Section 12 of the *Handbook of Optics* (Alpern, 1978) and in Smith (1966, chap. 5). For our purposes, consider a simplified model, the so-called "reduced eye," shown in Figure 9.13*b*. It treats the eye as a simple lens, 17 mm from the focal plane. There are two detector types: rods and cones. Near the center of the focal plane, the cones are about 1.5 μm in diameter and are tightly packed together. Near the edges the cones are 5 to 8 μm in diameter, and the rods have a diameter of 2 to 5 μm.

Assuming that the eye is focusing properly, how well can we see fine detail? That is, what is the resolution of the eye? Consider the two effects that limit the resolution of a well-focused system: the detector and the diffraction diameter. In the area on the retina that we use for detailed viewing, the cones have a diameter of about 1.5 μm. Since the image distance is 17 mm, the cones subtend an angle of about 0.1×10^{-3} rad, or about 0.3 minutes of arc (see Figure 9.14).

The diffraction limit can be calculated from Equation (9.17*a*). If we use an 8-mm iris diameter (about the largest opening possible) and a wavelength of 0.55 μm (the center of the visible band), we obtain a diffraction-limited angle of 0.17 mrad, (just under 0.6 minute of arc):

$$\theta = 2.44 \frac{\lambda}{D}$$

$$= 2.44 \frac{0.55 \times 10^{-3} \text{ mm}}{8 \text{ mm}}$$

$$\simeq 0.17 \times 10^{-3} \text{ rad}$$

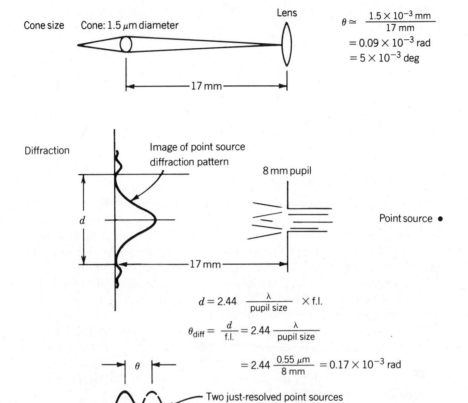

Cone size   Cone: 1.5 μm diameter   Lens

$$\theta \simeq \frac{1.5 \times 10^{-3}\ \text{mm}}{17\ \text{mm}}$$
$$= 0.09 \times 10^{-3}\ \text{rad}$$
$$= 5 \times 10^{-3}\ \text{deg}$$

|← 17 mm →|

Diffraction   Image of point source
                 diffraction pattern

                                           8 mm pupil

d

                                                      Point source •

|← 17 mm →|

$$d = 2.44\ \frac{\lambda}{\text{pupil size}} \times \text{f.l.}$$

$$\theta_{\text{diff}} = \frac{d}{\text{f.l.}} = 2.44\ \frac{\lambda}{\text{pupil size}}$$

$$= 2.44\ \frac{0.55\ \mu\text{m}}{8\ \text{mm}} = 0.17 \times 10^{-3}\ \text{rad}$$

θ

Two just-resolved point sources
angular separation $\geq \theta_{\text{diff}}/2$
$\simeq 0.01\ \times 10^{-3}\ \text{rad}$

Combined effect: can resolve point sources
  if separated by $\sim 0.1 \times 10^{-3}\ \text{rad}$

0.1×10^{-3}

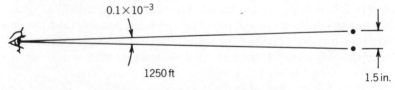

1250 ft   1.5 in.

Figure 9.14.   Resolution limits of the human eye.

Based on this analysis, we would predict the resolution of the eye to be a little under 1 minute of arc; this agrees well with measured values: Alpern (1978) states that in the construction of his eye charts, the Dutch opthamologist Snellen used 1 minute of arc as the minimum separable angle under optimum conditions. Winter (1971) indicates that the probability of detecting a target rises to 50% as the target subtends an angle of about 0.23 mrad, 0.8 minute of arc.

REFERENCES

Alpern, Mathew (1978). "The Eye and Vision," Section 12 in *Handbook of Optics*, Walter G. Driscoll, ed., William Vaughan, assoc. ed., McGraw-Hill, New York.

Bennet, Jean M., and Harold E. Bennet (1978). "Polarization," Section 10 in *Handbook of Optics*, Walter G. Driscoll, ed., William Vaughn, assoc. ed., McGraw-Hill, New York.

Jenkins, Francis A., and Harvey E. White (1957). *Fundamentals of Optics*, McGraw-Hill, New York.

Smith, Warren J. (1966). *Modern Optical Engineering*, McGraw-Hill, New York.

Winter, Donald C. (1971). "Matching an Image Display to a Human Observer," *Electro-Opt. Syst. Des.*, August, p. 40. (Cited by Khalil Seyrafi, (1973). *Electro-Optical Systems Analysis*, Electro-Optical Research Company, Los Angeles, p. 282.

SUGGESTED READING

Claus, Albert C. (1973). "Archimedes Burning Glass," *Appl. Opt.* 12, A14. Claus described an ancient account of Archimedes use of a "burning glass" to destroy attacking Roman ships and went on to explain how it might have been done (several hundred Greeks each reflecting sunlight with a flat, polished metal mirror). The initial article appeared in the October 1973 issue. It stimulated a series of other lively letters (February and May 1974) which included references to even earlier reports of focused light, published studies of the legend, estimates of the power levels possible, and suggestions that the designers of modern glass-walled hotels should consider the danger of frying passersby.

Defense Supply Agency (1962). *Optical Design* (MIL-HDBK-141), Defense Supply Agency, Department of Defense, Washington, D.C. This handbook combines many useful topics, including basic optics, functioning of the human eye, and IR optics. Because it is so old, its availability may be a problem.

Ditchburn, R. W. (1963). *Light*, 2nd ed. Wiley-Interscience, New York.

Driscoll, Walter G., ed., and William Vaughan, assoc. ed. (1978). *Handbook of Optics* (sponsored by Optical Society of America), McGraw-Hill, New York. Excellent reference for all aspects of optics and detection. Includes chapters on radiometry, photometry, sources, photosensitive materials, properties of optical materials, the eyes and vision, and even a table of standard machine screw dimensions.

Jenkins, Francis A., and Harvey E. White (1957). *Fundamental of Optics*, McGraw-Hill, New York.

Kuttner, Paul (1985). "Measuring the Critical Properties of Infrared Optical Materials," *Laser Focus/Electro-optics*, April, 90.

Moore, John H., Christopher C. Davis, and Michael A. Coplan (1983). *Building Scientific Apparatus*, Addison-Wesley, Reading, Mass. Chapter 4 of this book is an extensive practical summary of optical theory, components, and practice.

Savage, J. A. (1985). *Infrared Optical Materials and Their Antireflection Coatings*, Adam Hilger, Bristol, England.

Schawlow, A. L. (1965). "Measuring the Wavelength of Light with a Ruler," *Am. J. Phys.* 33, 922. The article describes a simple lab or lecture hall demonstration: the rulings on a steel ruler are used as a diffraction grating, and a laser produces interference patterns on a wall.

Walker, Jearl. "The Amateur Scientist," in *Scientific American*. This regular column often has applications of optics to everyday phenomena. A few examples: Apparatus to Measure the Speed of Light (October 1975); Optics from an Airplane (August 1988); Gyroscopic Eye Movement (November 1954); Polarizing Filters (December 1977); Optical Knife Edge Tests (July 1958); Effect of Humming on Vision (February 1984); Fly's Eye View of a Fisherman (March 1984); Afterimages; Bidwell's Ghost (February 1985); Shadows in a Pool (July 1988).

CATALOGUES AND HANDBOOKS FROM VENDORS

OCLI Infrared Stock Filter Catalogue and *OCLI Infrared Handbook: A Guide for . . . Filter Performance*. Optical Coating Laboratory, Inc. 2789 Northpoint Parkway, Santa Rosa, California 95041. OCLI sells both custom and off-the-shelf filters. Their off-the-shelf filters, primarily excess from prior special orders, are listed in this catalog. If you can find something that will fill your need in stock, it is much cheaper and faster than having a filter coated to your requirements.

Optics Guide 4, Melles Griot, Inc., 1770 Kettering Street, Irvine, California 92714. A handbook covering fundamental optics, MTF, materials, and a variety of more specialized optical topics. Free.

The following handbooks cover the optical properties of crystalline materials used for IR and UV windows, along with excellent tutorial material on reflection, transmission, and absorption in windows and the associated measurements. They include information on choice of window thickness and the cleaning and refinishing of various window materials. They are generally well documented and provide a convenient reference to the scientific literature.

Harshaw Optical Crystals (1966). The Harshaw Chemical Company, Crystal–Solid State Department. Harshaw is no longer in business, so this catalog is no longer available. If you can find a copy it is an excellent resource.

Kodak Irtran IR Optical Materials, Kodak Publication U72, Eastman Kodak Company, Rochester, NY 14650.

Optical Crystal Handbook, Optovac, Inc., North Brookfield, MA 01535.

Union Carbide Corp. (1988). *Optical Properties and Applications of Linde Cz Sapphire*, Union Carbide Corp, Crystal Products Department. 9320 Chesapeake Drive, San Diego, California 92123.

PROBLEMS

1. Carbon dioxide has a major absorption band at 2349.3 cm^{-1}. What is the corresponding wavelength, expressed in: (a) Nanometers? (b) Micrometers? (c) Inches.

2. If the width of the absorption band in Problem 1 is 25 cm^{-1}, what is it in μm?

3. Given a 120-in. (diameter) $f/3$ telescope used for the visible, determine the diffraction limit both in terms of angles and in linear distance at the focal plane.

4. A spot scanner has an $f/1.5$ system. What is the minimum useful spot size that one can obtain on an InSb detector ($\lambda \sim 5 \ \mu m$)?

5. Given a square detector 2 mils on a side and a peak wavelength of 5 μm, what can you say about the practicality of measuring detector spatial uniformity with the following spot scanner?

    1.5-in.-diameter lens

    0.020-in.-diameter blackbody aperture

    Blackbody-to-lens distance: 30 in.

    Lens-to-image distance: 3.0 in.

6. A traveling microscope is used to determine the clearance between a detector and the back side of a window. The window is CaF_2, 0.080 in. thick. The microscope was moved 0.092 in. as the focus was moved from the front of the window to the detector. What is the clearance between the window and detector?

7. The effective distance of a detector from the blackbody is less than the physical distance: We should use the optical thickness—not the

physical thickness—of all windows and filters. Since most windows and filters are thin compared to the normal detector–blackbody spacing, this effect is generally ignored. Show by an example that ignoring the optical thickness concept does not significantly affect the incidance calculation: Calculate the effective blackbody-to-detector distance if the physical distance is 37.3 cm and the dewar window and filter are both 2 mm thick and have an index of refraction of 2.0.

10

Electronics*

Signal processing for infared detectors is particularly challenging. Resistances can vary from 20-Ω HgCdTe to the 10^{12}-Ω-impedance low-background detectors. The noise contribution of the electronics must be low to minimize detectivity degradation. Frequency response is important: In many applications customers require dc response, or ac response beginning at 1 Hz and operating to the kHz and MHz frequencies. Sophisticated techniques are necessary to satisfy these requirements. For example, there are ways to minimize the effect of small stray capacitances that would otherwise limit the frequency of operation. There are ways to minimize the noise that will work with one detector but not with another.

As in other portions of this book, we will not discuss state-of-the-art techniques in any detail. Instead, we provide the fundamentals of IR detector electronics so that the reader will be in a position to seek out and learn from more specialized texts and articles and to "ask the right questions."

We begin with some basics—resistors and capacitors. Transistors, their use as amplifiers, and their noise mechanisms are discussed in Section 10.2; amplifiers for IR detectors are discussed in Section 10.3, and the grounding methods and test equipment used for laboratory testing, in Section 10.4.

10.1 PASSIVE COMPONENTS

The passive circuit elements are resistance, capacitance, and inductance. Normally, these are considered as discrete components, but in some applications we will need to consider unavoidable "stray" or "parasitic" resistance and capacitance associated with the circuit configuration and material. Inductance is not discussed since its effects are small at normal detector operating frequencies.

---

* This chapter was written from material prepared by Richard A. Chandos.

401

10.1.1. Resistors

Resistors are circuit components that obstruct the free flow of electrons in a conductor. These include not just resistors themselves, but also detector materials and even connecting wires and cables. Resistance in a circuit limits the current that can flow for a given applied voltage, making the current directly proportional to the applied voltage and inversely proportional to the resistance; this is sometimes called Ohm's law:

$$I = \frac{V}{R} \tag{10.1}$$

where I = current (amperes)
         V = voltage (volts)
         R = resistance (ohms)

The power dissipated by a resistor is the current times the voltage; by using Ohm's law, this can be written in three different ways:

$$P = I^2R = IV = \frac{V^2}{R} \tag{10.2}$$

where P = power (watts)
         V = voltage (volts)
         R = resistance (ohms)

Series and Parallel Combinations. A simple voltage divider with two resistors connected across a voltage source is illustrated in Figure 10.1. The

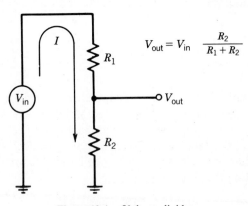

$$V_{out} = V_{in} \frac{R_2}{R_1 + R_2}$$

Figure 10.1.   Voltage divider.

same current flows through both resistors; they are said to be *in series*. Because of that, the voltage drop across each is proportional to the relative resistance of each.

For example, if $R_1 = R_2$, exactly half of the source voltage appears at the output. If an additional resistor is connected from the output to ground (as in Figure 10.2), some of the current is diverted from R_2 and the output voltage is reduced. R_2 and R_3 are *in parallel*, making the voltage across each the same, but causing the current to divide in inverse proportion to the ratio of the two resistors.

RESISTORS IN SERIES

$$R_{eq} = R_1 + R_2 + \cdots + R_n \tag{10.3}$$

RESISTORS IN PARALLEL

$$R_{eq} = \frac{1}{1/R_1 + 1/R_2 + 1/R_3 + \cdots + 1/R_n} \tag{10.4}$$

TWO RESISTORS IN PARALLEL

$$R_{eq} = \frac{R_1 R_2}{R_1 + R_2} \tag{10.5}$$

By successive application of the series–parallel equations (Eqs. 10.3–10.5), complex resistor networks can be reduced to their simplest forms.

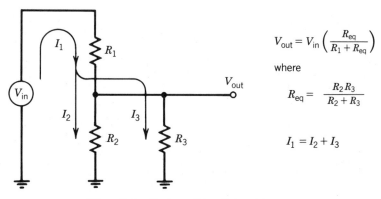

$$V_{out} = V_{in} \left(\frac{R_{eq}}{R_1 + R_{eq}} \right)$$

where

$$R_{eq} = \frac{R_2 R_3}{R_2 + R_3}$$

$$I_1 = I_2 + I_3$$

Figure 10.2. Series and parallel resistances.

For example, the circuit in Figure 10.1 can be reduced to an electrically equivalent form (Figure 10.3) which is identical in all respects except power dissipation within the network shown. The current, voltage, and power dissipated in a load connected to the output of the circuit in Figure 10.1 will be identical to that at Figure 10.3. This is an application of the Thévenin theorem.

Thévenin and Norton Equivalent Circuits. Any power source can be replaced, without loss of generality, by either of two simple circuits:

1. The *Thévenin circuit* is a voltage V in series with a resistance R, where V is the open-circuit voltage of the original source (the voltage across the terminals when no load is connected) and R is the terminal-to-terminal resistance.
2. The *Norton circuit* is a current source I in parallel with a resistance R, where I is the short-circuit current of the original source (the current from the terminals when they are shorted) and R is the terminal-to-terminal resistance.

These apply to both ac and dc sources.

A corollary: To characterize a power source, we need only know two of the following three parameters: the terminal-to-terminal resistance, the open-circuit voltage, and the short-circuit current. For example, consider a black box with an open-circuit voltage (V) of 200 mV and a short-circuit current (I) of 10 mA. The theorems state that the three circuits shown in Figure 10.4 are completely equivalent. It is easy to see that all three yield the same current when shorted and that all three yield the same voltage when open.

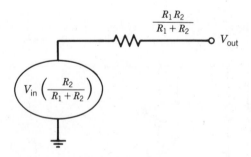

Figure 10.3. Simplified equivalent to divider.

We can also verify, for a few sample load resistances, that the last two yield the same voltages and currents:

For a load of 20 Ω, the current to the load is 5 mA and the voltage across it is 100 mV.

For a load of 80 Ω, the current to the load is 2 mA and the voltage is 160 mV.

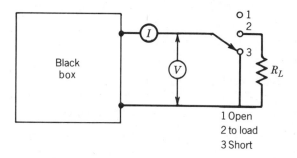

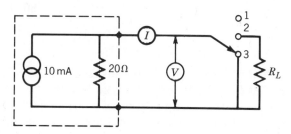

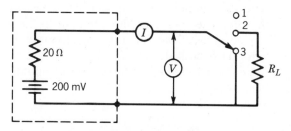

Figure 10.4. Thévenin and Norton equivalent circuits.

For Norton Equivalent

If $R_L = 20\ \Omega$ If $R_L = 80\ \Omega$

$I_L = \frac{1}{2}\cdot 10$ mA $R_{11} = \dfrac{20\cdot 80}{20\ +\ 80} = 16\ \Omega$

   $= 5$ mA $V_L = 10$ mA $\times$ 16 Ω

$V_L = 5$ mA $\times$ 20 Ω $= 160$ mV

   $= 100$ mV $I_L = 160$ mV/80 Ω

                                          $= 2$ mA

For Thévanin Equivalent

If $R_L = 20\ \Omega$ If $R_L = 80\ \Omega$

$I = 200$ mV/(20 Ω + 20 Ω) $I_L = 200$ mV/(20 Ω + 80 Ω)

   $= 5$ mA $= 2$ mA

$V = 5$ mA $\times$ 20 Ω $V_L = 2$ mA $\times$ 80 Ω

   $= 100$ mV $= 160$ mV

These equivalent circuits are useful because they can be chosen so that the resulting circuit is much easier to analyze than the original.

Several other methods of simplification of resistive networks by node analysis and current loops are possible (but not within the scope of this book) [see Fitzgerald et al. (1981) pages 50 to 58, for example].

Johnson Noise. Temperature-induced random electron movements in resistors cause a time-varying voltage to appear at the resistor terminals. This is called Johnson noise and is an ultimate limit to the performance of a circuit. In the absence of bandlimiting (reactive) components, the noise exhibits uniform spectral density from zero frequency to infinite frequency and is commonly called *white noise*. The magnitude of the noise voltage is described by

$$v_n = \sqrt{4kTR\ \Delta f} \qquad\qquad (10.6)$$

where k = Boltzmann's constant (about 1.38×10^{-23} J/K)
        T = absolute temperature (kelvin)
        R = resistance (ohms)
        Δf = noise equivalent bandwidth (hertz)
        v_n = rms noise voltage

It is often convenient to convert the noise voltage (v_n) to an equivalent noise current (i_n) as in Figure 10.5a and b, where

$$i_n = \frac{v_n}{R} = \sqrt{\frac{4kT\,\Delta f}{R}} \tag{10.7}$$

Figure 10.5a and b are equivalent circuits.

Noises do not add linearly. Instead, the noise *powers* add. The noise voltage output of Figure 10.5c equals the root squared sum (rss) of the two noise current generators, i_{n_1} and i_{n_2}, acting across the two parallel resistors R_1 and R_2:

$$v_{\text{out}} = \sqrt{(i_1)^2 + (i_2)^2}\;R_{\text{eq}} \tag{10.8}$$

where R_{eq} is the parallel combination of R_1 and R_2.

Noise values should be calculated by summing noise currents from individual resistors in a network [equation (10.8)]. If the resistors are at the same temperature, one can calculate the equivalent resistance for the network as described previously, then calculate the noise voltage [equation (10.6)].

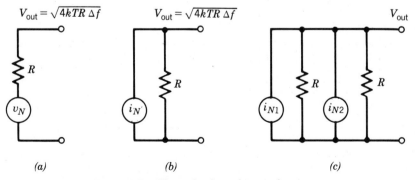

Figure 10.5.   Thermal noise voltage and current.

10.1.2. Capacitors

Capacitors are formed by conducting surfaces insulated from each other by a dielectric, and are capable of storing electrical charge. They are discrete electronic components included for ac signal coupling and filtering. Capacitance is also introduced by parasitic effects such as the biased diffusion layer of a semiconductor or even the wire, shield, and insulator of a shielded cable. The amount of charge held on a parallel-plate capacitor is proportional to the plate area, the dielectric constant of the insulator, and the voltage applied between the plates. It is inversely proportional to the insulator's thickness. All of these factors except the voltage are combined into a constant (for that device) called the *capacitance*, so that

$$Q = CV \tag{10.9}$$

where Q = charge (coulombs)
   V = voltage between the capacitor's terminals
   C = capacitance value (farads)

Since current is the rate at which charge flows, we can relate the voltage buildup on a capacitor to the current entering it:

$$V = \frac{I \, \Delta t}{C} \tag{10.10}$$

where I is the current (amperes) and Δt is the time the charging current is applied (seconds). This is shown in Figure 10.6a, where an amount of charge is formed by a current source I connected to capacitor C for time T, then disconnected. During the charging interval, the voltage rises linearly. At the end of the charging interval, the voltage remains constant because there is no current discharge path. The rate of voltage change (per unit time) during charging and the final voltage value on the capacitor are inversly proportional to the capacitance value. That is, for a given charging current a larger capacitor will cause a lower rate of voltage change during charging and smaller final voltage value.

Capacitors in Series and Parallel. For capacitors in parallel (Figure 10.6b), the charging current divides (as for resistors in parallel) because the terminal voltages of the capacitors are equal because they are connected. Therefore, the rate of voltage change during charging and the final voltage value are both less than in Figure 10.6a. For equal capacitors, the voltage will be exactly half of what it would have been with just one capacitor.

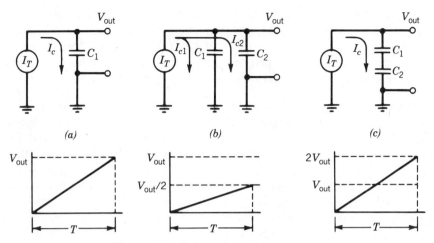

Figure 10.6. Series and parallel capacitors.

Referring to formula (10.10), the effective capacitance is larger than for just one capacitor. For the case of the series capacitors, in Figure 10.6, the same current charges both C_1 and C_2. At the end of the charging interval, each is charged to the same voltage as C_1 in Figure 10.6a, but since they are connected in series, the final voltage is twice that of C_1, so the equivalent capacitance is less than that of the components.

CAPACITORS IN PARALLEL

$$C_{\text{TOTAL}} = C_1 + C_2 + \cdots + C_n \tag{10.11}$$

CAPACITORS IN SERIES

$$C_{\text{total}} = \frac{1}{1/C_1 + 1/C_2 + 1/C_3 + \cdots + 1/C_n} \tag{10.12}$$

Capacitors store energy, measured in joules (or W·s). The energy W stored is a function of the square of the applied voltage V:

$$W_e = \frac{1}{2} CV^2 \tag{10.13}$$

As an example, use this equation to calculate the amount of energy delivered during a static discharge. The capacitance between a standing human and the floor is about 80 pF and can charge to a voltage of 20,000 V (or more). Formula (10.13) yields an energy of 16 mJ. If the discharge occurs in 1 ms, an average power of 16 W will be dissipated. Such a discharge is capable of destroying thin-film detectors, diodes (including photovoltaic detectors), and even large geometry transistors.

AC Coupling. Capacitors are particularly useful in conjunction with resistors for forming high-pass and low-pass filters as illustrated in Figure 10.7. While phasor notation and complex number theory are required for rigorous analysis, the fundamental concepts are illustrated here. See pages 119 to 136 of Fitzgerald et al. (1981) for a complete analysis.

Figure 10.7a and c illustrate the time response of an RC circuit to a pulse input waveform. The charge integration charactistic of the capacitor and current limiting characteristic of the resistor result in an exponentially rising and falling output. The rising waveform caused by a capacitance C charging through a resistance R is described by

$$V_{\text{out}} = V_{\text{in}} (1 - e^{-t/RC}) \tag{10.14a}$$

The falling waveform caused by a capacitor discharging is described by

$$V_{\text{out}} = V_{\text{in}}(e^{-t/RC}) \tag{10.14b}$$

where e = base of the natural logarithms
   t = time (seconds) after charging or discharging begins
   R = resistance (ohms)
   C = capacitance (microfarads)

The product RC has units of time and is often called the *time constant* of the circuit. It is usually represented by the symbol τ.

Figure 10.7b and d illustrate the sine-wave response of a single-pole RC low-pass and high-pass filter. In each case the 3-dB "corner" frequency is the intersection of the 20-dB/decade "roll-off" asymptote and the unattenuated "in-band" signal amplitude. (The dB notation is discussed in Appendix C.) The 3-dB frequency for a single-pole low-pass or high-pass circuit is defined as

$$f_{3\text{dB}} = \frac{1}{2\pi RC} = \frac{1}{2\pi\tau} \tag{10.15}$$

where $f_{3\text{ dB}}$ = frequency (hertz) for which the voltage is 0.707 of the unattenuated in-band value
   R = resistance (ohms)
   C = capacitance (farads)

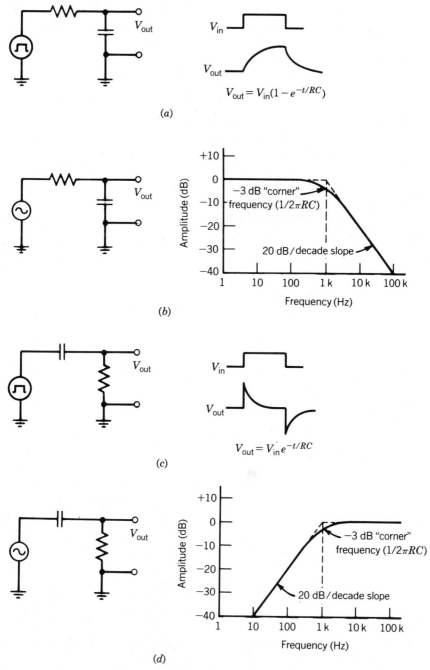

Figure 10.7. High-pass and low-pass filters.

In detector measurement, low-pass filters are particularly useful for limiting the system noise bandwidth. The equivalent noise bandwidth of a single-pole low-pass filter is

$$\Delta f = \frac{\pi}{2}\,(f_{3\text{dB}}) = \frac{1}{4\tau}$$ (10.16)

where Δf is the equivalent noise bandwidth and $f_{3\ \text{dB}}$ is the corner frequency calculated in equation (10.15).

Capacitors are generally considered to be noiseless devices; however, in extremely high-impedance detector circuitry the dielectric loss can sometimes limit detector performance. Dielectric loss (sometimes referred to as *loss tangent*) increases with frequency because it becomes more significant compared to the capacitive reactance. The noise current contributed by the loss tangent is described by

$$i_N = \sqrt{4kT{\cdot}2\pi fC\,\tan\delta}$$ (10.17)

where k = Boltzmann's constant
          T = temperature (kelvin)
          f = frequency (hertz)
          C = capacitive value (farads)
      $\tan\delta$ = loss tangent of the dielectric

10.2.   TRANSISTORS AND THEIR USE AS AMPLIFIERS

Three major types of transistors are of importance in detector electronics, either as discrete components or within integrated circuits. These are *bipolar junction transistors* (BJTs), *junction field-effect transistors* (JFETs) or *metal-oxide-semiconductor FETs* (MOSFETs). Each type has differences and similarities, which we discuss, as well as circuit configurations and noise characteristics.

10.2.1.   Bipolar Junction Transistors

Bipolar junction transistors are named "bipolar" because charges are carried by both holes and electrons, and "junction" because they are manufactured with two diode junctions. They are the most commonly used components in integrated circuits for preamplifiers, power supplies, and bias voltage regulators. Although this section deals with bipolar junction transistors, many principles are applicable to field-effect transistors as well and will not be repeated in those sections.

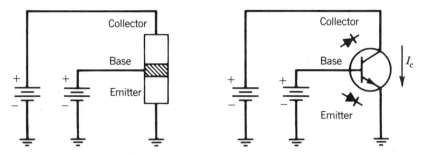

Figure 10.8. Bipolar junction transistor biasing.

Bipolar junction transistors are constructed primarily of doped silcon in the form of two junctions (see Figure 10.8); the one forming the base–emitter junction is operated forward biased, and the other, forming the collector–base junction is reversed biased. They are manufactured in both *npn* and *pnp* types, but for simplicity only the *npn* type is discussed here. The *pnp* types are similar with the exception that the diode junctions, power connections, and current flow are reversed. A detailed description of the physical mechanisms of charge movement is not possible here, but additional reading is suggested [see Holt (1978), for example].

The transistor exhibits current "gain": If a small amount of base current (I_b) is supplied by slightly forward biasing the base–emitter junction, a larger related current (I_c) flows from the collector to the emitter. The current gain (denoted by the symbol β) is a function of the transistor geometry and doping, as well as the operating current and temperature (see Figure 10.9). Even among transistors from the same wafer, β variations of more than a factor of 2 are not uncommon.

The base–emitter junction appears similar to any silicon diode, demonstrating a nominal 2.3 mV/°C negative temperature coefficient. But since the collector current must also pass through the base region, the diode resistance is a function of collector current. A good resistance approximation is

$$R_b = \frac{26 \ \Omega \cdot \text{mA}}{I_c} \tag{10.18}$$

where R_b is the base–emitter diode resistance (ohms) and I_c is the transistor collector current (milliamperes).

A simple method of demonstrating transistor operation is illustrated in Figure 10.10, where a 0- to 5-V logic signal is required to light a 100-mA 10-V lamp. Clearly, the logic signal is incapable of supplying either the

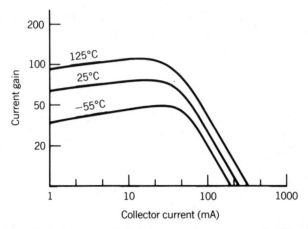

Figure 10.9. Bipolar junction transistor *b* versus collector current (at different temperatures).

required lamp voltage or current directly, so a transistor is inserted between the logic signal and the lamp to supply the necessary drive voltage and current. As the control signal rises above about 0.6 V (the V_{be} threshold), base current flows in the forward-biased transistor base–emitter junction, allowing a larger current to flow through the lamp and the collector–emitter terminals of the transistor. If sufficient base current is supplied, the transistor collector-to-emitter voltage will approach 0 V, and the entire +10-V supply voltage will be dropped across the lamp, causing it to turn "on."

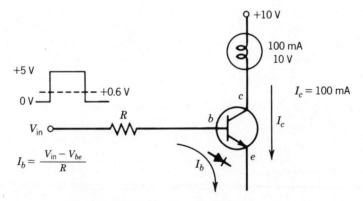

Figure 10.10. Bipolar junction transistor switch.

Knowing the transistor β and the lamp current requirements, the minimum required base current can be calculated.

$$I_{b\ (min)} = \frac{I_{lamp}}{\beta} \tag{10.19}$$

but

$$I_b = \frac{V_{in} - V_{be}}{R} \tag{10.20}$$

where R is the base current limiting resistor. Assuming a transistor β of 50, we can obtain the desired 100-mA lamp current with I_b of 2 mA, so we should select a maximum value for R of 2.2 kΩ.

Amplifier Circuits with Bipolar Junction Transistors. There are three connection configurations used for bipolar junction transistor amplifiers. These are common emitter, common collector, and common base. Each is named for the terminal that is made common to the power supply. Figure 10.11 shows all three as they might be used for detector signal amplification, although they are not all necessarily appropriate for that purpose. The unique properties and applications of each will be discussed.

Common Emitter Amplifiers. Common-emitter amplifiers have the same general configuration as the lamp driver circuit discussed previously, except that the lamp is replaced with a load resistor and a bias network is provided to ensure that the transistor is neither switched "on" or "off" but operates in the linear region between those states. Small changes in base current, caused by the signal to be amplified, produce larger signal swings at the collector (output) terminal. A set of collector curves is presented in Figure 10.12, with a superimposed load line equivalent to a collector load resistor of 5 kΩ. These curves represent the transistor collector voltage versus collector current for four different base currents (from 10 to 40 μA). The transistor β can be calculated by measuring the change in collector current between the base current step intervals. For this transistor, a 10-μA base current change produces about 0.5 mA of collector current change, making the β about 50. At the left of the figure, the nearly vertical stem of the curves represents a point at which the transistor is switched "on" (no longer in the linear operating region). The slope (V/I) of this line is a measure of the transistor's "on" resistance when used as a switch. For this transistor, that slope is about 0.6 V/2 mA, or about 300 Ω.

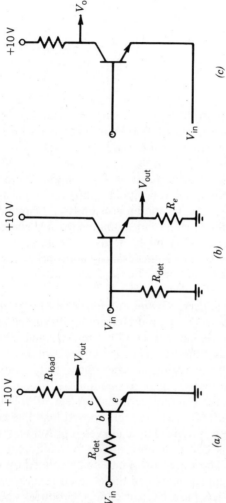

Figure 10.11. (*a*) Common-emitter, (*b*) common-collector, and (*c*) common-base circuits for bipolar junction transistors as they might be used to amplify detector signals.

416

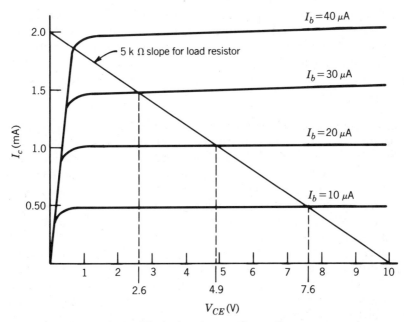

Figure 10.12. Bipolar junction transistor collector curves.

The circuit voltage gain (input/output voltage ratio) can also be calculated by comparing the intersections of the load line and two base current steps chosen about the desired quiescent operating current, with the corresponding V_{ce} voltages on the bottom (dashed lines) voltage scale. In this case, a base current change of 20 μA (from 10 to 30 μA) corresponds to a collector voltage change of 5 V (2.6 to 7.6 V). Using equation (10.18), the input base–emitter diode impedance can be calculated, to convert the 20-μA base current change to an equivalent input voltage. It is important to note, however, that the calculated value must be multiplied by β because both the base current and collector current flow through the transistor's base region. The gain equation is

$$\text{voltage gain} = \frac{\Delta V_{\text{collector}}}{\beta(R_{\text{base}})(\Delta I_{\text{base}})} \qquad (10.21)$$

For our example,

$$\text{voltage gain} = \frac{5\text{ V}}{(50)\,(26\ \Omega)\,(20\ \mu\text{A})}$$

$$= 192$$

Emitter degeneration is almost always used with common emitter amplifiers to stabilize both the dc operating point and the voltage gain. It is accomplished simply by inserting a resistor in the emitter lead of the transistor (see Figure 10.13). As the circuit gain is reduced by increasing the value of the degeneration resistor, it approaches the ratio of the collector and emitter resistors.

Common-Collector Configuration. The *common-collector* configuration (collector tied to the power supply) is also called an *emitter follower* (or *source follower* in the case of field-effect transistor implementation). Input signals are connected to the base terminal, and the emitter terminal is the output (see Figure 10.14). It is a totally degenerated circuit; that is, all of the transistor current gain is applied to create drive power (current gain) instead of voltage gain as in the common-emitter configuration. The voltage gain is slightly less than 1.

The circuit is useful for buffering high-impedance signals and for providing drive current for cables, focal plane readouts, or external circuitry. Most linear integrated circuits use it to provide a low-impedance output. The key to the circuit's performance is that the collector current and the base current both flow through the emitter resistor, making it appear β times larger to the input circuit (explaining the circuit's high input resistance) and making the impedance driving the circuit appear β times lower at the emitter terminal (explaining the circuit's low output resistance). This circuit configuration also operates to very high frequencies.

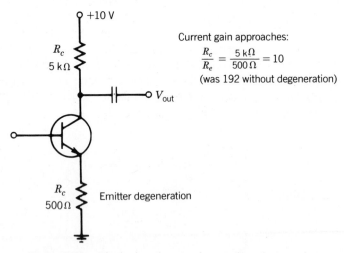

Figure 10.13. Bipolar junction transistor emitter degeneration.

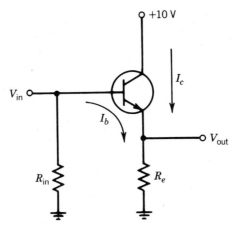

Figure 10.14. Bipolar junction transistor common collector. Input resistance $\approx \beta R_e$; output resistance $\approx Z_{in}/\beta$; voltage gain ≈ 1.

Common-Base Configuration. In the *common-base* configuration (Figure 10.15), input signals are connected to the transistor emitter terminal and output is taken from the collector. The input resistance is very low, approximately the resistance of the base–emitter diode, and the output resistance is quite high, approximately the value of the load resistor. For low-resistance inputs, the gain can be quite high, approaching the ratio of the load resistor divided by the driving source resistance.

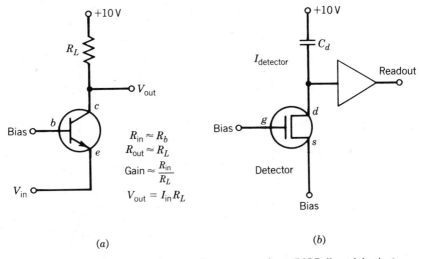

(a)   (b)

Figure 10.15. Biopolar junction transistor common base (MOS direct injection).

Common-base bipolar implementations are not particularly useful except for RF circuits, but the MOS transistor counterpart is useful as the input stage to focal plane readout chips (Figure 10.15b). It provides low input resistance to hold the detector bias constant while allowing the signal current to be integrated on the drain capacitance for subsequent readout.

Bipolar Junction Transistor Noise Sources. Three types of noise sources are associated with bipolar transistors: $1/f$ (thought to be associated with carrier generation/recombination in the transistor), thermal (associated with resistive properties), and shot noise, which is associated with the nonuniform current flow in the space charge regions. Shot noise is the result of nonuniform current flow in any conductor; the shot noise equation (Eq. 10.22) for transistors is identical to that for PV detecters.

SHOT NOISE

$$i_n = \sqrt{2Ie\,\Delta f}\qquad(10.22)$$

where i_n = rms noise current
  e = electron charge (about 1.6×10^{-19} C)
  I = current (amperes)
  Δf = noise equivalent bandwidth

The best noise performance can be obtained by operating the bipolar transistor from a very low source resistance, which minimizes the effects of shot noise, allowing the thermal noise (attributed primarily to the resistance of the base–emitter diode) to become dominant. $1/f$ noise, which increases with the square root of decreasing frequency, is always present but is usually not a significant part of the total noise above a frequency of a few hertz.

There are two standard conventions for characterizing transistor noise. The first convention models a noiseless transistor with a voltage noise or current noise generator at the base terminal, as illustrated in Figure 10.16a and b. These are useful for modeling low-resistance or high-resistance inputs. Typical noise spectral density plots for each case are presented in Figure 10.16c and d.

The second characterization method is more commonly used and involves measuring the total noise of the transistor–source resistance combination while varying the source resistance. The transistor's noise figure, F, is the increase in noise above that of the source resistance alone. It is

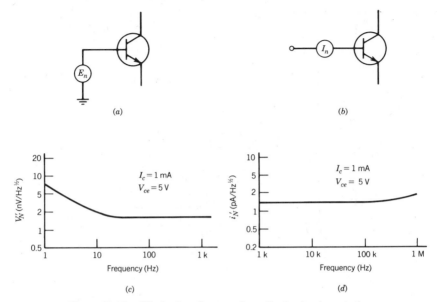

Figure 10.16. Bipolar junction transistor V_n, I_n circuit, and plot.

usually expressed in decibels. Figure 10.17a shows the frequency dependence of the device noise figure at a fixed collector current, and Figure 1.17b illustrates the dependence on collector current and source resistance.

10.2.2. Junction Field-Effect Transistors

Junction field-effect transistors (JFETs) are useful in detector signal processing because they are characterized by high input resistance and the best available noise performance when operating from high-impedance sources. They are also useful as almost-lossless analog switches and multiplexers. Like bipolar junction transistors, they are manufactured mostly of doped silicon and are available in both n-channel and p-channel configurations. For brevity, only the n-channel type is discussed, but by reversing diode polarities and supply potentials, the discussion is applicable to p-channel devices as well.

Device construction involves diffusing a single p (gate) junction into n-type silicon, with a source and drain connection at each end, as illustrated in Figure 10.18. The drain, gate, and source of the JFET are analogous to the collector, base, and emitter of the bipolar junction transistor. The amplifier circuits are called common source, common gate, and common drain.

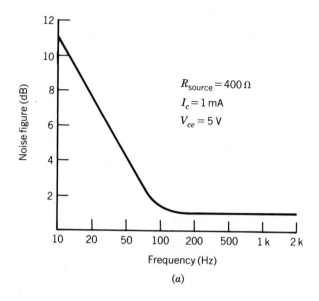

$R_{source} = 400\,\Omega$
$I_c = 1\,mA$
$V_{ce} = 5\,V$

(a)

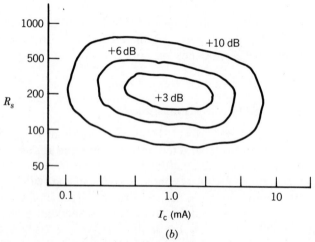

(b)

Figure 10.17.   Noise figure contours for bipolar junction transistor.

In normal operation the drain is biased at a positive potential and the gate from zero to a negative potential; it is always zero or reverse biased. At zero gate bias, the channel below the gate junction conducts, allowing current flow from the drain to the source, but as the gate bias is made more negative, a depletion layer under the gate junction thickens, closing

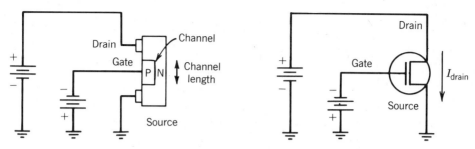

Figure 10.18. JFET construction and biasing.

off the channel to conduction. Device parameters are a function of the channel geometry, with high-power devices using wider channels (perpendicular to the page in Figure 10.18). Figure 10.19 illustrates a typical transfer function: drain current versus gate bias. From this curve the device transconductance (G_{fs}) can be calculated. Transconductance is expressed in inverse ohms or siemens, and is defined as

$$G_{fs} = \frac{\Delta I_{\text{drain}}}{\Delta V_{\text{gate-source}}} \tag{10.23}$$

The same calculation can be made from the drain curves presented in Figure 10.20. These are very similar to the collector curves for the bipolar junction transistor (Figure 10.12) except that the steps of base current become steps of gate voltage. A common-source load resistor is also plot-

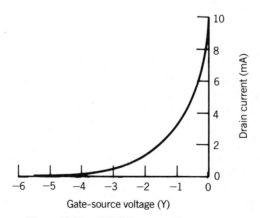

Figure 10.19. JFET I_d versus V_{gs} curve.

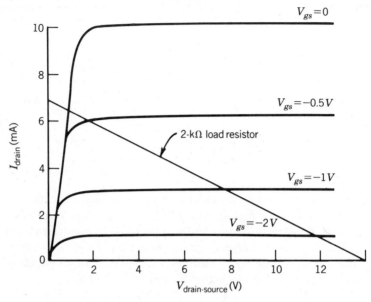

Figure 10.20. JFET drain curves.

ted. At a quiescent operating point of 4.5 mA, the G_{fs} is approximately 3 mA/0.5V, or 6 millisiemens.

JFET Amplifier Configurations. The three amplifier configurations listed for bipolar junction transistors have JFET analogs, called common source, common drain (or source follower), and common gate. The circuit gain expression for a common-source stage is

$$Av = G_{fs}R_L \tag{10.24}$$

where G_{fs} is the device transconductance at the quiescent operating point and R_L is the load resistor value (ohms). The common source gain is then 6 millimhos times 2 kΩ, or 12.0.

The common-drain (source-follower) configuration is similar to that of the bipolar transistor, characterized by a high input resistance (almost infinite) and low output resistance ($1/G_{fs}$). JFETs are seldom used in the common-base configuration because bipolar transistors have better noise figures when operating from low-impedance sources.

JFET Noise. The same three noise sources present in bipolar transistors are also present in JFETs, but in different proportions. $1/f$ noise is generated in the space charge layers by random generation and recombination of carriers and is visible in the noise spectrum below about 100 Hz. Thermal noise generated by the channel resistance ($\approx 1/G_{fs}$) is the dominant source, characterized by a flat spectral response. In extremely high-impedance circuits, shot noise can be seen which results from the leakage current of the gate diode diffusion. A typical noise plot is presented in Figure 10.21.

10.2.3. Metal-Oxide Field-Effect Transistors

Metal-oxide field-effect transistors (MOSFETs) are most commonly used on focal planes with cooled detectors because they operate well at extremely low temperatures (as opposed to bipolar transistors and junction FETs which cease to function below about 190 K). While they have inherently higher (and less predictable) noise than junction FETs and are easily damaged by static discharge, they have the highest input resistance of all semiconductors and are capable of interfacing with even the highest-impedance detectors.

MOSFETs, like bipolar junction transistors and JFETs, are manufactured in both n-type (n-channel) and p-type (p-channel) configurations. Again, only n-channel devices will be discussed here, with the understanding that p-channel devices operate identically if the bias potentials (and current flows) are reversed.

There are two channel configurations of MOSFETs; these are depletion mode and enhancement mode, illustrated in Figure 10.22. They require slightly different biasing and have different drain curves. The *depletion-*

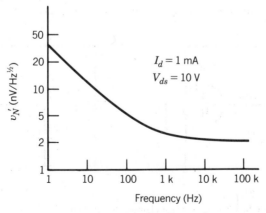

Figure 10.21. JFET voltage noise.

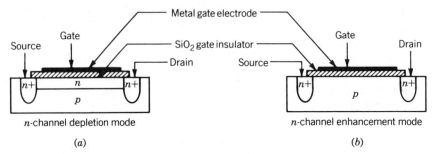

Figure 10.22. MOSFET structures: (*a*) depletion; (*b*) enhancement.

mode device (Figure 10.22*a*) consists of a *p*-type substrate into which are diffused two heavily doped n^+ source and drain electrodes, connected by a more lightly doped *n*-type diffusion, which forms the channel. An oxide layer (insulator) above the channel provides the substrate onto which the gate electrode metal is deposited.

With no gate bias applied ($V_{gs} = 0$) the channel conducts, allowing drain current to flow to the source electrode. As the drain potential becomes more positive (Figure 10.23*a*), conduction continues to increase also until channel saturation occurs, which limits the current flow. The saturation current is a function of channel dimensions (width/length) and doping density. When a negative gate bias is applied, electrons are forced onto the gate electrode (similar to a capacitor plate), attracting positive charges in the channel and depleting it of carriers, thereby inhibiting the current flow from the drain to the source. Depletion-mode drain curves are shown in Figure 10.24*a*.

The **enhancement-mode** device is the type most commonly used in high-density integrated circuits, focal plane readout chips, and consumer prod-

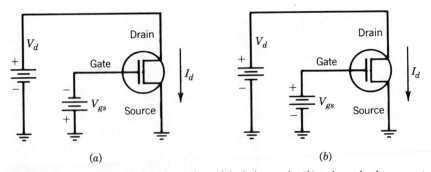

Figure 10.23. MOSFET biasing: (*a*) *n*-channel depletion mode; (*b*) *n*-channel enhancement mode.

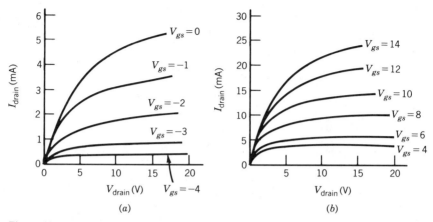

Figure 10.24. MOSFET drain curves: (a) n-channel depletion mode; (b) n-channel enhancement mode.

ucts such as watches and calculators. Like the depletion-mode MOSFET, it consists of a p-type substrate with two highly doped n-type source and drain diffusions, as shown in Figure 10.22b. There is no channel diffusion, however, and the substrate itself serves as the channel. The gate configuration is similar to the depletion-mode device. The biasing is shown in Figure 10.23b. With a positive drain bias and a zero gate potential, the drain diffusion is reverse biased, so there is no conduction between the drain and source and the device is "off." When a positive gate bias is applied, negative carriers are generated, forming an inversion layer under the gate, converting the p-type substrate to n-type, forming a conductive channel. If the positive gate bias is increased, the inversion layer becomes thicker and conduction is enhanced. As with the depletion-mode device, the channel width/length ratio determines the maximum amount of channel conduction. Typical enhancement-mode drain curves are illustrated in Figure 10.24b. MOSFET circuit configurations (common source, common drain, and common gate) are similar to those of the bipolar junction transistor and the junction field-effect transistor already described and so are not discussed here. The noise mechanisms, however, are somewhat different.

MOSFET Noise. For most source impedances, MOSFET noise performance is not as good as that of junction FETs. As with JFETs the minimum noise level is determined by the thermal noise of the channel resistance, but for MOSFETS this level is almost never achieved. The dominant noise mechanism is generally attributed to charge "traps" in

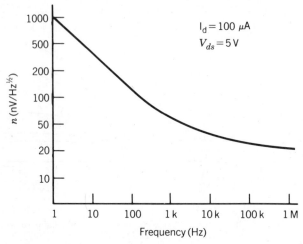

Figure 10.25. MOSFET noise voltage versus frequency.

the gate oxide, which attract and release electrons in a random fashion, thereby changing the channel conduction. This noise has a $1/f$ spectral shape with corner frequencies as high as tens of kilohertz.

Noise characteristics are highly variable even among devices of the same type, and are particularly sensitive to static damage caused by improper handling. The noise performance can be severely degraded by static exposure even if operational damage is not apparent. A typical noise voltage plot is presented in Figure 10.25.

10.3. AMPLIFIERS

Amplifiers are used in detector signal processing for amplifying and band-limiting the low-level signal from the detector. They must provide a stable predictable gain and contribute as little distortion and extraneous noise as possible. In this section we describe some of the principles of amplifier gain, negative feedback, and noise performance, as well as standard amplifier circuit configurations applicable to high-impedance, moderate-impedance, and low-impedance detector signal amplification.

10.3.1. Feedback Theory and Stability

As we have seen from previous pages, the semiconductor devices from which amplifiers are constructed are temperature and voltage sensitive and their gains are not predictable within a factor of 2 or 3 at the time of manufacture. Without a gain stabilization mechanism, every detector amplifier would require individual adjustment to achieve the desired gain. Component replacement, temperature changes, and voltage changes would all make gain and offset readjustment necessary.

Negative feedback provides a powerful stabilization tool that makes amplifier circuits insensitive to component parameter changes. It also enhances both input and output characteristics and improves linearity.

Figure 10.26 illlustrates the general case of a high-gain inverting amplifier $(-A)$ with a feedback network that returns a fraction β of the amplifier output signal and a summing network for combining the input signal with the feedback signal. For negative feedback, the general gain expression is

$$g = \frac{V_{out}}{V_{in}} = \frac{A}{1 + \beta A + \beta} \simeq \frac{A}{1 + \beta A} \qquad (10.25)$$

where A = amplifier (open-loop) gain
β = transfer ratio of the feedback loop = R_{in}/R_f
g = circuit (closed-loop) gain

If the open-loop gain is large, the gain of the circuit (the closed-loop gain) will approach the inverse of the feedback ratio, and be insensitive to large changes in the open-loop gain. The feedback loop can be made of stable resistors and provide gain stability of fractions of a percent.

A disadvantage of negative feedback is that it introduces the possibility of amplifier oscillation. Because components are not purely resistive, frequency dependent phase shifts occur between the input and output signals. This phase shift—if large enough—can cause the sign of the feedback to reverse from negative to positive, reinforcing the amplifier input and causing oscillation or "rail to rail" saturation. In general any phase

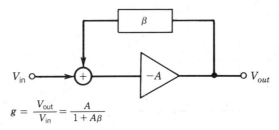

Figure 10.26.   Feedback model. For $A = 100 \times \beta$, gain error = 1%. For $A = 1000 \times \beta$, gain error = 0.1%.

shifts greater than 90° tend to allow oscillation and/or overshoot in the pulse response.

To prevent oscillation, most general-purpose amplifiers are designed with a single *RC* low-pass characteristic, that reduces the amplifier gain at 20 dB/decade of frequency (6 dB/octave, V ∝ 1/*f*) above a few hertz. Figure 10.27 shows a typical gain versus frequency plot. If the 20-dB/decade roll-off is maintained through unity gain, the maximum phase shift that the amplifier can have is 90°, ensuring that any feedback ratio (β) can be used with no danger of oscillation. For a single-pole amplifier, the frequency point at which the gain crosses unity is called the *gain–bandwidth product*. It is useful for predicting how much open loop gain an amplifier will have at frequencies less than the unity gain frequency. Using the gain–bandwidth product, the open-loop gain *A* can be found for any frequency by

$$A = \frac{\text{gain–bandwidth product}}{\text{frequency}} \qquad (10.26)$$

This is important because it must be used in conjunction with the feedback equation (10.25) to determine the maximum usable frequency for an amplifier system.

10.3.2. Operational Amplifiers

Operational amplifiers are the basic amplifier building blocks from which most detector processing circuitry is constructed. They are high-gain differential amplifiers, amplifying the difference between their inverting (−)

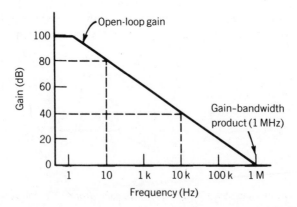

Figure 10.27. Gain versus frequency, showing gain–bandwidth limit.

and noninverting ($+$) inputs. Typical open-loop voltage gains range from 10^4 (80 dB) to 10^6 (120 dB) and typical gain–bandwidth products from 1 to 10 MHz.

There are two common types of gain configurations for operational amplifiers, inverting and noninverting, illustrated in Figure 10.28. The simplest gain expression for the inverting configuration is

$$g \simeq \frac{R_f}{R_i} \tag{10.27}$$

where g = circuit voltage gain
R_f = feedback resistor value (ohms)
R_i = input resistor value (ohms)

The assumptions necessary for this simplified expression are:

1. The amplifier (open-loop) gain is much larger than the closed-loop gain.
2. No appreciable current flows into (or out of) the amplifier input terminals.
3. R_i includes the impedance of the driving source.

The rationale for the gain expression can easily be understood by assuming that the ($-$) amplifier input remains at 0 V, which is the same as assuming that the open-loop amplifier gain is very large. The circuit input current then becomes V_{in}/R_{in}, which equals the feedback current $V_{out}/$

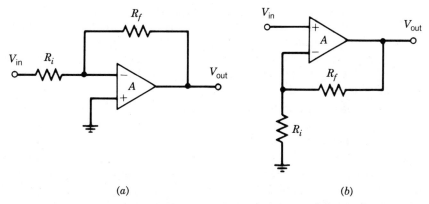

(a)   (b)

Figure 10.28. (*a*) Inverting and (*b*) noninverting operational amplifier circuits. For $A \gg \beta$, $g = R_f/R_i$ for inverting amplifier; $G = 1 + (R_f/R_i)$ for noninverting amplifier.

R_f, making the gain (V_{out}/V_{in}) equal to the feedback resistor divided by the input resistor.

The noninverting configuration simplified gain expression also requires the same assumptions. The circuit operates similarly except that the feedback loop, instead of holding the ($-$) input at virtual ground, causes it to follow the signal present at the ($+$) input; hence

$$g \simeq 1 + \frac{R_f}{R_i} \qquad (10.28)$$

The input resistance for this configuration is very high, equaling that of the operational amplifier, while the input resistance for the inverting configuration is equal to R_{in}, which can be quite low.

The previous gain expressions have assumed ideal amplifier performance, but since operational amplifiers are constructed of nonideal semiconductors, there are other effects to be considered:

Input offset voltage is the consequence of imperfectly matched base–emitter or gate–source voltages of the transistors at the amplifier ($+$) and ($-$) inputs. The result is an unwanted offset voltage at the amplifier output, which is a function of the circuit gain as determined by R_f and R_i.

Input bias current is the base current or gate leakage current required to bias the amplifier input transistors. It also produces an unwanted offset voltage at the amplifier output, which is a function of both the circuit gain and also the resistance values of R_f and R_i. Higher resistance causes a greater output offset voltage.

Slew rate is the maximum rate of voltage change per unit time that an amplifier output is capable of following. It is independent of the gain–bandwidth product previously discussed, and limits the maximum high-frequency undistorted output voltage swing. Slew-rate limitations cause waveform distortion and overshoot of pulse-type inputs. Usually, the input voltage level or the gain must be reduced to eliminate the distortion.

Finite gain results in the departure of the amplifier from the feedback determined gain [equation (10.25)] or inadequate bandwidth to pass the required signal frequency [equation (10.26)].

Equivalent input voltage and *current noise* reduce the system signal-to-noise ratio by introducing noise into the amplified signal—these are discussed in more detail in the following pages. Imperfect power supply rejection allows ripple and noise from the amplifier power supplies to be introduced into the amplified signal. The amount of signal contamination is a function of the circuit gain and usually increases with frequency.

10.3.3. Amplifiers for High-Impedance Detectors

Photovoltaic Detectors. High-impedance photovoltaic detectors operate best when interfaced to a transimpedance preamplifier as shown in Figure 10.29. The *transimpedance preamplifier* (TIA) is similar to the inverting operational amplifier discussed previously, except that the input resistor is replaced by the detector, so that detector signal current is converted to an output voltage by the amplifier and feedback resistor. A TIA is sometimes called a *current mode amplifier*. The TIA is characterized by the ratio of input current to output voltage. This ratio is called the *transfer impedance*, or *transimpedance*. Low-voltage biases may be applied simply by biasing the noninverting amplifier input (see Figure 10.29). The $(-)$ input is forced to the same potential by the amplifier gain via the feedback resistor. This provides a low-impedance bias source to the detector that will not change with detector signal current. Since the input current must flow through the feedback resistor (as with the inverting operational amplifier), the output voltage is proportional to the feedback resistance:

$$V_{out} = I_{det}R_f \qquad (10.29)$$

where I_{det} is the detector current and R_f is the feedback resistor value.

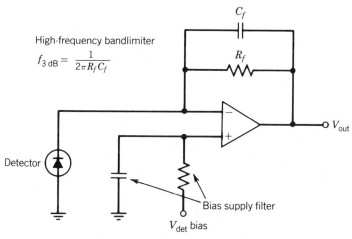

Figure 10.29. High-impedance PV detector with a TIA.

For best noise performance, the detector resistance must be much larger than the feedback resistance, making the circuit transfer function nearly unity for feedback resistor thermal noise, detector bias supply noise, amplifier equivalent input voltage, and current noise. Since the feedback resistor thermal voltage noise increases as the square root of the resistance value [equation (10.6)] but the amplifier transimpedance is a linear function of feedback resistance [equation (10.29)], the circuit signal-to-noise ratio is improved by making the feedback resistor as large as possible.

Circuit input capacitance, whether it is caused by capacitance from the detector, interfacing cable, or amplifier input, has a detrimental effect on preamplifier noise performance. Input capacitance has the effect of diverting the high-frequency feedback that normally counteracts the amplifier equivalent input voltage noise (see Figure 10.30). This causes the amplifier noise voltage to appear "boosted" at the amplifier output, increasing at 20 dB per decade from the R_f/C_{in} corner frequency. A typical composite noise plot for a 100-MΩ feedback resistor JFET TIA is presented in Figure 10.31. At frequencies above 800 Hz, the amplifier performance is degraded by "boosted" amplifier TIA voltage noise. Reducing the amplifier voltage noise or the input capacitance would allow better performance at higher frequencies.

High-frequency bandlimiting for the detector signal can easily be accomplished by a capacitor across the feedback resistor. This attenuates high-frequency detector signals (and resistor thermal noise) at 20 dB per decade above the $R_f C_f$ corner frequency.

Transimpedance preamplifiers are susceptible to microphonics from vibration-induced changes in capacitance. The most severe problems are observed when detectors are remotely located from the preamplifier and

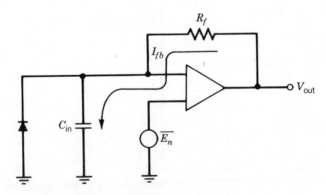

Figure 10.30. Input capcitance in a TIA circuit.

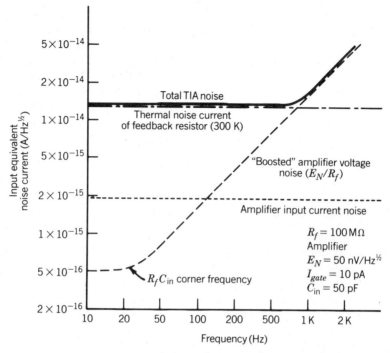

Figure 10.31. Input equivalent noise current from a TIA circuit.

interfaced with shielded cable. Motion of the cable causes capacitance fluctions that inject current into the TIA summing junction. The current is similar to the detector signal and is amplified by the preamplifier in the same manner. While plated shield printed circuit cables demonstrate less microphonic change in capacitance than conventional shielded cables, the best solution is to move the detector close enough to the TIA that shielded cable is not required. When very high value feedback resistors in the range 10^9 to 10^{12} Ω are used, even a rigid printed circuit board in relatively low-vibration environments will allow enough motion of the preamplifier to modulate the stray capacitance at the summing junction and induce microphonics. It is often difficult to distinguish these microphonics from 60-Hz power-line pickup because they are often caused by line-powered fan motors or humming transformers in nearby test equipment. In some cases it is necessary to cushion the complete dewar–amplifier assembly to reduce the microphonics to an acceptable level.

Photoconductor Detectors. The TIA circuit described for high-impedance PV detectors (Figure 10.29) could be used for high-impedance PC detec-

tors, but the bias voltages required are generally higher than can be applied in that way. One option is to bias the opposite end of the detector,
as shown in Figure 10.32, leaving the TIA end of the detector at ground.

Another option is to use a source-follower FET between the detector
and the TIA, as shown in Figure 10.33. The FET can be cooled to low
temperatures, so it can be mounted very close to the detector. Since its
output resistance is relatively low, the precautions needed with the high
resistances can be confined to a small space around the detector and FET.
Dereniak et al. (1977), Goebel (1977), and Sloan (1978) described early
versions of this system; a more recent source is Dereniak and Crowe
(1984).

10.3.4. Amplifiers for Intermediate Impedance Detectors

Many photoconductors—the lead salts and high-background doped germanium and silicon—have impedances between 500 kΩ and 5 MΩ. Their
thermal noise alone, even at 100 or 200 K, is considerably greater than
the equivalent input voltage and current noise of modern FET input operational amplfiers, making the preamplification task relatively simple.
Developing a low-noise detector bias voltage is more difficult. Some detectors require bias voltages of greater than 50 V. The bias must be applied
to the detector from a high-impedance source to allow the detector voltage
to change in response to the detected signal. Semiconductor current
sources typically have output resistances of 250 to 300 kΩ, which is low
enough to attenuate the detected signal significantly. A better alternative

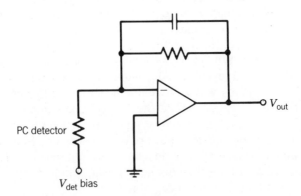

Figure 10.32. High-impedance PC detector with a TIA. Compare the biasing scheme with
that in Figure 10.29.

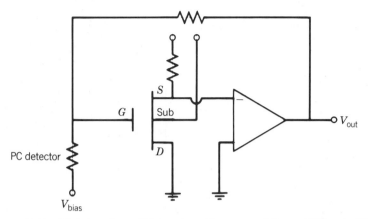

Figure 10.33. High-impedance PC detector with a source-follower FET and a TIA.

is to apply the bias voltage with a high-value wire wound or metal film resistor, if possible two or more times larger than the detector resistance. It is also advisable to operate the bias resistor at the detector temperature to reduce its thermal noise contribution.

Bias supply ripple and noise are attenuated by the ratio of the bias resistor and the detector impedance and amplified by the preamplifier, just as the detected signal. This means that the bias source must be extremely well filtered so that only the detector noise is observed at the preamplifier output. Paper or polycarbonate dielectric capacitors are usually best for bias supply bypassing and input signal coupling because of their low dielectric loss. Plastic film dielectric capacitors are physically smaller but can introduce noise pulses resulting from dielectric imperfections that allow microscopic (self-healing) shorts between the capacitor plates.

It is also important that the bias supply voltage be unreferenced to other preamplifier voltage supplies ("floating") and the "low" side connected as close as possible to the preamplifier input signal return along with the detector common. A typical preamplifier application is illustrated in Figure 10.34.

As with high-impedance detectors, microphonic cable effects are also a problem for intermediate-impedance detectors. In addition, microphonic effects of the ac coupling capacitor and bias supply bypass capacitors can degrade the detector signal. In some vibration environments, soft mounting of these components is required to achieve acceptable performance.

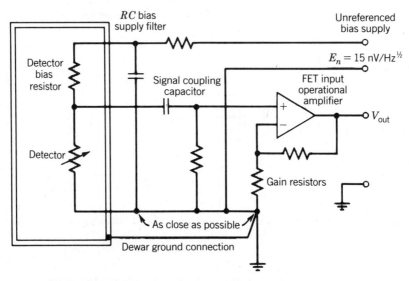

Figure 10.34. Connections for an intermediate-impedance detector.

10.3.5. Amplifiers for Low-Impedance Detectors

Low-impedance photoconductive detectors such as HgCdTe typically range from 50 to 500 Ω.* They are the most difficult to amplify because there are not many semiconductors with noise levels adequate for such low-source resistances. Custom discrete component preamplifiers are almost always required; however, some operational amplifiers with equivalent input voltage noises of about 1 nV/Hz$^{1/2}$ have recently become available.

The lowest-noise amplifier input stage for these detectors is a common-base or common-emitter bipolar junction transistor operating at a collector current of greater than 1 mA so that the base resistance (and its thermal noise contribution) is low. Further noise improvement can be obtained by paralleling N input transistors, which results in a square root of N decrease in equivalent input voltage noise at the expense of a square root of N increase in equivalent input current noise. These two sources must be traded to determine the optimum configuration for a particular detector impedance.

---

* Section 10.3.4 described amplifiers for 500 KΩ to 5 MΩ detectors. Amplifiers for 500 Ω to 500 KΩ must be designed on a case-by-case basis; neither the intermediate-impedance nor the low-impedance schemes is perfectly suitable.

Detector biasing is somewhat easier than for intermediate-impedance detectors because semiconductor current sources made from low-noise bipolar junction transistors or junction FETs, if properly designed, can contribute less noise than the detector thermal noise. Their output resistance is high enough that the detector signal is not attenuated.

Detector–preamplifier interfacing is also a particularly difficult problem for low-impedance detectors because the temperature sensitivity of bipolar junction transistors requires that they remain at ambient temperature and therefore cannot be located close to the cooled detector. Because of the limited cooling capacity of most dewars, the detector must be connected to the preamplifier by leads with high thermal impedance. Such leads have electrical resistance almost as high as the detector. Thermal noise from the cable resistance must be considered. In the case of multielement arrays, interchannel crosstalk caused by returning the current from more than one element through a resistive return conductor must also be considered. Biasing each detector with a high-impedance constant-current source helps to minimize the signal return crosstalk, but making the return conductor as low resistance as possible is imperative.

Output signals from low-impedance detectors are extremely small and usually require preamplifier gains of more than 1000 (60 dB). Gain-determining resistors must be small enough in value that their thermal noise does not degrade the detector performance. Dc amplification is difficult because offsets generated by detector bias or temperature change will cause the preamplifier to saturate. In most applications it is easier to ac couple the detector signal, passing only the frequencies of interest. High-quality coupling capacitors such as those recommended previously for bias supply filtering are also recommended here, although aluminum electrolytic types have been used successfully. In most ac-coupled applications, preamplifier output noise can be seen to increase at frequencies below the 3-dB corner frequency of the ac coupling network. It is often confused with detector $1/f$ noise (which is also present) but is really due to the amplifier equivalent input current noise developing a noise voltage across the increasing source impedance created by the capacitive reactance of the ac coupling capacitor by

$$v_{n1} = (i_n)(R_{\text{det}} + X_c) \tag{10.30}$$

where v_{n1} = input noise voltage (V/Hz$^{1/2}$) at the preamplifier input caused by the amplifier input current noise
i_n = preamplifier input current noise (A/Hz$^{1/2}$)
R_{det} = detector impedance (ohms)
X_c = capacitive reactance of coupling capacitor (ohms)

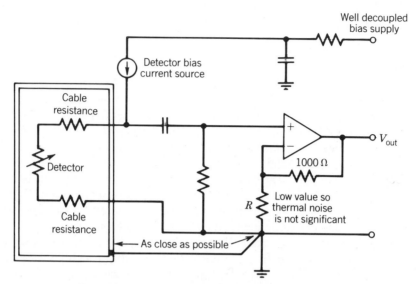

Figure 10.35. Connections for a low-impedance detector.

This low-frequency boosted noise must be removed by high-pass filtering the detector signal after the preamplification process.

Cable-induced microphonics are not usually a problem with low-impedance detectors because the microphonically induced currents are shunted by the very low detector impedance. Detector thermal fluctuations caused by boiling cryogens are often confused with microphonics. These minute temperature fluctuations cause the detector resistance to change and because of the detector bias current, they result in a false detector output signal. A typical low-impedance preamplifier configuration is illustrated in Figure 10.35.

10.4. LABORATORY TESTING

In laboratory testing it is sometimes difficult to achieve repeatable correct measurements, particularly if grounds are not established properly or if test equipment is not used properly. Operating manuals for all test equipment should be studied carefully, not just to learn the mechanics of operating the controls, but also to gain an understanding of the measurement principles that the equipment employs and the applicability of those principles to the types of signals encountered in detector testing.

This section includes a brief discussion of system grounding, static safety, and a short description of some of the operating principles of laboratory equipment useful for making detector measurements.

10.4.1. Grounding

Laboratory testing of detectors is difficult primarily because of the grounding constraints placed on the test setup by the line-powered test equipment. In addition, safety laws require that line-powered test equipment have the chassis connected to the third wire (safety ground) of the power line. Often the instrument chassis ground is also the instrument input signal return. It is difficult to remove all measurable traces of the 60-Hz power-line interferance, so making accurate noise measurements near 60 Hz and its harmonics should not be attempted unless they are specifically required. If they are required, a completely shielded battery-powered setup should be used.

A test setup grounding diagram is illustrated in Figure 10.36, showing the central signal ground at the preamplifier input return fanning out to the various circuit components. This signal ground point is not chosen arbitrarily. The preamplifier will amplify the difference between its input and input return even if that difference is the result of the voltage drop across a wire. Therefore this is the only point at which the current loops from the detector, power supplies, and preamplifier output should be connected.

The preamplifier power supply should be well isolated from the power line even if that means using batteries. Measurement equipment such as oscilloscopes and meters should be connected to the circuit being tested with a shielded cable, and should be placed several feet away from the low-level circuitry so that their power-line cords and power transformers do not interfere with the low-level detector signals. Some preamplifiers will require a small isolation resistor between their output and the shielded cable so that the capacitive loading will not cause oscillation. The line-powered measurement equipment should be connected to the same 60-Hz outlet box so that chassis grounds will be at the same potential.

All of the above recommendations are consistent with testing at a static-safe workstation. Static-safe workstations usually connect the working surface and the test operator to a central power system ground via a high resistance (about 1 MΩ) connection. This allows static charges generated by movement or clothing to be safely dissipated, but limits the amount of current that can flow between the test setup and the workstation so that there is no interference between the two systems. It also minimizes electrical hazards to the operator since it limits the current flow from a 110-V source to about 100 μA. Bypassing antistatic provisions of the

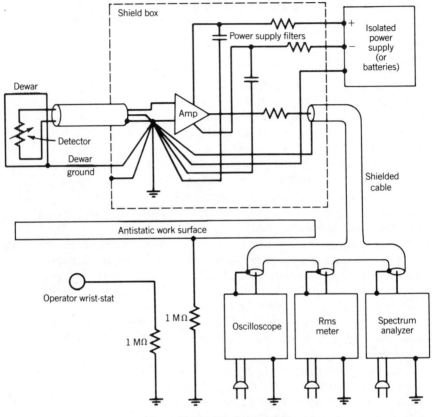

Figure 10.36. Ideal grounding.

workstation is not only unnecessary but unwise because it jeopardizes the very detector parameters being measured.

10.4.2. Laboratory Equipment

Oscilloscopes. Oscilloscopes present a voltage versus time display of their input on a cathode ray tube. They are useful for observing preamplified detector signals and making rough measurements of noise and frequency response. They are also useful as an aid to optical alignment and to ensure that the detector signal is not contaminated with unwanted signals, such as 60-Hz pickup. Most oscilloscope inputs have a resistance of 1 MΩ and capacitance of about 50 pF, approximately the same capacitance as 2 ft of shielded cable. Low-level oscilloscope preamplifiers are capable of displaying microvolt signals, and some have adjustable high-pass and low-pass filters. Most oscilloscopes also have an output jack at the rear that provides a low output impedance for connecting other equipment, such as spectrum analyzers or rms meters. The oscilloscope gain from the input to the output jack is a function of the vertical sensitivity and must be calibrated using a known signal to be useful for making absolute measurements.

Lock-In Amplifiers. Lock-in amplifiers are useful for extracting periodic signals from random noise by narrowband filtering. They are often used in detector signal measurement when the detector radiation source is a chopped blackbody. A reference signal from the chopper can provide a frequency/phase reference for controlling the center frequency of the lock-in amplifier's narrowband filter. The filter bandwidth can be as narrow as fractions of a hertz, making it possible to measure signals accurately well below the broadband noise level. As with any narrowband measurement instrument, care must be taken that the input preamplifier does not saturate on high noise peaks, making the output reading invalid. Most instrument manuals specify an absolute voltage limit for each input range setting.

Preamplifiers. Commercial preamplifiers are available in both voltage-mode and current-mode (transimpedance) configurations. It is useful to calibrate the noise performance and bandwidth of such amplifiers before trusting any detector performance readings based on their use. For voltage-mode preamplifiers, the equivalent input voltage noise can be measured by shorting the amplifier input and measuring the output noise versus frequency. For each frequency of interest, the input voltage noise can be calculated by dividing the output noise reading by the amplifier gain at the same frequency. The equivalent input current noise is somewhat more difficult to measure. A known resistor is connected across the amplifier input, and the noise versus frequency is again measured. In this case the input voltage noise and the input termination resistor thermal noise contributions are "root subtracted" from the measured output. The input current noise is then calculated by dividing the resultant output voltage for each frequency by the amplifier gain and the input resistor value. Some experimentation with the input resistor value is required to allow the current noise term to be large enough to be measured accurately.

For transimpedance amplifiers it it not possible to make input shorted measurements because the amplifier output will saturate ($R_f/R_i = \infty$). The measured output noise with the input open should agree very closely with the thermal noise calculated for the feedback resistor.

Spectrum Analyzers. There are two major types of spectrum analyzers for displaying the electrical spectrum of the detector output signals. The first creates the spectral plot by tuning a narrowband filter across the spectrum of interest and evaluating the filter output at intervals along the spectrum. The disadvantage of this approach is that only the portion of the spectrum where the filter is presently tuned is being measured. This can be particularly troublesome when very low-frequency spectral plots are made which require that the filter be tuned very slowly. During the tuning interval, detector temperature or detector stimulus can change making the spectral plot invalid.

The second type contains a powerful microcomputer and digitizer that rapidly sample the input signal and perform a Fourier transform, converting the impulse response to a spectral plot. Successive plots are then averaged for the display. This type is generally preferrable for detector measurement because the data acquisition can be operator triggered when detector and stimulus temperatures are stable. The equivalent noise bandwidth is greater than the "bin" or resolution width; see the operators manual for details.

RMS Meters. Rms meters are useful for measuring the root-mean-square value of the broadband noise after filtering. Filtering and precise knowledge of the filter shape are necessary to interpret the broadband noise reading. There are two general types of rms meters. The first is called *average responding*. This type of meter responds to the average voltage value (not the true power value) of an ac input, and applies a gain factor to correct the reading for the rms of a pure sine wave. This means that for any waveform except a pure sine wave, the reading is in error. For random noise, average responding meter readings must be multiplied by 1.128 to obtain a correct true rms reading. The second type, called *true rms responding*, actually measures the power value of the input signal irrespective of the waveform shape. The measurement requires no correction factor for noise measurement.

This brief list in no way covers all the necessary detector test equipment and is not intended as a substitute for reading the operating manuals and specifications.

REFERENCES

Dereniak, Eustace L., and Devon G. Crowe (1984). *Optical Radiation Detectors*, Wiley, New York.

Dereniak, Eustace L., R. R. Joyce, and R. W. Capps (1977). "Low-Noise Preamplifier for Photoconductive Detectors," *Rev. Sci. Instrum.* 48, 392.

Fitzgerald, A. E., David E. Higginbotham, and Arvin Grable (1981). *Basic Electrical Engineering*, McGraw-Hill, New York.

Goebel, John H. (1977). "Liquid Helium-Cooled MOSFET Preamplifier for Use with Astronomical Bolometer," *Rev. Sci. Instrum.* 48, 389.

Holt, Charles A. (1978). *Electric Circuits—Digital and Analogue,* Wiley, New York.

Sloan, William W. (1978). "Detector-Associated Electronics," Chapter 16 in *The Infrared Handbook*, William L. Wolfe and George J. Zissis, eds., Environmental Research Institute of Michigan.

SUGGESTED READING

Dereniak, Eustace L., and Devon G. Crowe (1984). *Optical Radiation Detectors,* Wiley, New York.   An excellent reference for the electronics associated with IR detector operation.

Fitzgerald, A. E., David E. Higginbotham, and Arvin Grable (1981). *Basic Electrical Engineering*, McGraw-Hill, New York.   Basic engineering text; the circuit analysis sections are particularly good.

Holt, Charles A. (1978). *Electric Circuits—Digital and Analogue*, Wiley, New York.   Electrical engineering text; the sections on semiconductors are especially good.

Horowitz, Paul, and Winfield Hill (1980). *The Art of Electronics*, Cambridge University Press, Cambridge.   General electronics text in a down-to-earth style.

Keithley, J. F., J. R. Yeager, R. J. Erdman (1984). Low Level Measurements, Keithley Instruments, Inc., Cleveland, Ohio.

Moore, John H., Christopher C. Davis, and Michael A. Coplan (1983). *Building Scientific Apparatus,* Addison-Wesley, Reading, Mass.   Includes a chapter on practical electronics.

Morrison, Ralph (1967). Grounding and Shielding Techniques in Instrumentation, Wiley, New York.

Sloan, William W. (1978). "Detector-Associated Electronics," Chapter 16 in *The Infrared Handbook*, William L Wolfe and George J. Zissis, eds., Environmental Research Institute of Michigan.   A review of electronics applied to infrared detectors.

PROBLEMS

1. For the resistor network shown in Figure 10.37, what will the voltmeter read? How much power will be dissipated by the 820 Ω resistor?

2. A 10-MΩ resistor is operated at a temperature of 200 K. What RMS Johnson noise voltage is expected if the measurement device has a noise measurement bandwidth of 1 kHz?

3. For the DC network shown in Figure 10.38, what will the oscillo-

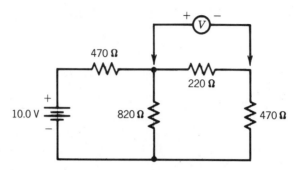

Figure 10.37. Circuit for Problem 1.

scope waveform look like if the sweep is triggered by the circuit input voltage step? What will the capacitor voltage be 10 ms after the voltage step is applied? If the voltage step source is replaced by a 0–1 milliampere positive current step source, what will the capacitor voltage be 10 ms after the current step is applied?

4. A variable frequency sine wave generator is connected to the circuit of Figure 10.39 and its amplitude is adjusted to 5 V RMS. To what frequency must the generator be set in order for the RMS meter to read 3.54 V?

5. A silicon transistor circuit is constructed as shown in Figure 10.40. If the operating Vbe is 0.6 V, and the transistor β is greater than 200 at the operating current, approximately what dc voltages are expected at points A and B measured with respect to the circuit ground? If a 0.1 V rms sine wave voltage source is ac-coupled to the transistor base terminal, what rms voltage is expected at point A measured with respect to circuit ground?

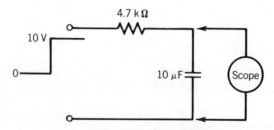

Figure 10.38. Circuit for Problem 3.

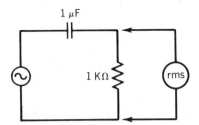

Figure 10.39.   Circuit for Problem 4.

6. A 10-MHz gain bandwidth operational amplifier is connected as shown in Figure 10.41. What rms output voltage is expected for a 0.5 V RMS, 1 KHz sine wave input?

7. If the amplifier of problem 6 is connected as shown in Figure 10.42, what rms output voltage is expected for a 0.5 V RMS, 1-kHz sine wave input? What output voltage is expected if the input frequency is increased to 20 kHz?

8. Two random noises, measured with the same true rms meter, read 2.46 mV and 3.16 mV. If these sources are summed together (with a lossless summing circuit), what will the rms meter read if it is connected to the summing point?

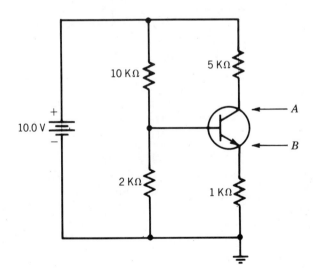

Figure 10.40.   Circuit for Problem 5.

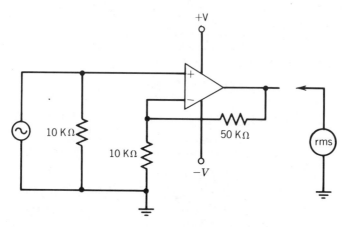

Figure 10.41. Circuit for Problem 6.

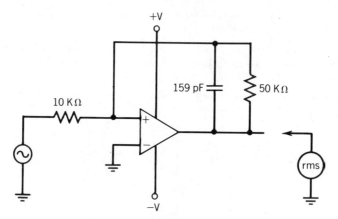

Figure 10.42. Circuit for Problem 7.

9. A 100 Ω photoconductive HgCdTe detector is biased with a 10-kΩ resistor connected to a 10.0 V dc bias supply. The detector signal is amplified by a voltage mode preamplifier with a gain of 1000. If the bias supply has 1.0 millivolt of ripple, how much ripple will be observed at the preamplifier output?

Appendixes

Appendix A

Symbols and Abbreviations

a	Cross-sectional area
A	Sensitive area of detector; open-loop gain of feedback amplifier
A	Ampere
ac	Alternating current
AR	Antireflection (coating)
BB	Blackbody
BLIP	Background limited infrared photon detector
c	Speed of light in vacuum; specific heat capacity
c_1	First radiation constant
c_2	Second radiation constant
C	Heat capacity; blackbody-to-peak conversion factor; conductance; capacitance
C	Coulomb
cm	Centimeter
d	Diameter; distance
D*	Specific detectivity
D	Distance
dc	Direct current
e	Electronic charge; the number 2.718 . . .
E	Electric field; incidance
EMF	Electromotive force
f	Focal length; frequency
ft	Foot
F	Farad
g	Voltage gain
G	Photoconductive gain; transconductance of a JFET
GR	Generation–recombination

h	Planck's constant
H	Heat of vaporization; enthalpy
i	Current (ac)
I	Current (dc); intensity
in.	Inch
IR	Infrared
JFET	Junction field-effect-transistor
J	Joule
J	Current density
k	Kilo- (10^3)
k	Boltzmann's constant; thermal conductivity; spatial frequency
K	Thermal conductance
K	Kelvin
l	length
L	Liter
L	Sterance; Lorenz ratio
LSB	Least significant bit
LSF	Line spread function
m	Meter; mill– (10^{-3})
mi	Mile
mm	Millimeter
M	Mega– (10^6)
m	Mass
M	Exitance; modulus of rupture
MF	Modulation factor
MLI	Multilayer insulation
MTF	Modulation transfer function
n	Nano–(10^{-9})
n	Index of refraction; gas particle density; carrier density; noise spectral density
N	Noise
NA	Numerical aperture
ND	Neutral density
NEI	Noise equivalent incidance
NEP	Noise equivalent power

NEP′	Noise equivalent power per root bandpass
NE ΔT	Noise equivalent temperature difference
P	Pico–(10^{-12})
P	Pressure; power
PC	Photoconductor
PRT	Platinum resistance thermometer
PSF	Point spread function
PV	Photovoltaic
Q	Radiant photon incidance (when used with detector current, signal, or responsivity formulas); outgassing rate; throughput, power in a thermal system
r	Radius; reflectance from a single surface
rms	Root mean square
rss	Root of the sum of the squares
R	Resistance; reflectance from a window
$\mathscr{R}$	Responsivity
RGA	Residual gas analyzer
s	Standard deviation of the sample
S	Signal; pumping speed
s	Second
S/N	Signal-to-noise ratio
t	Time
T	Temperature; transmittance
TC	Thermocouple
v	Velocity
V	Potential difference ("voltage")
V	Volt
W	Watt
$\bar{x}$	Average or mean value of x
Z	Impedance
α	Absorbtion coefficient
β	Current gain of a transistor; transfer ratio of a feedback loop
δx	Increment or small part of x; variation in parameter x
Δf	Equivalent noise bandwidth
Δx	Increment or small part of x
$\Delta \lambda$	Spectral bandwidth (of a spectral filter)

ϵ	Emissivity
η	Quantum efficiency, "sticking probability" of gas molecules
θ	Plane angle
λ	Wavelength; mean free path
μ	Mobility; micro– (10^{-6})
ν	Frequency; wavenumber
π	The number 3.14 . . .
ρ	Resistivity, density
σ	Stefan–Boltzmann constant; electrical conductivity; standard deviation
σ_q	Photon analog of the Stefan–Boltzmann constant
τ	Time constant; lifetime; transmittance
Θ	Temperature
Φ	Flux
Ω	Projected solid angle; Ohm
ω	Solid angle

Appendix B

Glossary

In this glossary we define a few words encountered in the IR business that may be unfamiliar to the newcomer. Some of the words are discussed more thoroughly in the text.

Bias. Application of an electrical voltage or current to make a detector or transistor operate in a desired way (see Chapter 2).

Blackbody. A source of infrared whose output can be predicted accurately by Planck's law (see Chapter 3).

BLIP. Background-limited infrared photon detector. BLIP performance results when all noise sources are negligible compared to the background-generated noise. A detector cannot exceed the BLIP limit (see Chapter 2).

Chevron. One or more pairs of detectors oriented at an angle with respect to each other (see Figure B.1). When scanned with a known speed across a target, the location of the target in two dimensions can be determined. The in-scan location is determined from the time at which detector 1 sees the target, and the cross-scan location is determined from the interval between the signals from the two detectors.

Cryogen. A liquid whose boiling point is below normal room temperature. Cryogens are used to cool detectors to suitable operating temperatures (see Chapter 7).

D* ("D-star", Specific detectivity). A figure of merit for infrared detectors. For many conditions D^* is independent of detector area and electrical bandwidth, so it is a convenient way to specify and compare performance (see Section 1.8).

Decibel (dB). A "unit" of electrical gain or attenuation (see Appendix C).

Dewar. A container with vacuum-insulated, low-emissivity walls, used to store cryogens (see Chapter 7).

Exitance. The rate at which energy or photons leave a surface of unit area (see Chapter 3).

Incidance. The rate at which energy or photons reach a surface of unit area (see Chapter 3).

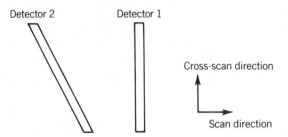

Figure B.1. Chevron detector array.

Modulate. To periodically interrupt or modify the amplitude of the radiometric or electrical signal. Modulated signals can often be amplified more easily or accurately than can unmodulated signals.

Multiplex. To sequentially sample or observe the signal from a series of detectors. Multiplexing allows the use of many detectors with only a few electrical leads leaving the detector plane. This reduces the heat load to the cooler and simplifies cable and feed-through construction.

PC. Photoconductive detector; a detector whose conductivity is modified when illuminated (as opposed to photovoltaic detectors) (see Chapter 2).

Photon. A ''particle'' of light, infrared, or other electromagnetic energy. Also called quanta. Results of some experiments can best be explained by treating electromagnetic energy as ''particles'' or photons, as opposed to continuous waves (see Chapter 1).

PV. Photovoltaic detector; a detector that generates a voltage when illuminated (as opposed to photoconductors, which are passive) (see Chapter 2).

Reticle. A thin mask with an applied pattern. The reticle could consist of an opaque pattern on a transparent substrate or a stencil-like pattern etched in a thin, opaque plate.

Spatial frequency.   The number of cycles of a pattern or target per unit length. Spatial frequency is usually stated in cycles per millimeter ($\sim$/mm). The spatial frequency k is the reciprocal of the pattern ''wavelength'' λ: $k = 1/\lambda$.

Appendix C

Decibel Convention

The gains or attenuation of electrical circuits are often expressed in decibels, (abbreviated dB) as

$$\text{gain (dB)} = 10 \log_{10}(P_{\text{out}}/P_{\text{in}})$$

where P_{in} and P_{out} are the input and output power from the circuit.

For most applications, power varies as the square of the voltage or current, so we can write

$$\text{gain (dB)} = 20 \log_{10}\left(\frac{V_{\text{out}}}{V_{\text{in}}}\right) = 20 \log_{10}\left(\frac{I_{\text{out}}}{I_{\text{in}}}\right)$$

Table C.1 Decibel Convention

Gain (dB)	Power Ratio		Voltage Ratio	
60	10^6		10^3	
50	10^5			
40	10^4		10^2	
30	10^3			
20	10^2		10	
10	10		3.162	$= \sqrt{10}$
9.03		8	2.828	$= \sqrt{8}$
6.02		4	2.0	$= \sqrt{4}$
3.01		2	1.414	$= \sqrt{2}$
0	1	1	1	1.0
-3.01		0.5	0.707	$= \sqrt{1/2}$
-6.02		0.25	0.50	$= \sqrt{1/4}$
-9.03		0.125	0.3535	$= \sqrt{1/8}$
-10	0.1		0.3162	$= \sqrt{1/10}$
-20	0.01		0.1	

Examples:

1. $P_{out}/P_{in} = 1000 = 10^{+3}$, $\log_{10}(P_{out}/P_{in}) = 3$, so gain = 30 dB.
2. $P_{out}/P_{in} = 10 = 10^{+1}$, $\log_{10}(P_{out}/P_{in}) = 1$, so gain = 10 dB.
3. $P_{out}/P_{in} = 3 \simeq \sqrt{10} = 10^{0.5}$, $\log_{10}(P_{out}/P_{in}) = 0.5$, so gain = 5 dB.
4. $P_{out}/P_{in} = 0.01 = 10^{-2}$, $\log_{10}(P_{out}/P_{in}) = -2$, so gain = -20 dB.

Evidently, a power ratio of less than 1 yields a negative gain when expressed in dB; both represent an attenuation.

The statement "the gain is 13 dB" is complete and unambiguous. You do *not* need to say "the *voltage* gain is 13 dB" or "the *power* gain is 13 dB." In fact, to do so indicates that you are a little fuzzy about how the decibel convention works.

Amplifier gain and filter attenuations are often expressed as a function of frequency. An octave means a factor of 2 in frequency, and a decade means a factor of 10 in frequency. Thus the expressions "the signal rolls off at 3 dB per octave," ". . . at 10 dB per decade," and "voltage varies

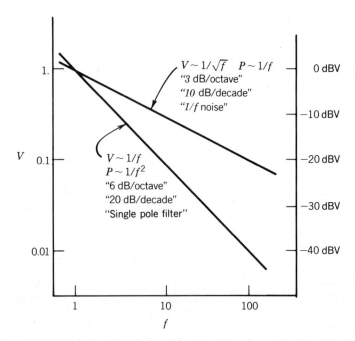

Figure C.1. Equivalent descriptions of two common frequency dependencies.

as 1 over the square root of f" all mean the same thing. (This roll-off is characteristic of "1-over-f" noise.) These relationships are all illustrated in Figure C.1.

The dB convention can be used to express absolute voltages (as opposed to ratios, or gains): -40 dBV means that the desired voltage is 40 dB below one volt—10 mV.

Appendix D

Characterization of Semiconductor Materials

The material from which infrared detectors are made is normally characterized much like semiconductors for any other electronic application. The band gap, mobility, and carrier density are derived from resistance and Hall voltage measurements made as a function of temperature. Lifetime is determined from the response to a pulsed source, or from the roll-off frequency of the noise spectral density. The measurement of resistance is described first, along with determination of the band gap from resistance versus temperature data. Hall voltage measurements are then described, and the determination of carrier concentration. Lifetime determination is described last.

1. RESISTANCE MEASUREMENTS

1.1. Rectangular Bar

The most direct method to measure resistivity is to construct a bar of constant cross section, apply a voltage to the ends, and measure the resulting current. The resistance is then V/I, and the resistivity is the resistance times the cross-sectional area a, divided by the distance l between contacts:

$$R = \frac{V}{I} \tag{D.1a}$$

$$\rho = R\frac{a}{l} = \frac{V}{I}\frac{a}{l} \tag{D.1b}$$

1.2. Four-Lead Method

In many situations the lead resistance may be high enough to affect measurement accuracy. This will be the case if the sample resistance is low,

461

or if long, high-impedance leads must be used, as in cryogenic applications. The effect of lead resistance on sample resistance determination can be eliminated by use of the four-wire technique. The current is applied through two leads, and the voltage is measured through two others (see Figure 7.11). Since the current flow through a voltmeter is small, there is negligible IR drop in the voltmeter part of the circuit, and the voltage measured is essentially that across the sample.

1.3. Hall Bar

Solder joints between the sample and the leads are another source of error. The solder can introduce effects that are difficult to interpret: Impurities may alter the carrier concentration we are attempting to measure and may introduce nonohmic voltage drops. The effect of this contact contamination can be eliminated if the sample is cut to the shape shown in Figure D.1. Current leads are attached to the ends of the bar, and the voltage leads are attached to the "arms," which serve as contact probes made of the sample material. Since no current flows through the arms, the resistance of the solder contacts there does not affect the voltages measured.

Thermal EMFs can still be present, so one generally measures the voltages twice: once with the current flowing in each direction, and averages the results.

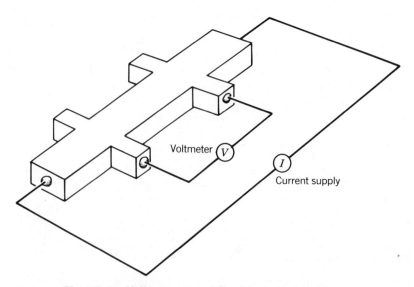

Figure D.1. Hall bar connected for resistance measurements.

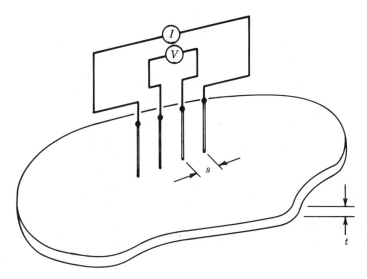

Figure D.2.   Resistance measurement using four probes on a wafer.

Only two arms are needed for resistance measurements, but the others are useful if the device is used for mobility measurements—to be described soon.

2.   WAFER MEASUREMENTS

If four equally spaced in-line probes contact a wafer of dimensions much larger than the space s between the probes (Figure D.2), the resistivity is given by Dunlap (1959):

$$\rho = 2\pi s \frac{V}{I} \tag{D.2}$$

where V and I are the current and voltage. Corrections for other geometries are available. For more detailed information on the four probe method, see Valdes (1954) and Smits (1958).

3.   HALL EFFECT

When charged carriers move in the presence of a magnetic field, they are acted upon by a magnetic force; this will generate a voltage that is pro-

Table D.1 Equivalent Magnetic Units*

Field	System of Units	
	Gaussian	MKS
H	10^4 oersted	$\dfrac{1}{4\pi} \times 10^7$ amp/meter
B	10^4 gauss	$1 \text{ tesla } = 1\,\dfrac{\text{weber}}{\text{meter}^2} = 1\,\dfrac{\text{newton}}{\text{amp·meter}}$

* There are three magnetic field vectors: the magnetic field strength H due to "free" (conventional) current; the magnetization M due to "effective" (molecular) currents; the magnetic intensity B due to the combination of H and M. Since B and H represent different physical phenomena, 10^4 oersted does not *equal* 10^4 gauss, but if H equals 10^4 oersted, then (for nonmagnetic materials) B equals 10^4 gauss, which equals 1 tesla. Most semiconductor materials are nonmagnetic and we can use Table D.1 to convert between different units.

portional to their velocity. From that voltage we can calculate the carrier concentration. By combining that with the resistivity, we can determine the carrier mobility. The experiment is normally done with a sample cut or sandblasted, then etched, to the shape of Figure D.3. This is called a *Hall bar*. Current is driven through the length of the bar and the voltage

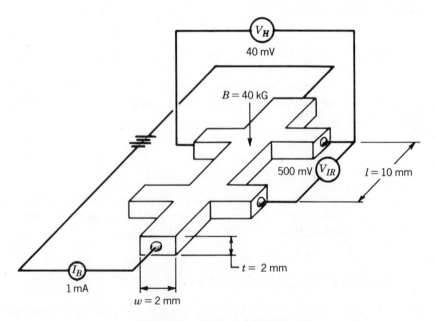

Figure D.3. Hall bar and typical data.

difference between two arms on the same side of the bar allows calculation of the sample resistivity. The Hall voltage is generated perpendicular to the sample length and can be measured between either set of opposing arms.

It is common to compute an intermediate parameter—the Hall coefficient R—from experimental data, then to use the Hall coefficient and the resistivity to determine the carrier concentration and mobility. Before writing the equations, note that the computations are complicated by the variety of units used for magnetic fields; Weber/m^2, tesla, gauss, or oersteds may all be encountered. Although it may require an extra step, the least confusing method is to convert sample dimensions to meters and magnetic fields to tesla (Weber/m^2) when calculating the Hall coefficient from Hall data. Then convert the Hall coefficient to centimeters prior to calculating carrier density and mobility. Equivalence of the different magnetic units is shown in Table D.1.

CARRIER CONCENTRATION FROM HALL DATA

$$n \ (\mathrm{cm}^{-3}) = \frac{1}{eR_\mathrm{H}(\mathrm{cm}^3/\mathrm{C})} \qquad (\text{D.3})$$

where

$$R_\mathrm{H}(\mathrm{cm}^3/\mathrm{C}) = R_\mathrm{H}(\mathrm{m}^3/\mathrm{C}) \times 10^6 \ \mathrm{cm}^3/\mathrm{m}^3 \qquad (\text{D.4}a)$$

$$R_\mathrm{H}(\mathrm{m}^3/\mathrm{C}) = \frac{V_\mathrm{H}}{I_\mathrm{B}B} t = \frac{E_\mathrm{H}}{J_\mathrm{B}B} \qquad (\text{D.4}b)$$

V_H = observed Hall voltage (volts)
I_B = bias current (amps)
B = magnetic intensity (weber/meter^2)
t = sample thickness (meters)
e = charge on the electron (about 1.6×10^{-19} C)

The mobility can be calculated directly from the resistivity ρ and either the Hall coefficient or carrier concentration, in CGS units, as seen in Equations D.5, D.6, and D.7.

MOBILITY FROM HALL DATA AND RESISTIVITY

$$\mu = \frac{R_H \ (\text{cm}^3/\text{C})}{\rho \ (\Omega \ \text{cm})} \tag{D.5}$$

MOBILITY FROM CARRIER CONCENTRATION AND RESISTIVITY

$$\mu = \frac{1}{ne\rho} \tag{D.6}$$

where

$$\rho = \frac{V_B}{I_B} \frac{w \times t}{l} = \frac{E_B}{J_B}$$
$$= \text{resistivity} \ (\Omega \cdot \text{cm}) \tag{D.7}$$

Example: Apply the method outlined here to the data in Figure D.3:

$$B = 40 \ \text{kG} \times \frac{1 \ \text{w/m}^2}{10^4 \ \text{G}} = 4 \ \text{w/m}^2$$

$$R_H(\text{m}^3/\text{C}) = \frac{40 \times 10^{-3} \ \text{V}}{10^{-3} \ A \times 4 \ \text{w/m}^2} \times 2 \times 10^{-3} \ \text{m}$$

$$= 0.02 \ \text{m}^3/\text{C}$$

$$R_H(\text{cm}^3/\text{C}) = 0.02 \ \text{m}^3/\text{C} \times 10^6 \ \frac{\text{cm}^3}{\text{m}^3}$$

$$= 2 \times 10^4 \ \text{cm}^3/\text{C}$$

$$n = \frac{1}{1.6 \times 10^{-19} \text{C} \times 2 \times 10^4 \ \text{cm}^3/\text{C}}$$

$$= 3 \times 10^{14} \ \text{cm}^{-3}$$

$$\rho = \frac{500 \times 10^{-3} \ \text{V}}{10^{-3} \ A} \times \frac{0.2 \ \text{cm} \times 0.2 \ \text{cm}}{1.0 \ \text{cm}}$$

$$= 20 \ \Omega \cdot \text{cm}$$

$$\mu = \frac{1}{3 \times 10^{14} \ \text{cm}^{-3} \times 1.6 \times 10^{-19} \ \text{C} \times 20 \ \Omega \ \text{cm}}$$

$$= 10^3 \ \text{cm}^2/(\text{V} \cdot \text{s}) \qquad \blacksquare$$

4. ENERGY LEVEL AND BAND GAP DETERMINATION

The band gaps and energy levels for semiconductor materials appear in the formula for carrier density as a function of temperature:

$$n \propto \exp\left(\frac{-E_g}{2kT}\right) \qquad \text{for the intrinsic case}$$

$$n \propto \exp\left(-\frac{E}{kT}\right) \qquad \text{for the extrinsic case}$$

Once we know carrier concentration versus temperature, we can use it to extract the band gap or energy level.

It is also possible, although slightly less accurate, to obtain the band gap or energy level from the resistance data. This is possible since the mobility is relatively independent of temperature, so the carrier concentration dominates the temperature dependence of the resistance:

$$R = \text{constant} \times \exp\left(\frac{+E_g}{2kT}\right) \quad \text{or} \quad \exp\left(+\frac{E}{kT}\right)$$

Thus the desired information can derived directly from the temperature dependence of either the Hall data or the resistance vs temperature data. In either case, it is usual to plot resistance or carrier concentration vs $1/T$ (or $1000/T$) on semilog paper.

The band gap or energy level is proportional to the slope of the resulting straight line; see Equation D.8.

BAND GAP FROM LOG R VERSUS $1/T$ (INTRINSIC REGIME)

$$E = \frac{2\ln(10)k \, \log(R_2/R_1)}{e(1/T_2 - 1/T_1)} \tag{D.8a}$$

$$= (3.97 \times 10^{-4} \text{ eV/K}) \frac{\log(R_2/R_1)}{1/T_2 - 1/T_1} \tag{D.8b}$$

ENERGY LEVEL FROM LOG R VERSUS $1/T$ (EXTRISIC REGIME)

$$E = \frac{\ln(10)k \, \log(R_2/R_1)}{e(1/T_2 - 1/T_1)} \tag{D.8c}$$

$$= (1.98 \times 10^{-4} \text{ eV/K}) \frac{\log(R_2/R_1)}{1/T_2 - 1/T_1} \tag{D.8d}$$

---

5. LIFETIME

Carrier lifetimes can be determined by pulsing a detector with a fast source of electromagnetic radiation, and observing the resulting signal growth or decay. An alternative method is to observe the noise spectral density with an amplifier circuit whose frequency response is much faster than the lifetimes of interest. In this case the roll-off in the noise spectral density is due to the material lifetime, and the lifetime can be calculated from the 3-dB frequency using the relationship $\tau = 1/(2\pi f_{3\text{-dB}})$.

REFERENCES

Dunlap, W. Crawford, Jr. (1959). "Conductivity Measurements on Solids," Section 7.2 in *Methods of Experimental Physics*, Vol. 6, *Solid State Physics, Part B: Electrical, Magnetic, and Optical Properties*, Lark-Horovitz and Vivian A. Johnson, eds., Academic Press, New York.

Smits, F. M. (1958). "Measurements of Sheet Resistivities with the Four Point Probe," *Bell Syst. Tech. J.* 37, 711.

Valdes, L. B. (1954). "Resistivity Measurements on Germanium for Transistors," *Proc IRE* 42, 420.

Index